Palaeo-Environmental Change and the Persistence of Human Occupation in South-Western Australian Forests

Joe Dortch

BAR International Series 1288
2004

Published in 2016 by
BAR Publishing, Oxford

BAR International Series 1288

Palaeo-Environmental Change and the Persistence of Human Occupation in South-Western Australian Forests

ISBN 978 1 84171 638 1

BAR Publishing is the trading name of British Archaeological Reports (Oxford) Ltd. British Archaeological Reports was first incorporated in 1974 to publish the BAR Series, International and British. In 1992 Hadrian Books Ltd became part of the BAR group. This volume was originally published by Archaeopress in conjunction with British Archaeological Reports (Oxford) Ltd / Hadrian Books Ltd, the Series principal publisher, in 2004. This present volume is published by BAR Publishing, 2016.

Printed in England

BAR
PUBLISHING

BAR titles are available from:

BAR Publishing
122 Banbury Rd, Oxford, OX2 7BP, UK
EMAIL info@barpublishing.com
PHONE +44 (0)1865 310431
FAX +44 (0)1865 316916
www.barpublishing.com

Abstract

This study investigates hunter-gatherer responses to environmental changes in south-western Australian forests. It examines human reactions to terminal Pleistocene and early Holocene expansions of Karri (*Eucalytpus diversicolor*) tall open-forest, a forest type identified as difficult to occupy due to its dense understorey and limited resources. Archaeological and historical evidence for hunter-gatherer occupation of different forest types shows that hunter-gatherers occupied entire forested regions throughout many vegetational changes. They did this by following small geographical shifts in favourable habitats, and by controlling the extent of unfavourable habitat with fire. The extent to which hunter-gatherers controlled vegetation in the past is assessed here from palaeo-environmental and human occupational records at four archaeological sites dated from the last millennium to 47,000 BP.

The study area, the Leeuwin-Naturaliste Region, south-western Australia, is known for its deep limestone cave floor deposits and its intricate mosaic of forest communities. Previous excavations at Devil's Lair and Rainbow Cave revealed hearths, stone artefacts and biotic remains in well-stratified contexts. These records are replicated by new investigations at Tunnel Cave and Witchcliffe Rock Shelter. Devil's Lair and Tunnel Cave are located in present-day Karri forest, Witchcliffe Rock Shelter and Rainbow Cave in coastal woodland and scrub.

Interpretation of radiocarbon assays and stratigraphy indicates that at Devil's Lair, human occupation continued intermittently from 47,000 BP until the entrance collapsed, sometime after 12,000 BP. At Tunnel Cave, there are six hearth complexes built between 20,000 and 12,000 BP, traces of occupation up to 8,000 BP, and a hearth built at 1,400 BP. Witchcliffe Rock Shelter and Rainbow Cave were each occupied by people building hearths between 800 and 400 BP.

Analysis of stone artefact attributes indicating conservation of stone raw material during manufacture or use, which is thought to vary with changes in raw material supply and site occupation patterns, suggest no changes in occupation intensity at any site (although raw material conservation is more evident during terminal Pleistocene marine transgression, thought to have removed access to some sources). The only change in occupation intensity suggested by artefact studies is at Tunnel Cave during the height of the last glacial, when a thick hearth layer was built up at 17,000 BP. Cold and wind perhaps encouraged more frequent cave occupation at this time, but weather conditions alone probably did not cause people to abandon Tunnel Cave as a campsite at 8,000 BP. A more likely factor in cave occupation/abandonment in the Holocene was change in the vegetation surrounding cave sites.

Changes in vegetation structure or habitat are indicated by changes in the proportions of mammal species identified from bone fragments, even after allowing for possible biases created by contributions of hunter-gatherer prey animals in occupation layers. The major prey animals are identified from their stratigraphic associations with artefacts and other evidence of human occupation. The abundance of non-prey fauna varies in ways consistent with their present-day habitat preferences and putative terminal Pleistocene increases in rainfall, which probably encouraged understoreys to become denser. Cessation of human firing, indicated by the cessation of human occupation at Tunnel Cave, could have also enabled understorey regrowth.

Taxonomic abundances of identified charcoal fragments indicate concurrent changes in the floristic composition of the forest canopy as well. Preliminary study of changes in vessel size in the charcoal fragments suggests that these changes coincided with increased uptake of water by Jarrah (*Eucalyptus marginata*) trees. Considering the continental evidence for climatic change, it appears that Jarrah forest or woodland gave way to Karri forest at 10,000 BP due to increases in rainfall, rather than a cessation of firing. Karri-dominated forest encroached totally at Tunnel Cave by 8,000 BP, the same time that people abandoned the site.

These results show primarily that hunter-gatherer site occupation did alter in response to vegetational change. However, research cited in this study shows that people did not abandon the Leeuwin-Naturaliste Region when Karri forest encroached over some parts, and people did use some sites in Karri forest. As in historical time, forest occupation probably involved flexible, short-term site occupations, and therefore localised vegetational shifts had little impact on regional occupation patterns. These findings apply particularly to all regions where vegetation forms a mosaic. Hunter-gatherers prioritised their use of different vegetation communities, and phases of site occupation and abandonment reflect the persistence of a regional occupation pattern that followed the geographic shifts of those communities.

CONTENTS

List of chapters

List of figures

List of tables

Appendices

Figures in Appendices

Tables in Appendices

Please note that the compact disc referred to above and within the text has now been replaced with a download available at www.barpublishing.com/additional-downloads.html

Acknowledgments

This study, carried out as doctoral research in the Centre for Archaeology, University of Western Australia, benefits immeasurably from the help of many people and institutions. I acknowledge them gratefully, with the qualification that errors or inadequacies that may be found are my responsibility alone.

The Department of Anthropology and the Centre for Archaeology, UWA, funded the first excavations and all subsequent laboratory maintenance costs. The Australian Institute of Aboriginal and Torres Strait Islander Studies, Canberra, granted funds for a one year research stipend, two field seasons, numerous radiocarbon determinations, and a laboratory assistant. The South-West Development Commission, Bunbury, funded further radiocarbon age determinations. The Research School of Pacific and Asian Studies at the Australian National University paid a three-month research stipend and other expenses, and the Electron Microscopy Unit, ANU, supported long periods of scanning electron microscope use.

Many people outside the academic field allowed my work to begin. I thank members of the Aboriginal community of Busselton, who approved and supported the work. I particularly thank Wayne and Toni Webb, George Webb, and Mike Hill. Members of the Western Australian Speleological Group directed me to potential sites in Leeuwin-Naturaliste Region caves and rock shelters: Barry Loveday, Rauleigh Webb, Rob Fouldes, Peter Bell, and Peter Wood are foremost. I am much in debt to Barry Loveday for the use of his incredibly detailed map of Leeuwin-Naturaliste Region caves, and for discussion of feasible locations for sites. Staff of the Department of Conservation and Land Management, South-West Capes District, and the Augusta-Margaret River Tourist Association supported me in many ways in the field. I acknowledge the help of Roger Banks, Peter Bell, Rob Klok, and the late Keith Tritton in this respect. Rob Reynolds and Peter Randolph, Department of Indigenous Affairs, are thanked for advice on excavation permits.

For assistance on excavations I am indebted to Caroline Allen, Peter Bell, Bill Bennell, Nelson Boundry, Wendy and Jeff Brown, Manfred and Jesse Bruggler, Shane Burke, Katherine Crisp, Kier Davis, Sandy Davis, Robyn Dennis, Scott Dethlefsen, Adrian Di Lello, Peter Dorrington, Charlie Dortch, Martin Gibbs, Sarah Grimes, Loraleen Kelly, Andrew and Mary Hancock, Fiona Hook, Ryan Hovingh, Jo Jordan, Julie McKay, Sally McGann, Jenny Potts, Jenn Richman, John Selby, Robin Stevens, Emma Suerz, Bruce Veitch, Jackie Watt, and Peter, Anne, Kate, Carlen and Michelle Wood. For assistance in many hours of laboratory sorting, I thank Caroline Allen, Shane Burke, Kier Davis, Adrian Di Lello, Charlie Dortch, Rebecca Gagliardi, Amy Gardos, Wayne Glendenning, Sarah Grimes, Ryan Hovingh, Melissa Johnston, Justin MacDonald, Julie McKay, Sally McGann, Christine Martin, Monique Pasqua, Kathryn Robinson, Erica Rowley, Ian Ryan, Robin Stevens, and Emma Suerz. For logging stone artefact attributes, I thank Christiana Jones, Ben Marwick, and Ian Ryan.

For their hospitality during fieldwork, I thank Wendy and Jeff Brown, Manfred and Robbie Bruggler, Christa Cameron, David Newstead, David Samson, and Peter and Anne Wood, and for accommodating me during further study in Canberra, I thank John Chappell.

I am perhaps most indebted to those colleagues who provided the intellectual environment without which this study would never have materialised. Charlie Dortch, my mentor and fellow-student, encouraged me to begin this study and always gave sound advice. My thesis supervisors, Jane Balme, Sandra Bowdler, David Bulbeck, Ian Lilley, and Sue O'Connor, fostered rigorous thinking about my topic. I owe a considerable debt to many other colleagues for their stimulating comments: Jim Allen, Shane Burke, John Chappell, Zoe Chappell, Richard Cosgrove, Nick Dolby, David Frankel, Amy Gardos, Jack Golson, Richard Gould, Tom Higham, Peter Hiscock, Alan Hogg, Simon Holdaway, Rhys Jones, Jo Kamminga, Duncan Merrilees, Cait Mitchell, Al Patterson, Jenn Richman, Mike Smith, Moya Smith, Robin Torrence, Lynley Wallace, and Peter White. I especially thank Jo Kamminga, who instructed me in macroscopic use-wear analysis of stone artefacts, and Mike Smith, for suggesting that I investigate the potential of charcoal fragments as palaeontological evidence. For allowing me to cite unpublished age determinations, I thank Paul Abell, Gifford Miller, Bert Roberts, Mike Smith, Chris Turney, and Bill Wilson.

Alex Baynes, George Kendrick, and Liz Pickett are owed special thanks for their informed and constructive comments on all aspects of south-western Quaternary research. For advice on palaeontological matters generally, I thank Ken Aplin, Julian Ash, Bob Baird, Doreen Bowdery, John Chappell, John Dodson, Joe Gentilli, Geoff Hope, Richard Hubbard, Geoff Hunt, John Long, Ernie Lundelius, Mike MacPhail, Jim Mead, Giff Miller, Jane Newsome, Brett Smith, Ian Thomas, and Karl-Heinz Wyrwoll. Brenton Knott and Jamie O'Shea answered zoological questions; Glen Kelly, Craig Macfarlane, Gabe Magyar, and David Robertson directed me to references on forest ecology and plants; and John Glover gave me geological advice. For help with technical aspects of charcoal and wood identification, I thank Kingsley Dixon, Phil Evans, Ian Godfrey, Jugo Ilic, Bill Loneragan, Emer O'Gara, Ian Rotheram, and Lins Vellen; for practical assistance with electron microscopes and photography, I am indebted to Frank Brink, Tammy Hasselfeldt, Roger Heady, Sally Stowe, and David Vowles.

I sincerely thank Mary Dortch, whose careful comments improved the first drafts of this work immeasurably. Finally, I thank my wife, Dirima Cuthbert, who I mention last only to mark her unique contribution. Without her optimism and unflagging support, I would not have written the same work, and I dedicate it to her.

Chapter 1 Introduction

This study investigates hunter-gatherer responses to environmental change in south-western Australian forests. It examines how hunter-gatherers reacted to terminal Pleistocene and early Holocene expansions of Karri (*Eucalytpus diversicolor*) tall open-forest, a forest type identified as difficult to occupy (Hallam 1975, Ferguson 1985). The putative hunter-gatherer reaction requires careful assessment because past hunter-gatherers could have continued to occupy forested areas by using many different habitats within forests and controlling the extent of unfavourable habitats by firing. I assess the issue by reviewing ecological and archaeological research in south-western and south-eastern Australian forests and analysing archaeological evidence for occupation in various types of forest. The study region is the Leeuwin-Naturaliste Region, extreme south-western Australia (Figure 1.1).

This region is chosen because palaeo-ecological studies there indicate terminal Pleistocene changes in vegetation (Balme *et al.* 1978, Merrilees 1984), and because its limestone cave floor deposits contain detailed evidence of Aboriginal occupation (C.E. Dortch 1979a, J. Dortch 1996, Lilley 1993). These cave floor deposits are stratified and include large accumulations of well-preserved archaeological and palaeontological material. They record changes in campsite occupation and local forest type extending from the last millennium to 47,000 years BP (Dortch and Dortch 1997, Turney *et al.* in press). I interpret these accumulations in terms of their geographical positions within changing forest habitats and hunter-gatherer site occupation patterns.

The wider importance of this study is that it assesses apparently conflicting truisms of environmental archaeology: hunter-gatherers manipulate the environment, and they are constrained by it. These assertions have featured in the archaeology of small-scale societies since J. Iversen's palaeo-environmental reconstructions in the 1930s (Bryant 1996). The present-day study of landscape archaeology, or landscape socialisation, sees society and environment as influencing one another (Gosden and Head 1994; Head 1989, 1994). Thus hunter-gatherers can perpetuate their use of a region's habitats, and only drastic environmental changes upset the existing social and economic order. In this study, I show that hunter-gatherers continued to occupy a region in much the same way, even though they abandoned small parts of that region. It is critical to distinguish regional and local scales. On the regional scale, human occupation in forests persisted throughout vegetational changes. On the local scale, campsites in some forested locations were abandoned as the forest type changed around those locations. How these two phenomena, continuity and abandonment, reflect the same use of the environment is therefore the widest possible subject of this study.

Research background

Following other environmentally-oriented archaeological studies, I assume that hunter-gatherer societies must have reacted when climatic changes induced alterations in the configuration or accessibility of plant and animal resources (cf. Allen and O'Connell 1995, Cosgrove *et al.* 1990, Straus *et al.* 1990). Human reactions to losses of foraging area and relocation of littoral or periglacial zones due to marine transgression or glaciation are not in dispute. Less evident, however, are the possible reactions to vegetational change. I shall consider forest expansion here.

The view that forest expansion caused people to abandon regions makes apparent sense (Cosgrove *et al.* 1990, Driver *et al.* 1996, Ferguson 1985, R. Jones 1971). Hunter-gatherer groups probably cannot subsist wholly within the forest of any region. Other vegetation types generally provide more food, since the productivity of forest is concentrated in inedible woody growth and the relatively inaccessible canopy (Bailey *et al.* 1989, Byrne 1983, Head 1989, Headland 1987, McBryde 1978). Forest distribution does appear to determine human population distribution. Moreover, there is archaeological evidence for hunter-gatherers abandoning and re-colonising forest regions following world-wide glacial expansion, *c.*22-18,000 BP, and glacial contraction, *c.*15,000-5,000 BP. Forest expansion in the early to mid-Holocene is interpreted as causing people to abandon sites in south-western Australia, Tasmania, and high latitude boreal forests in North America (Bailey *et al.* 1989, Cosgrove *et al.* 1990, Driver *et al.* 1996, Lourandos 1983a, Ferguson 1985, R. Jones 1971, B.D. Jones 1994).

However, two factors suggest that people did not react to vegetational change simply, and that human responses to vegetational change should be argued in detail. Firstly, the activity of modern hunter-gatherers suggests that past hunter-gatherers fired vegetation to suit their needs (Head 1989, Jochim 1996, Lourandos 1985, Ross 1985). Firing vegetation would have promoted growth of food plants and improved access to plants and animals (Clark 1975, Dimbleby 1961, Gould 1971, Hallam 1975, R. Jones 1969, Lewis 1973, Mellars 1976, Mellars and Reinhardt 1978, Patterson and Sassaman 1988, Simmons *et al.* 1981, A.G. Smith *et al.* 1989, Thomas 1994). Palaeo-ecological research suggests that Australian forest structure changed swiftly following European settlement, even before widespread clearing and regrowth, suggesting that the primary cause was the demise of traditional Aboriginal firing patterns (Bowman and Brown 1986, Head 1989). The questions therefore raised are, if Aboriginal hunter-gatherers' environmental control was so far-reaching, could it have slowed or contained environmental change? Could it have allowed occupation of sites to continue wherever forest environments changed?

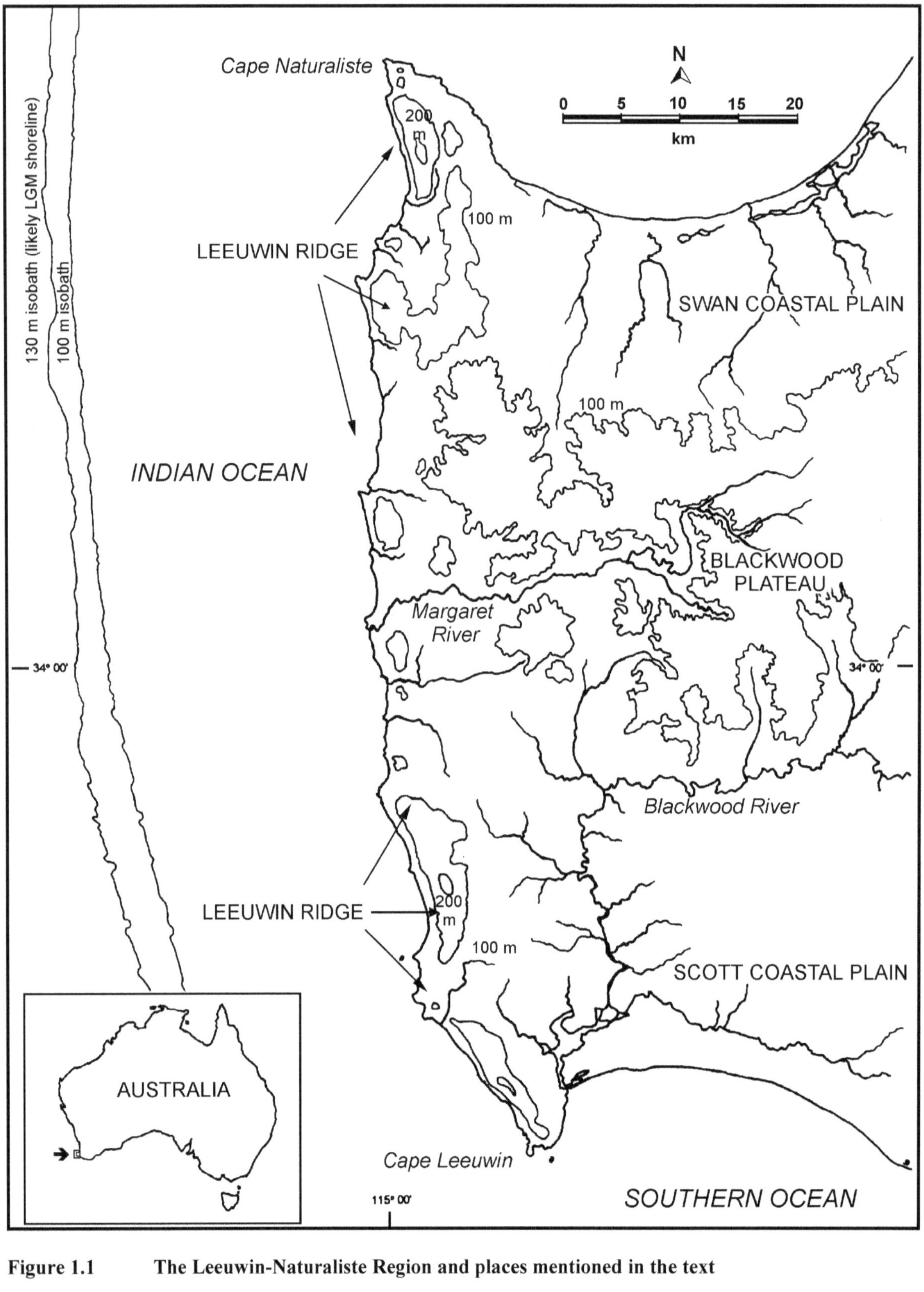

Figure 1.1 **The Leeuwin-Naturaliste Region and places mentioned in the text**

Secondly, arguments for vegetational change limiting human populations ignore ecological complexities. Certainly, foraging people may distribute themselves according to their differential use of a region's various habitats (cf. Deacon 1995, Byrne 1983, Head 1989, Thomas 1993). A corollary assumption is that some habitats, especially forests, severely limit or even exclude human occupation (Bailey *et al.* 1989, Cosgrove *et al.* 1990, Cosgrove 1995, Driver *et al.* 1996, Ferguson 1985, Hallam 1975, Headland 1987, R. Jones 1971, B.D. Jones 1994, McBryde 1978). However, in south-western Australia at least, forests are sufficiently varied to allow hunter-gatherers to find, in any locality, large patches of more favourable vegetation, *e.g.*, woodland or open-forest (see Chapters 2, 4). Annual ranges of hunter-gatherer groups could have passed through or around various interfingering forest types depending on their relative attractions. The question of forests being difficult to occupy is partly answered if the most difficult forest types have limited extent. South-western Australian forest types are also internally heterogenous. The region's plants are highly adapted to small differences in soils, topography, and climate, resulting in varied associations of plants, including edible species, that change every hundred metres or so. This diversity and heterogeneity is assessed in the next chapter.

To summarise, I examine the proposition that hunter-gatherers occupied forests in ways that varied according to how difficult they found each type of forest. The scale or intensity of forest occupation, among other things, certainly appears to have varied from one forest type to another. In any region, hunter-gatherers may identify, on one hand, forests that are attractive, full of resources, and easy to manage, and on the other, forests that are extremely limited in terms of resources, accessibility, or manageability. The scale or extent of site occupation would have depended on the forest type around sites, but in south-western Australia at least, this should not be assumed to be a simple or wholly deterministic relationship. I therefore identify the scale and duration of Leeuwin-Naturaliste Region site occupations, compare these characteristics with the evidence for types of forest at each period of occupation, and assess these results against ecological, palaeo-environmental, and archaeological evidence from the wider region of south-western Australia.

Approach

In this thesis, the vegetational changes of interest are putative Holocene expansions of Karri (*Eucalyptus diversicolor*) tall open-forest, which now extends over southern parts of south-western Australia (Beard 1981, Churchill 1968). Hallam (1975) and Ferguson (1985) propose that Aboriginal hunter-gatherers avoided Karri forest. The period of interest is the terminal Pleistocene and the Holocene, when in some localities, woodland may have given way to Karri tall open-forest as a result of regionally important climatic changes.

As mentioned, the limestone cave sites in the Leeuwin-Naturaliste Region are chosen for study because few other sites in south-western Australia offer well-preserved stratigraphy and biotic remains (Dortch and Dortch 1997). The fine stratigraphy, deep deposit, and well-documented archaeological and palaeo-environmental record at one cave site, Devil's Lair (Balme *et al.* 1978, Dortch 1979a), provided at least one site to infer a long and detailed record of human occupation in forest. Its presence suggested that similar sites exist in the region, and here I report my surveys and test-excavations aimed at locating sites for comparison to the Devil's Lair record (Chapter 5; J. Dortch 1996).

The region is also known for its "intricate mosaic" of diverse vegetation communities (Beard 1981: 194), enabling comparison of occupational records at sites that are located close together in the same limestone belt. Devil's Lair and a Pleistocene site that I located at Tunnel Cave (Figure 5.2) are located in present-day Karri forest, therefore offering the opportunity to infer site occupation and abandonment in what Hallam (1975) and Ferguson (1985) identify as the most difficult forest type in the region. Two other sites, Witchcliffe Rock Shelter and Rainbow Cave, offer the opportunity to study, as comparison, hunter-gatherer occupation in coastal woodland and scrub (J. Dortch 1996, Lilley 1993).

In these four sites, concentrations of artefacts, items that were carried only by people, and hearths indicate human occupation. Absence of these items is negative evidence, difficult to assess in test-excavations representing small parts of large sites. However, trends in stone artefact types and debitage characteristics can indicate changes in intensity of human occupation.

Stone artefacts are generally produced or maintained more carefully during large or long site occupations, when stone raw material for making artefacts is not replenished by visits to stone quarries (Bamforth 1986, Callow 1986, Dibble 1987, Hiscock 1996a, Rolland and Dibble 1990, Roth and Dibble 1998, Walsh 1998). In this investigation, I search for changes in stone artefact production and maintenance and raw material conservation. I address probable changes in physical access to raw material, caused by marine transgression. Allowing for changes in physical access, analysing trends in artefact characteristics can indicate trends in site occupation.

By comparing evidence for human occupation, and changes in intensity of human occupation, with evidence for changes in vegetation structure and composition, I show whether people occupied different forest types and hence infer their extent of occupation in those types. I infer the types of vegetation growing at the time of those layers' formation, by identifying mammals and forest trees whose remains are abundant in these deposits in the form of bones and charcoal. Vegetation structure (openness of canopy, understorey, and other layers) and composition are two major characteristics. Mammal remains indicate changes in mammal habitats (on the assumption that present mammal habitat preferences have not altered) and hence general vegetation structure (Balme *et al.* 1978). Charcoal fragments indicate forest trees, or composition.

With the variety of forest types across south-western Australia, one might expect that people avoided the less favourable, but I shall present some evidence that they did not, after all. From this, and a discussion of ecological and palaeo-environmental factors, I conclude that people were more capable of controlling forest than has been supposed. Decisions to abandon sites were not regionally significant, and are consistent with the continuation of the same pattern of forest occupation.

Chapter outline

Chapter 2 provides an ecological background to these propositions by reviewing the ecology of Karri and other forests, describing the Leeuwin-Naturaliste Region environment, and the habitat preferences of south-western Australian mammals. Chapter 3 summarises climatic changes across Australia, from the Pleistocene to the Holocene. I discuss evidence for these changes in south-western Australia and suggest how they may have affected the environment in the Leeuwin-Naturaliste Region. Chapter 4 appraises archaeological research in south-eastern and south-western Australian forests and shows that the question of hunter-gatherer occupation in forests is complex and deserves detailed analysis, as attempted here.

The subsequent analysis is based on detailed examination of archaeological and biotic remains from Devil's Lair, Tunnel Cave, Witchcliffe Rock Shelter, and Rainbow Cave. Methods for identifying and excavating archaeological deposits are described in Chapter 5.

In Chapter 6, I argue that artefacts and hearths indicate a human presence or occupation in the site during the formation of layers they are found in. On stratigraphic evidence, I show which layers can be grouped together, and present arguments for combining or distinguishing combined radiocarbon age estimates. In Chapter 7, I identify periods of intensive site occupation. Analysis of stone artefacts indicates variations in the conservation of stone raw material, relating to the duration of occupations and physical access to raw material.

In Chapters 8 and 9, I analyse faunal and botanic remains to infer the vegetation growing near the site during the formation of each layer. In Chapter 8, I allow for the different contributions of animals from various predators that deposit bones, and on the argument that certain marsupials require open or closed vegetation (e.g., for shelter or access), I infer palaeo-vegetation structure. In Chapter 9, I assess the effect of other types of depositional bias on the representation of charcoal taxa, and identify dominant trees growing at the time of charcoal deposition. In Chapter 10, the conclusion, I summarise my analyses and assess their implications.

Chapter 2 Forests in south-western Australia

This chapter first of all details south-western Australian forest types. It brings out the differences between them that could emerge as significant to Aboriginal hunter-gatherers - differences in their food and other resources, their accessibility, and the ease or otherwise with which they could be controlled by fire.

Secondly, this chapter describes how the forest vegetation develops in relation to widespread and long-term climate change and to environmental conditions such as precipitation.

Thirdly, this chapter shows how the vegetational changes of the past can be inferred from fossil remains. These fossils are of the animals whose present-day ranges are restricted to certain vegetation formations, and of the plants themselves. In conclusion, the chapter summarizes the influence that humans could have had on vegetation.

Ecology of south-western Australian forests

This section reviews ecological studies, which indicate whether various forest types are unattractive environments for human occupation, compared with other vegetation formations in south-western Australia. The ecological literature suggests constraints on human activity in the forests, such as the quantity or quality of potential food resources, the ease of movement in them, and their susceptibility to control or alteration by firing. This review begins with the continentally-applied definitions of Australian forests (Groves 1994).

Australian forests

Forests in Australia are either "closed" or "open", according to the amount of sunlight that is blocked by foliage, or *projective foliage cover* (pfc; Groves 1994). Closed forests, also known as rain-forests, have no bearing on this study, because the virtual lack of southern Australian rain-forest taxa such as *Nothofagus* or *Podocarpus* (represented by one species of shrub) suggests that these forest types have been absent from south-western Australia for millions of years (Christensen 1992: 47, Ashton and Attiwill 1994: 167).

In Australia, all open-forests and woodlands are sclerophyllous, that is, the plants have hard, leathery leaves. The characteristic sclerophyllous forest tree in Australia is the eucalypt (*Eucalyptus*), a genus[1] containing 700 species and virtually endemic to New Guinea and Australia. Other important sclerophyll taxa are *Acacia* and *Casuarina*. Sclerophylly is an adaptation to nutrient-poor soils and frequent drought (Groves 1994). Australian sclerophyll forests include the inland and coastal forests covering much of eastern Australia, many parts of Tasmania, and all of south-western Australia; examples of sclerophyll forests outside Australia are the Northern Hemisphere, high latitude boreal forests. Generally, sclerophyll forests grow on nutrient-poor soils where annual rainfall is between c. 500 and 1500 mm (Groves 1994).

One can further divide Australian sclerophyll forests, on the basis of the understorey shrub layer, into wet sclerophyll and dry sclerophyll (Groves 1994, Specht 1970). In wet sclerophyll, the shrub understorey is generally dense and fast-growing, and it can include non-sclerophyllous or soft-leaved plants. In dry sclerophyll, the shrub understorey is largely sclerophyllous and it is open and slow-growing. Both types exist within large regions. Local variations in soils, rainfalls, and topography determine whether the vegetation is wet or dry sclerophyll forest.

These same factors determine the height of the trees, leading to the current use of Specht's (1970) system based on physiognomy. His terms "tall open-forest" and "open-forest" relate to wet and dry sclerophyll forests respectively. Tall open-forest refers to forest where the dominant trees are greater than 30 m height; open-forest to forest where the dominant trees are between 10 m and 30 m height (Ashton and Attiwill 1994, Gill 1994). In both types of open-forest, the pfc is 30-70%.

Sclerophyll plants conserve resources by growing slowly and having hard leaves resistant to solar and biological damage (Groves 1994). Nutrient cycling is important but it takes place slowly since the hard leaves and twigs decompose slowly. Aids to nutrient cycling are the rapid life cycles of understorey plants, which may have a symbiotic relationship with trees, and the occurrence of fire (Hingston *et al.* 1989).

Fire is common in most forests, but perhaps most frequent in drier, sclerophyll vegetation (Ashton and Attiwill 1994, Bell *et al.* 1989, Christensen and Abbott 1989, Gill 1994, Gill *et al.* 1981). Lightning-strikes, frequent in the southern Australian summer, are the major natural cause of fires (Ashton and Attiwill 1994: 169, 187; Gentilli 1989: 36). The scale of adaptations to fire among sclerophyll forest plants indicates that fire has recurred in forests since they spread across Australia in the late Tertiary (Kemp 1981). Some researchers deduce from rich charcoal horizons in pollen cores that forest fires intensified after human arrival in Australia (Kershaw 1986, Singh *et al.* 1981). A major element in this deduction is the much-vaunted relationship between humans, vegetation, and fire known as the "fire-stick

---

[1] Some botanists, not all, have elevated the several sub-genera of *Eucalyptus* to genera. Thus, *E. calophylla* is also known as *Corymbia calophylla* (Brooker and Kleinig 1990). As the older classification is understood by many, I have used it in this study.

farming" hypothesis, expressed by researchers in ecology (Jackson 1968), ethnography (Tindale 1959), archaeology (R. Jones 1969), and palaeontology (Merrilees 1968). Put simply, this hypothesis is that Aboriginal people enhanced natural food production by firing vegetation.

The fire-stick farming hypothesis simplifies complex and incompletely known human and natural factors (Bowman and Brown 1986), but finds broad support from ethnographic and ecological research (e.g., Bowman 1998, Bowman and Panton 1993, Gould 1971, Thomas 1994). Some plants in open-forest and tall open-forest depend on heat or smoke to stimulate seed release and germination (Bell *et al.* 1989, Dixon *et al.* 1995). Other plants colonise fire-cleared areas rapidly, attracting grazing animals, including human prey. Both the fire-dependent and the rapid colonising plants include human foods. Fire is also a tool for removing inflammable vegetation around pockets of closed vegetation containing certain plant foods and providing shelter to prey animals (Bowman 1998). Used carefully, fire therefore protects potential resources until the time comes to use them, when it is then the means of flushing out prey. Finally, firing dense undergrowth removes serious obstacles to the easy movement of people.

The dependency of vegetation on human firing, rather than natural ignitions, is seriously questioned (Horton 1982). This question is addressed in Chapter 4. In either case, whether applied by people or sparked by lightning, fire is a factor in the growth of tall open-forest and open-forest. In tall open-forest, the success of eucalypt seedlings depends partly on fire effects such as the creation of gaps in the canopy, the volatilisation of nitrogen and phosphorous from heating of the soil, and the death of soil microbes which retain nutrients and attack seedlings (Ashton and Attiwill 1994). The forest's recovery from fire is enhanced by nitrogen-fixing, colonising understorey plants (legumes and *Acacia* species). Rapid growth of these plants prevents the loss by erosion of volatilised nutrients and returns the nutrients in the form of leaf litter (Attiwill and Leeper 1990).

Plants in tall open-forest have various sensitivities to fire. Legumes and *Acacia* species usually regenerate from soil-stored, fire-resistant seeds (Ashton and Attiwill 1994). However, these plants produce seed only after three or four years' growth, so fires recurring over shorter periods eliminate these plants (Christensen and Abbott 1989). Other plants produce no fire-resistant seed and recolonise burnt areas from adjacent unburnt patches, which are preserved according to the influences of topography and wind direction during the fire (Attiwill and Leeper 1990). Given that certain seasonal winds promote fire, some locations are more fire-prone than others. Some experience no fire for decades while others burn after only one or two successional stages. In a well-established tall open-forest which experiences fire, there is a structural and floristic range in the understorey and canopy communities. There are sharp, fire-influenced boundaries between patches of different plant communities with different fire sensitivities or dependencies (Ashton and Attiwill 1994: 181).

In tall open-forest, fires are often intense and short-lived (Ashton and Attiwill 1994). Seeds of many tall eucalypts are released and germinate only after scorching by fires large enough to touch the canopy and yet too short-lived to destroy the seeds. For example, more than 30 seconds of exposure to the very high temperatures of a fire spreading into the canopy is probably fatal to an *E. regnans* seed inside its seed case (Ashton and Attiwill 1994: 181). A canopy fire can kill the parent tree and still stimulate seed germination (Attiwill and Leeper 1990: 185).

Such intense fires are naturally short-lived in well-established tall open-forests, as the range of structure and floristics creates a range of fuel quantity and quality (Attiwill 1994). There is a range of fire frequencies and intensities corresponding to the patches of different fire-sensitive plant communities. Eliminating diverse fire regimes has tested this proposition. When late 19th and 20th C forest managers attempted to introduce a uniform fire regime, i.e. no fires at all, the result was forest devastation over large areas.

All over Australia, deliberate exclusion of fire from settled regions allowed native vegetation to grow vigorously and densely. Other factors were probably also involved (Horton 1982), but lack of frequent fire was a proximate cause (Bowman and Brown 1986). Where there had been grassy and open understoreys, thick scrub appeared decades after European settlement (Gell *et al.* 1993, Howitt 1890, Thomas 1994). Throughout the late 19th C and much of the 20th C, catastrophic wild-fires resulted from massive accumulation of litter and logging waste, growth of understoreys, and the outbreak of fire from settlements and railways (Gill and Moore 1997, Lang 1997). In the summer of 1960-61, despite the introduction of controlled burning eight years before, fires ignited by lightning burnt 140,000 hectares in south-western Australia and razed the towns of Dwellingup and Karridale (Gill and Moore 1997: 25, 33). As late as 1983, a single fire in Victoria burnt 1.4 million ha, or 13% of the state (Ashton and Attiwill 1994: 169).

If there were wild-fires as large as these in the prehistoric past, people might have been at some risk. Clearly they did escape them, but a large extent of such fires implies a large extent of dense, inaccessible understoreys. This does not appear to have been the case when Europeans first saw southern Australian forests. In south-western and south-eastern Australia, numerous historical documents attest to open understorey in tall open-forest becoming dense after long periods of European settlement and fire-exclusion (Attiwill 1994). In any case, in the absence of rigorous fire suppression, it seems likely that large regions would have experienced enough lightning-ignited fires to have produced some variety in the geographic extent of various tall open-forest structures. A regime of few or no fires, as was actively sought in the first half of the 20th C, seems unlikely in any prehistoric situation.

If one extreme fire regime, a total lack of fires, is in the long run catastrophic, the other extreme is also undesirable, ecologically. Very frequent fires destroy legumes and vulnerable saplings of the large eucalypts, and deplete soil nutrients, removing them by convection and post-fire erosion (Attiwill and Leeper 1990, Jackson 1968). If frequent fires were extensive, soil would be mobilised over large areas.

In tall open-forests, frequent burning encourages the most fire-dependent plants, which promote fire by virtue of their rapid growth and natural flammability. Even though many plants seed only after fire, few tolerate very frequent fires - the eucalypts mentioned above, for example, only produce viable seed from age 10-15 years. Thus frequent burning over very large areas converts tall open-forest into dense scrub composed of "fire-weed" shrubs (Attiwill 1994, Attiwill and Leeper 1990). Dense scrub could be eliminated by extremely frequent burning - every year or two - to promote grasses (cf. Ward 1998, on Jarrah open-forest), but this frequency would seem difficult for people to maintain in high-rainfall areas where in some years the soil dampness prevents any fire (Burke 1997, Burrows 1987).

In open-forests, where soils and litter are drier, and summers are longer and still promote thunderstorms (Gentilli 1989), fires are naturally more frequent. Summer drought encourages another important set of adaptation towards conserving nutrients. Trees withdraw nutrients from leaves that are dying (Gill 1994). Nutrient deficiency encourages diversity of understorey plants (Beard 1983, Bell and Heddle 1989, Marchant 1973, Tilman 1983) and slow growth and efficient cycling of nutrients in litter (Hingston *et al.* 1989). These adaptations are also adaptations to fire, based around conserving nutrients and *resisting* destruction by fire (Dell *et al.* 1986, Bell *et al.* 1989). Many understorey plants regenerate after fire from underground organs called lignotubers (Bell *et al.* 1989). Trees resist burning with thick bark that scorches but protects the trunk (in contrast, tall open-forest trees have thin, inflammable bark). If fire damages their crowns, they sprout leaves from the trunk (epicormic growth). The effect of a fire in open-forest is less of a transformation than in tall open-forest, since in the former, plants seem to protect their investment in precious resources (Dell *et al.* 1986). Thus after fire they either continue to grow or re-sprout from trunks or lignotubers.

In assessing forests' dependency on fire, Attiwill and Leeper (1990) concede that both tall open-forest and open-forest trees can grow for many decades without fire. For example, eucalypts' fine (adventitious) roots can absorb nutrients from unburnt forest litter. But the role of fire is vital - it regenerates trees in tall open-forest, some species seed only after fire, and patchy firing promotes a mosaic structure and hence diversity. Fire is variable and influential in the growth patterns of sclerophyll forests. It controls the long-term destiny of large forest areas, as summarised by Attiwill and Leeper (1990):

As ecologists, we therefore view fire as an inherent part of the development and maintenance of the Australian flora. ... By its controlled use, we can regenerate forests over large areas or within small clear-felled areas ('coupes'). We can to some extent avert the major forest fire by burning the fuel accumulated in the litter and understorey at regular intervals. There are major arguments for and against such practices, and many of them speculative since the effects of fire on many organisms and processes are not known. Furthermore, 'fire' is variable; we should argue in terms of fire regimes, *a term proposed by Gill (1975) to include the* frequency *of fire, the* intensity *of fire, and the* season *in which the fire occurs.*

The choice of fire regime has a direct influence on the ecology of the forest. The influence of excluding fire might be just as great as the influence of allowing fire. A few examples illustrate the point:

(1) the effect of frequency of fire is illustrated by plant communities in the Florentine Valley, Tasmania (Gilbert 1963):

at a frequency of 350-400 years, rain-forest forms; the frequency is close to the life-span of most eucalypts

at a frequency of 100-350 years, eucalypt forest is retained with an understorey of rain-forest species [=tall open-forest]

at a frequency of a few years to 100 years, eucalypt forest is retained with an understorey of Pomaderis, Olearia, *and* Acacia *[=open-forest]*

*at a frequency of less than a few years, buttongrass plains are maintained. Buttongrass (*Gymnoschoenus sphaerocephalus, *Family Cyperaceae) is actually a sedge and tends to grow on shallow impoverished soils; Tussock Grass (*Poa australis*) occurs where the soil is deeper*

*(2) Alpine Ash (*Eucalyptus delegatensis*) in the higher altitudes of Tasmania may decline in vigour and die. In the absence of fires, rain-forest species develop in the understorey, as in (1) above. Where this understorey was felled and burned, dieback of the mature* E. delegatensis *was markedly reduced and growth was enhanced (Ellis* et al. *1980).*

*(3) High-intensity fire in Jarrah forest (*E. marginata*) in Western Australia changes the proteaceous understorey to leguminous. Although the trees are completely scorched by fire, their growth after the fire is enhanced and it has been suggested that legumes decrease the*

susceptibility of E. marginata *to the dieback fungus* Phytophtora cinnamoni *(Shea* et al. *1979)*
(4) The growth of herbaceous species following surface fires in E. regnans *forest is often luxuriant (Ashton 1981: see Plate 12.9). It is possible that with more frequent fires of lower intensity in the past, the understorey over substantial areas of these forests may have been herbaceous and grassy rather than woody and shrubby. (Attiwill and Leeper 1990: 186)*

South-western Australian forests

The above quotation indicates a potential for change in wet and dry sclerophyll vegetation in south-western Australia. For example, supporting the suggestions made above, Ward (1998) cites historical evidence that annual or biennial burning in Jarrah open-forest and woodland once helped maintain a grassy understorey. The following section assesses human and environmental influences on south-western Australian vegetation, and its attractiveness for human occupation.

In south-western Australia, open-forests, tall open-forests, and woodlands are the most extensive formations (Figure 2.1). They cover a 40,000 km² area (Dell *et al.* 1989) known as the Darling Botanical District (Beard 1981). This district includes four sub-districts that are convenient for illustrating the distribution of resources used by people. The sub-districts are Drummond, the coastal woodlands and scrub on the west coast; Dale, the northern, drier open-forest; Menzies, the southern, wetter open-forest; and Warren, a combination of the wettest and southernmost open-forests, all the tall open-forest, and coastal woodland, heath, scrub, and sedgelands. The Warren Sub-District extends 300 km south-east from Cape Naturaliste to King George's Sound (Albany). At its western end it includes the Leeuwin-Naturaliste Region, the focus of this study.

The Darling Botanical District has a Mediterranean climate, with hot dry summers and cool wet winters. Plants in Mediterranean climates are adapted to seasonal drought, fire, and poor soils. Of the five Mediterranean climate regions (the Mediterranean basin, California, Chile, south-west Africa, and south-western Australia), the soils in south-west Africa and south-western Australia are the poorest (Lamont 1995), yet the trees in south-western Australia are the tallest and the forests there most extensive. Such dry, nutrient-poor forests would seem to pose more difficulties for human occupation than forests in other regions. However, the paradox of an environment with poor nutrients is botanical diversity (Tilman 1983). This paradox is significant to people, as shown below.

In south-western Australia, plants must be highly adapted to cope with the combination of Mediterranean climate and poor soils. These conditions, in combination with periods of extreme aridity and isolation of the region in the geological past, have encouraged c. 500 plant genera known for south-western Australia to diversify into 3611 recorded species, of which 2841 (79%) are endemic (Beard 1981: 88-91). High diversity arises from poor conditions because of competitive pressure and because no single species can command enough of a given nutrient to dominate a locality (Flannery 1995, Krebs 1985, Tilman 1983). Small differences in soils and climate are significant to any new varieties that can take advantage of them. The aggregate of thousands of plant species' various responses to subtly different niches even within a single forest type means that the south-western vegetation is diverse, although not as diverse as vegetation on equally impoverished soils in more rugged Mediterranean regions, such as South Africa (Beard 1981). The highly endemic south-western vegetation helps define a botanical region defined by Beard (1981) as the South-West Botanical Province. The Darling Botanical District is the south-westerly and forested part of this Province.

The following section addresses the ecology, particularly the fire ecology, of two of the major south-western Australian forest types. It will be shown below that the ecologies have implications for the foods that people could have obtained from each of the forest types and how easily they could have moved in them. For reference, the several south-western Australian forest types indicated in Figure 2.1 are listed in Table 2.1, below.

The Jarrah forest

The Jarrah forest is the most extensive forest in the Darling Botanical District. It is dominated by Jarrah (*Eucalyptus marginata*) and Marri (*E. calophylla*; Beard 1981, Dell *et al.* 1989). Since Marri trees co-dominate with Jarrah in this forest, its name could equally be the Jarrah-Marri forest, but according to general usage, the name used here will be Jarrah forest.

The Jarrah forest extends 400 km north-south and 300 km east-west, thus featuring significant rainfall, temperature, and seasonality gradients in both directions (Dell *et al.* 1989). It covers a triangular area with apices at Toodyay, Cape Leeuwin, and Albany (Figure 2.1). The substrate is largely nutrient-deficient laterites and unconsolidated quartz sands. Across the Jarrah forest, the summer drought varies from four to six months, and winter rainfall ranges from 600 to 1,000 mm, with rainfall higher and more reliable in the south and west (Gentilli 1989). The southern Jarrah forest, corresponding to the Menzies Botanical Sub-District, probably includes the optimum physical environments for Jarrah trees, since it is here that Jarrah trees are most dominant and tallest.

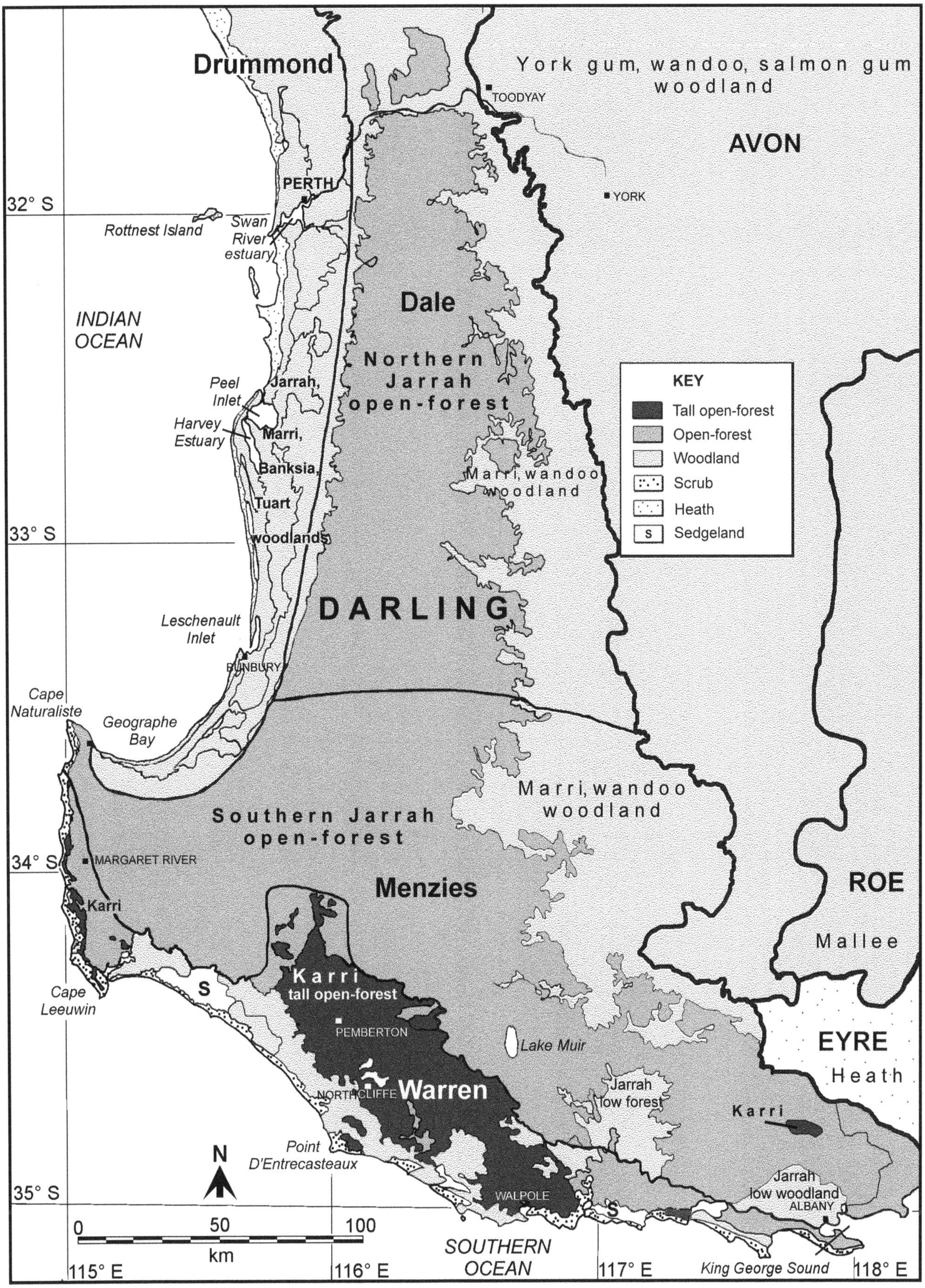

Figure 2.1 **Map of the major vegetation associations, botanical districts and sub-districts, and other locations, in the most south-westerly part of the South-West Botanical Province (after Beard 1981).**
DARLING = Botanical District ; Warren = Botanical Sub-District.

Table 2.1 Summary list of south-western Australian forests and woodlands and dominant trees. The general usage of common tree names (several are Anglicised Nyoongar names) eliminates confusion caused formerly by the use of descriptive names such as blue gum, white gum, box, mahogany, red gum, etc (Beard 1981). They are therefore used here as a shorthand, thus Karri forest, Tuart woodland, etc.

Structural types	*Dominant canopy trees (common examples)*
Tall open-forest	Karri; Karri-Marri
Mixed open-forest	Karri-Jarrah; Karri-Jarrah-Marri; Jarrah-Marri; Jarrah-Yarri
Open-forest	Jarrah; Marri; Jarrah-Marri
Tall woodland	Tuart
Woodland	Marri; Tuart; Marri-Wandoo; Jarrah; Wandoo; York gum; Salmon gum
Low woodland	Jarrah-Banksia; Banksia; Peppermint; She-oak

Common name	*Scientific names*
Karri	*Eucalyptus diversicolor* F. Muell.
Marri	*E. calophylla* R. Br. (*Corymbia calophylla*)
Jarrah	*E. marginata* Sm.
Yarri (Blackbutt)	*E. patens* Benth.
Tuart	*E. gomphocephala* DC.
Wandoo	*E. wandoo* Blakely
York gum	*E. loxophleba* Benth.
Salmon gum	*E. salmonophloia* F. Muell.
Peppermint	*Agonis flexuosa* (Spreng.) Schau.
She-oak	*Allocasuarina fraseriana* (Miq.) L. Johnson
Karri she-oak	*Allocasuarina decussata* (Benth.) L. Johnson
Bull banksia	*B. grandis* Willd.
Candle banksia	*B. ilicifolia* R. Br.
Holly-leaf banksia	*B. attenuata* R. Br.

Jarrah has an adaptable form, from mallee (copse-like) to tall tree. It combines with many other trees and shrubs, as do most of these plants with one another throughout south-western Australia. In a study of only 75 of the tree and shrub species (10% of the estimated total) in the northern Jarrah forest (Dale Sub-district), Havel (1975) identifies 20 structural combinations - in the canopy and mid-layers alone. Understorey shrubs and herbs also combine in probably hundreds of ways, according to local variations in topography, soil type, and soil moisture. The Jarrah forest contains about 800 of the plant species recorded in south-western Australia and most of these are found throughout all vegetation associations of the south-west (Bell and Heddle 1989). Thus some structural combinations are very restricted, yet they are occupied by species that have a wide distribution.

The diversity of plants provides many potential plant resources for people. Yet the same condition of diversity indicates poor nutrients, and individual productivity is poor. Plants produce small fruits, flowers, and seeds (Beard 1981: 93-94). By bulk, tubers probably dominated human plant foods (which still included the fruits, flowers, and seeds; Meagher 1974). Poor nutrients are significant for animals also. Because foods are limited, there are fewer and less diverse mammals in south-western Australian forests, relative to other forests (Nichols and Muir 1989: 134-135). Animals tend to be small - only two species weigh more than 10 kg (Emu, *Dromaius novaehollandiae*; and Western Grey Kangaroo, *Macropus fuliginosus*) and few of the remaining vertebrates weigh more than 1 kg (Nichols and Muir 1989). Some south-western mammals subsist only on resources that are found in small quantities over large areas, thus limiting their numbers and body size. The Honey-possum (*Tarsipes rostratus*), weighing 20 g, subsists entirely on nectar and pollen (Strahan 1995).

Plant defences also limit animal numbers. Sclerophyll shrubs have hard, indigestible leaves. Species adapted to the poorest soils produce more phenolic substances, i.e., toxic compounds that affect digestion (Cork and Catling 1996). About 30 Jarrah forest species, of the genera *Gastrolobium* and *Oxylobium*, contain flouroacetate, a toxin for animals lacking a digestive technique called pre-gastric fermentation (Nichols and Muir 1989). This technique, common to south-western mammals, allows them to eat a range of plants and fungi, at the cost of lowering reproductive capacity and metabolic rates.

As in all open-forests, fires, whether naturally or artificially induced, influence Jarrah forest ecology. Jarrah forest is fire-resistant, that is, hard to dislodge from its state that existed before the disturbance posed by fire (Attiwill 1994). Most Jarrah forest plants have fire-resistant exteriors or subterranean lignotubers that re-sprout after fire (Dell *et al.* 1989). During the summer fire season, Jarrah trees withdraw nutrients from leaves that may burn (Attiwill 1994, Christensen and Abbot 1989). After fire, surviving plants put on a growth spurt, and seeds germinate, apparently encouraged by smoke particles in the topsoil (Dixon *et al.* 1995).

Jarrah forest animal populations recover quickly after fire (Christensen 1992, Christensen *et al.* 1985). Macropods and possums are common in recently burnt areas (Christensen and Kimber 1975). The recovery of animal populations after introduced predators are eliminated suggests that species are adapted to disturbance in this naturally periodically stressed environment (Fox and Fox 1986, Nichols and Muir 1989). Small mammals, reptiles, and invertebrates are exposed to predators and therefore scarce after a fire, but populations recover if patches of thick understorey survive (Nichols and Muir 1989). For example, the requirements for shelter and food of the Tammar (*Macropus eugenii*) and the Brush-tailed bettong (*Bettongia penicillata*) include firing of habitat and foraging areas at different intervals (Christensen 1980).

Very frequent burning in Jarrah forest appears to encourage grasses (Ward 1998), as happens in other open-forests (Gell *et al.* 1993). Grasses attract large grazing animals such as wallabies and kangaroos (Cork and Catling 1996). Aboriginal hunter-gatherers using the Jarrah forest may have sought this result, although they probably did not achieve it, or seek to achieve it, throughout the whole forest.

Analysis of Jarrah tree-rings and *Xanthorrhoea* (grass-tree) trunks that are two to three centuries old suggests that before European settlement, fires were frequent in Jarrah open-forest (Burrows *et al.* 1995, Ward and van Didden 1997). Jarrah trees are deeply scorched, or scarred, by very hot fires. Scarring fires probably result from fuel-loads that accumulate after five years or more of no fire (Burrows *et al.* 1995: 12, 14). Scars are visible in transverse cuts of tree trunks, enabling accurate dating of hot fires. Relatively few fire-scars formed in Jarrah trees before AD 1850, which is the approximate date of two events that probably limited Aboriginal ability to fire forests. These events were the first major measles epidemic in the Aboriginal population and a government edict forbidding Europeans and Aborigines to set fires in summer, with penalties of fines or flogging (Ward 1998). As the seasonally dry Mediterranean climate and the high frequency of lightning cause summer fires naturally, the absence of fire scars for periods averaging 80 years in parts of Jarrah trees older than 1850 suggests that before that date, Jarrah trees did experience fires, but these fires were so mild that they did not scar the trees. After 1850, the number of fire scars per decade recorded in Jarrah trees increased ten-fold.

In a similar study, Ward and van Didden (1997: Figure 15) show that the number of fire scars per decade in *Xanthorrhoea* trees, which in this genus are obtained in both in mild and hot fires, decreased steadily from three to four per decade in the period 1750-1850, to less than one per decade by the period 1980-1997. These analyses support extensive historical evidence amassed by Hallam (1975) and Ward (1998), that Aboriginal people set fires in many parts of the Jarrah forest (and coastal woodlands) very regularly, perhaps every two to three years. The historical observations suggest that Aboriginal people burnt the forest frequently to create cool, controllable fires (Bell *et al.* 1989), regenerate plants, preserve fire-sensitive species (with cool fires, some areas can escape burning), and promote a grassy or herbaceous understorey, now rare or absent, that probably attracted grazing (folivore) prey animals such as kangaroos and large wallabies (Ward 1998).

In sum, because of the limiting soil and climatic conditions, the wide geographical and environmental range, and irregular disturbances from fire, the total Jarrah forest is floristically and structurally heterogenous. The biological heterogeneity and the climatic and geological variation inherent in the large total area means that the Jarrah forest can be considered as an assemblage of many ecosystems of varying extent whose differences grade subtly into one another and cannot be easily defined (Dell *et al.* 1989, cf. Wardell-Johnson and Horwitz 1996). Many flora and fauna (but not all) are highly adapted to seasonal fires and recolonise areas quickly.

Potential human plant foods are widely distributed although edible parts of plants are small (see below). There are few large animal species, though they are conspicuous (Nichols and Muir 1989: 135). Human movement through the Jarrah forest is generally easy because of the natural tendency for frequent, low-intensity cool fires, removing dense vegetation (Hallam 1975). These fires encourage plant regrowth and regeneration of animal habitats and forage plants. In the Jarrah forest, a fire-regime of frequent fires, as could be implemented by hunter-gatherers, would improve the quantity and diversity of food species, limited as they are.

The Karri forest

In contrast to the Jarrah forest, the Karri forest covers less than one eighth of the Darling Botanical District. Nonetheless, Karri is the main tree forming tall open-forest in Western Australia (Beard 1981). Karri trees grow only in the Warren Sub-district, where rainfall exceeds evaporation for eight or more months of the year, and are restricted to soils with moisture-retaining properties (Beard 1981, 1983). Karri trees are recorded as reaching 80 m in height, making the species one of the tallest flowering plants in the world (Ashton and Attiwill 1994). They form tall open-forests in the southern part of the Leeuwin-Naturaliste Region (covering an area here of less than 500 km²) and across a much larger area (2,000 km²) of the south coast and its hinterland (Christensen 1992). This second area includes the towns of Pemberton and Walpole, which receive 1400-1450 mm of rain annually, and have a nine or ten month rainy season

(Bureau of Meteorology 1975). The tallest Karri trees and most extensive Karri stands grow in this area.

The soils in the Karri areas include sands and laterites, which do not support Karri, but the laterite plateau is deeply incised, exposing the granite-gneiss country rock in valleys, and providing suitable substrate for Karris. Mid-slope, red earths have developed from granite-gneiss or other rock, and the valley bottoms contain alluvium. These soils support Karri because they retain moisture and have a physical structure capable of supporting very large trees (Beard 1981). However, nutrient levels differ little from Jarrah forest soils (Attiwill and Leeper 1990).

Like Jarrah, Karri associates with many of the taxa that are distributed across the south-west. It can form tall open-forest with Marri or mixed open-forest with combinations of Jarrah, Marri, or other eucalypts. Because soil boundaries are well-defined, there are abrupt transitions from Karri tall open-forest to Marri or Jarrah open-forest (Beard 1981, Gill 1994, McArthur 1991). The resultant mosaic is largely a function of the dendritic drainage pattern in the Warren Sub-District, with Jarrah on ridge-tops, Marri and Karri mid-slope, and pure Karri in the valley bottoms. In the Leeuwin-Naturaliste Region, an added topographic effect derives from the Leeuwin Ridge, a coastal dune complex that protects the Karri forest from sea winds and salt spray. Where rainfall favours Karri, that is, in the southern two-thirds of the Leeuwin-Naturaliste Region, the Karri forest extends along the lee of the ridge and up to the ridge crest, where it borders coastal heath, scrub, or woodland. Some of the Karri forest in the Leeuwin-Naturaliste Region grows on a brown earth derived from calcarenite, which is evidently an adequate substitute for the red earths derived from granite-gneiss in other Karri regions (Beard 1981, Northcote *et al.* 1967).

Like other tall open-forests, Karri forest experiences intense fires because of the high fuel loads. It is fire-resilient, returning quickly to its condition before disturbance, and fires seem to promote rapid regeneration of forest and dense understorey (Ashton and Attiwill 1994, Attiwill 1994). Fuel accumulates in Karri forest two to three times faster than in Jarrah forest, so for a given fire-free interval more severe fires are experienced in Karri forest (Bell *et al.* 1989). Karri forest is likely to have long fire-free periods because the wetter soil keeps fuel damp too (Burrows 1987). Nonetheless, it experiences fire more often than other tall open-forests, because its understorey regenerates faster (Grove and Malajczuk 1985) and unlike other tall open-forests, it has the same Mediterranean climate as the Jarrah forest, with dry, lightning-prone summers (Attiwill 1994). Because of local topography and prevailing weather conditions, lightning-strikes are probably more successful at lighting fires in some locations than in others, leading to a range of fire regimes and structural configurations across the whole extent of forest (Underwood 1978).

Thus, vegetation is adapted to severe, infrequent fires. As in other tall open-forests, destruction of large trees and dense understorey in wild-fire makes light, water, and nutrients available to new plants (Christensen and Annels 1985, Rotheram 1983). Karri forest understorey shrubs (*Trimalium floribundum*, *T. spatulatum*, *Bossiaea laidlawiana*, *Acacia pentadenia*, *Chorilaena quercifolia*) regenerate from soil stored seed, and Karri seedlings grow faster in ash-beds thanks to the release of nitrogen and phosphorous from burnt litter (Christensen and Abbott 1989, Loneragan and Loneragan 1964). Four to nine years after fire, leguminous shrubs produce seeds and ensure their propagation in the next fire (Skinner 1984); at age seven years Karri saplings are resistant to mild fires (Christensen and Annels 1985). In Karri forest, fires set more frequently than once every six years would probably deplete nitrogen to levels below the needs of Karri saplings (Christensen and Abbott 1989). Frequent fires would eliminate the leguminous understorey, young Karri trees, and favour Jarrah and Marri trees, other specialised sclerophyll species, and grasses (Bradshaw 1985, McCaw 1986, Ward 1998).

At the same time, there is evidence for at least moderately frequent fires in Karri forest. Rayner (1992) reports analyses of Karri tree-rings which suggest that in Karri stands dating from AD 1620, fire-scars pre-dating 1850 are rare or absent. Like Burrows *et al.* (1995) working on Jarrah tree-rings, Rayner (1992) concludes that the best explanation for the absence of fire-scars in trees growing in a seasonally dry, fuel-producing environment is that fuel loads were low when fires burnt, and therefore, in the stands sampled, that fires were frequent and milder than present. The implication for vegetation is that the "Karri" forest was in places composed of large, mature trees, of mixed ages and species. In other places, the results of infrequent, intense fires would have been Karri stands composed purely of Karri trees, all the same age.

In the latter type of Karri forest, the pure stand, flora and fauna are less diverse or scarcer because of the rapid growth of a limited number of understorey species and the dominance of one canopy species. In the former, mixed type of Karri forest, sustained by varying fire frequencies, plants and animals were probably more diverse, and animal populations perhaps larger (Christensen and Abbott 1989, Christensen *et al.* 1985). Today, this effect can result from a fine mosaic of vegetation structures (Christensen and Kimber 1975). Vigorous herb and shrub layers in recently burnt, open Karri forest attract small birds, emus, and kangaroos; thickets are habitat for small macropods, potoroids (rat-kangaroos), and dasyurids. A group of adjacent stands, resulting from fire-free periods of less than six to more than 30 years, would create the greatest floristic and structural diversity, and the greatest number of animal habitats (Williams *et al.* 1994).

In the Karri forest, a variety of fire regimes would probably promote the most variety of human foods, although little precise data exists with which to compare productivity in Karri and Jarrah forests. However, historical data relating to ethnographic Aboriginal foods allows one to assess the number of known human foods from each of these forests.

Recorded Aboriginal foods in Jarrah, Karri, and other forests

The food potential of different forests can be assessed from listing historically and ethnographically recorded Aboriginal plant foods, and a large group of "potentially edible" plants (Bindon 1996, Daw *et al.* 1997). These plants are included here as possible human foods, though no study of their practicality as such has been made. However, all the food animals listed have been historically documented. Tables 2.2-2.3, showing the distributions of food species, are based on the total list of published ethnographic food plants and animals, plus the potential ones (Appendix 1).

Both lists are limited because historical observers did not identify some foods to species, especially the less conspicuous, smaller animals, so their occurrence in open-forest or tall open-forest cannot be surmised by reference to ecological data. Distribution of such foods is given as uncertain in the tables.

A greater limitation is that the food species record largely comprises 19th C Europeans' observations of species eaten by Aboriginal (Nyoongar) people living in woodlands near the coastal settlements of Perth and Albany. Ethnographic investigations are only recently extending the list to other regions and vegetation (Bird and Beeck 1988, Daw *et al.* 1997). While many of the species historically recorded as Aboriginal food plants and animals do occur in inland forests as well in coastal woodlands, one can still expect the observers' bias to be important. Woodland species would perhaps prefer open-forest to tall open-forest, since the former are more similar to woodland in structure and composition. Moreover, it is very likely that coastal plains actually provided the largest number of food species, thus making the food resources of inland forests even more obscure. The importance to Nyoongar people of large-scale estuarine fishing and wetland exploitation is well substantiated in 19th C records (Anderson 1984, C.E. Dortch 1997, Gibbs 1987, Hallam 1987). Large wetlands as found on the coastal plains are absent in the interior Jarrah and Karri forests (Christensen 1992, Halse and Blyth 1992). Thus there is probably both a preponderance of both actual and recorded food species in coastal districts and woodlands, making it hard to compare open-forests and tall open-forests.

However, the limitation is perhaps less significant for food animals, or vertebrates at least, since few vertebrates are endemic to open-forest or tall open-forest (Christensen 1992). Those animals that inhabited tall open-forest or open-forest *were* found elsewhere, and 19th C observers could have seen Aboriginal people hunting them. These observers' bias can be assessed to some extent by comparing the records for coastal and inland botanical sub-districts (Warren and Drummond are coastal, Dale and Menzies are inland).

The bias of ethnographic observers' locations, mentioned above, is apparent. More food plants and animals are found in the coastal sub-districts (Drummond and Warren), even if the large number of aquatic animals found mainly in estuaries is excluded from the total. However, the bias may not be the only reason for the great disparity between food plants in tall open-forest and in other formations, since Warren Sub-district which contains all the tall open-forest also has the most food plants (it has a great range of vegetation types: Figure 2.1; Williams *et al.* 1994). Therefore, the tall open-forest probably does contain significantly fewer food plants than other formations.

In contrast, many recorded food animals live in tall open-forest, and observers' bias is possibly the main reason for any difference between tall open-forest and open-forest. Vertebrate adaptability is confirmed by analysis of animal habitats, below. Given the uncertainties with this analysis, tall open-forests potentially contain about as many food animal species as open-forests. The question of relative productivity therefore returns - which formation contains the greatest mass of food animals? Again, the answer probably depends partly on fire regime.

Biological surveys show that relative to open-forest, tall open-forest contains fewer plant species (confirming ethnographic records) and smaller vertebrate populations, partly because of the lower diversity and higher density of the understorey, which is in turn partly a product of the fire regime (Christensen 1992). This is the most difficult factor to assess in the use of tall open-forests. How easily could hunter-gatherers fire these forests to produce a vegetation mosaic, and improve diversity and productivity? This factor can potentially be assessed by long-term experiments with forest plots (Bowman and Brown 1986), but given the absence of such experiments or of detailed historical records of Aboriginal firing in the tall open-forests and southern open-forests (Hallam 1975, Tingay 1985), no-one can identify the probable fire regimes once imposed by south-western Aboriginal people. However, one can assess the extent of hunter-gatherer use of tall open-forest, and the impact of such use on vegetation, by investigating detailed archaeological and palaeontological records. A key part of this type of research is assessment of past environmental conditions, which can be inferred partly from present-day physiography, soils, climate, and their relationship to vegetation.

Table 2.2 Distribution of plant and animal resources in south-western Australian vegetation formations

Number of plant species that were:
Recorded in tall open-forest and nowhere else - 0
Recorded in tall open-forest, open-forest, and nowhere else - 6
Recorded in tall open-forest, open-forest, and other formations - 4
Total ever recorded in tall open-forest - 10

Recorded in open-forest and nowhere else - 2
Recorded in open-forest, woodland, and nowhere else - 20
Recorded in open-forest, woodland, and other formations - 27
Total ever recorded in open-forest - 55

Recorded in woodland and nowhere else - 19
Recorded in woodland and other formations - 56
Total ever recorded in woodland - 75

Recorded in scrub, scrub-heath, heath, and nowhere else - 16
Recorded in scrub, scrub-heath, heath, and other formations - 29
Total ever recorded in scrub, scrub-heath, or heath - 45

Recorded in sedgeland, swamps, and nowhere else - 4
Recorded in sedgeland, swamps, and other formations - 7
Total ever recorded in sedgeland or swamps - 11

Total plant species - 103

Number of animal species that:
Inhabited tall open-forest and nowhere else - 0
Inhabited tall open-forest, open-forest, and nowhere else - 0
Inhabited tall open-forest, open-forest, and other formations - 24
Total that ever inhabited tall open-forest - 24

Inhabited open-forest and nowhere else - 0
Inhabited open-forest, woodland, and nowhere else - 2
Inhabited open-forest, woodland, and other formations - 29
Total that ever inhabited open-forest - 31

Inhabited woodland and nowhere else - 0
Inhabited woodland and other formations - 32
Total that ever inhabited woodland - 32

Inhabited scrub, scrub-heath, heath, and nowhere else - 1
Inhabited scrub, scrub-heath, heath, and other formations - 28
Total that ever inhabited scrub, scrub-heath, or heath - 29

Inhabited sedgeland, swamps, and nowhere else - 1
Inhabited sedgeland, swamps, and other formations - 1
Total that ever inhabited sedgeland or swamps - 2

Habitat uncertain (because species is not identified) - 24

Inhabited fresh-water streams - 4
Inhabited estuaries - 18
Inhabited open-shore marine waters - 5
Total aquatic fauna - 27

Total animal species - 88

Table 2.3 Distribution of plant and animal resources in sub-districts of the Darling Botanical District

Number of food plant species that were:
Recorded in Drummond and nowhere else - 7
Recorded in Drummond, Dale, and nowhere else - 4
Recorded in Drummond, Menzies, and nowhere else - 0
Recorded in Drummond, Warren, and nowhere else - 10
Recorded in Drummond, Dale, Menzies, and nowhere else - 1
Recorded in Drummond, Dale, Warren, and nowhere else - 0
Recorded in Drummond, Menzies, Warren, and nowhere else - 1
Recorded in Drummond, Dale, Menzies, and Warren - 51
Total ever recorded in Drummond - 74

Recorded in Dale and nowhere else - 2
Recorded in Dale, Menzies, and nowhere else - 1
Recorded in Dale, Warren, and nowhere else - 0
Recorded in Dale, Menzies, Warren, and nowhere else - 0
Total ever recorded in Dale - 59

Recorded in Menzies and nowhere else - 1
Recorded in Menzies, Warren, and nowhere else - 5
Total ever recorded in Menzies - 60

Recorded in Warren and nowhere else - 12
Total ever recorded in Warren - 79

Distribution in Darling Botanical District unknown or uncertain - 8

Total plant species - 103

Number of food animal species that:
Inhabited Drummond and nowhere else - 1
Inhabited Drummond, Dale, and nowhere else - 0
Inhabited Drummond, Menzies, and nowhere else - 0
Inhabited Drummond, Warren, and nowhere else - 23
Inhabited Drummond, Dale, Menzies, and nowhere else - 1
Inhabited Drummond, Dale, Warren, and nowhere else - 0
Inhabited Drummond, Menzies, Warren, and nowhere else - 1
Inhabited Drummond, Dale, Menzies, and Warren - 33
Total that ever inhabited Drummond - 59

Inhabited Dale and nowhere else - 0
Inhabited Dale, Menzies, and nowhere else - 0
Inhabited Dale, Warren, and nowhere else - 0
Inhabited Dale, Menzies, Warren, and nowhere else - 0
Total that ever inhabited Dale - 34

Inhabited Menzies and nowhere else - 0
Inhabited Menzies, Warren, and nowhere else - 2
Total that ever inhabited Menzies - 35

Inhabited Warren and nowhere else - 0
Total that ever inhabited Warren - 57

Distribution in Darling Botanical District unknown or uncertain - 25

Total animal species - 86

Leeuwin-Naturaliste environmental conditions

Physiography

The Leeuwin-Naturaliste Region extends from Cape Leeuwin to Cape Naturaliste and from the Indian Ocean coast to the Swan and Scott Coastal Plains and the Blackwood Plateau (Figure 2.2a). It features a geological unit called the Leeuwin Complex (Geological Survey of Western Australia 1990, Lowry 1967). This complex is formed of the Precambrian granite-gneiss Leeuwin Block and overlying Quaternary dune sands and Tamala limestone, a calcarenite or dune limestone that fringes much of the Western Australian coast (Figure 2.2b). It is bounded on the east by the Dunsborough Fault, which separates it from the Mesozoic sediments of the Perth Basin. Most of the Leeuwin Complex is emergent above sea level. The Quaternary deposits rise to 200 m ASL, forming the Leeuwin Ridge.

Surface sediments

Sands deposited during successive Quaternary high sea-level stands form a three-part series, progressively more leached and re-worked on the inland and older side. The oldest and innermost sands are podzolic sands comprising mainly quartz grains (McArthur 1991, McArthur and Bettenay 1974, Northcote *et al.* 1967). The middle sands in the age series are yellow or brown, due to iron-staining of the quartz grains, and are also heavily leached. These yellow sands interfinger with the calcareous Tamala Limestone. The youngest sands in the series are mobile, shelly dunes along the coast, deposited during the present high sea-level stand, that is, since 6,000 BP.

Stream valleys dissecting the Leeuwin Ridge, such as Ellen Brook, Margaret River, and Boodjidup Brook (Figure 2.3a), contain red earths developed on granite-gneiss, and alluvium derived from the Blackwood Plateau (McArthur 1991). East of the ridge, brown earths have developed on weathered calcarenite; where calcarenite is absent, weathering of granite gneiss has produced a red earth or "Karri loam", so-called after the vegetation that it supports. Other areas of granite-gneiss are mantled by acid grey earths, sometimes containing lateritic gravel, and some sandy yellow-mottled soils.

Cave development in calcarenite is important to the search for detailed archaeological records in limestone cave floor deposits. Water entering the porous rock eventually creates underground streams whose channels grow by a series of collapses (Bastian 1964). Some of these break the surface to form dolines. Overhangs around dolines, formed by a lip of hard limestone "cap rock" or "calcrete" (Lowry 1979), and entrances to caves inside dolines, are potentially suitable for human occupation. Notable doline occupation sites include Devil's Lair and Tunnel Cave, but because of the angle of dune bedding in aeolian calcarenite, most cave entrances in dolines incline steeply, making them less accessible to humans. Other Leeuwin-Naturaliste Region caves were created by entry or exsurgence of streams into or from the calcarenite dunes (Williamson 1979). These caves have relatively easy access and the entry or exit chamber is sometimes perhaps large and flat enough to provide shelter (*e.g.*, Mammoth Cave, Quininup Lake Cave, Witchcliffe Rock Shelter, and Witchcliffe Cave). Human occupation potential of the recorded Leeuwin-Naturaliste Region sinkholes, caves, rock shelters, and cliffs is discussed further in Chapter 5.

Climate

The Leeuwin-Naturaliste Region climate is classed as moderate Mediterranean, having a high winter rainfall and warm dry summers (Beard 1981). In summer, the sub-tropical high pressure belt descends southward so that high pressure anticyclones are centred in the Australian Bight, bringing hot, dry north-easterlies from the interior to south-western Australia, and deflecting low pressure disturbances from the south-west. Thus changes in the seasonality of the sub-tropical high pressure belt's southward movement may have been critical for rainfall and hence vegetation (see Chapter 3). As the anticyclones move east, occasional onshore south-westerlies bring coastal rain, reducing summer evaporation (Gentilli 1989).

Since this light summer rainfall comes from the south-west, the dry season "water deficit" (Gentilli 1989) is less towards the south of the Leeuwin-Naturaliste Region. This is demonstrated by the number of dry season rain days at four weather stations in the Leeuwin-Naturaliste Region (Table 2.6, Figure 2.3b). Weather station data also show that the summer drought lasts five months at Busselton and Cape Naturaliste, and four months at Margaret River and Cape Leeuwin, presumably because of the north-south, seasonal movement of the sub-tropical high-pressure belt (Beard 1981, Gentilli 1989).

In winter, the sub-tropical high pressure belt returns north to about 28-32° S and moist westerlies influence south-western Australia. Periodic and frequent cold fronts from the south-west bring heavy rains to the lower west coast, increasing over hills and ranges such as the Darling Scarp and the Blackwood Plateau, and then declining eastwards. Rainfall is higher at Cape Leeuwin than at Cape Naturaliste, and higher still at Margaret River, on the Blackwood Plateau's western margin. Further east, rainfall declines steadily.

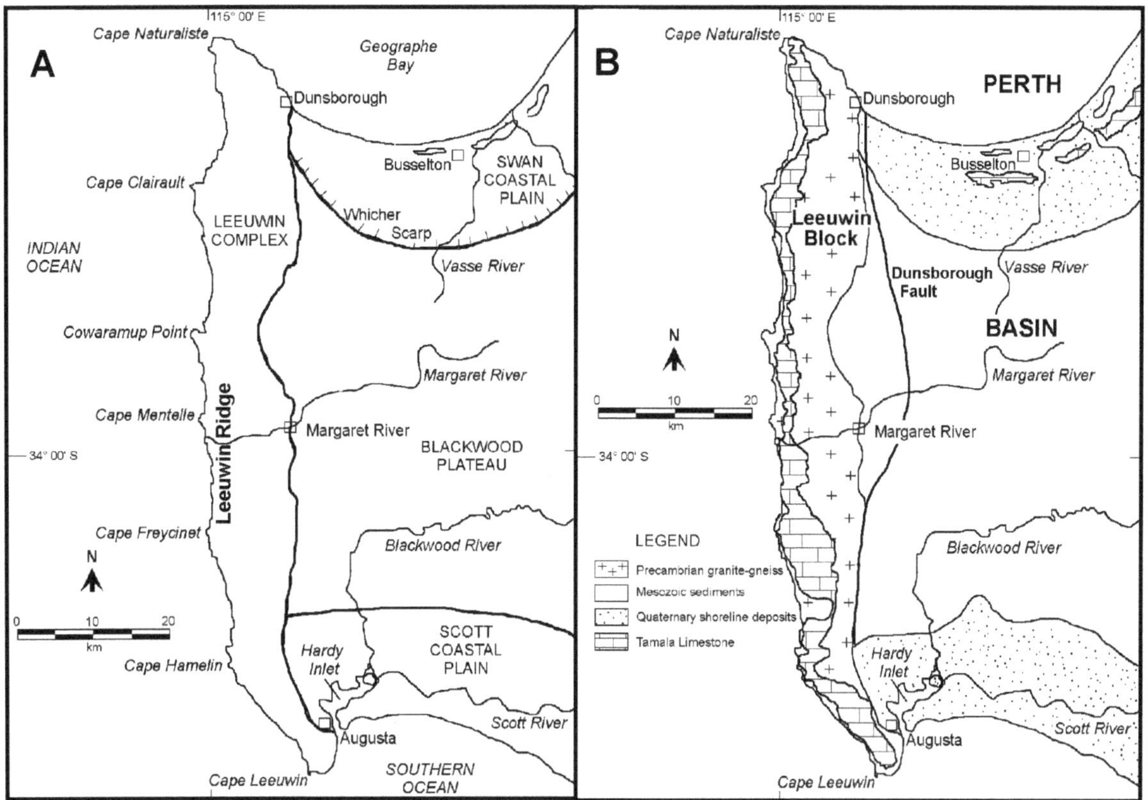

Figure 2.2 Physiography (Figure 2.2a) and surface geology (Figure 2.3b) of the Leeuwin-Naturaliste Region (Geological Survey of Western Australia 1990, Glover 1984).

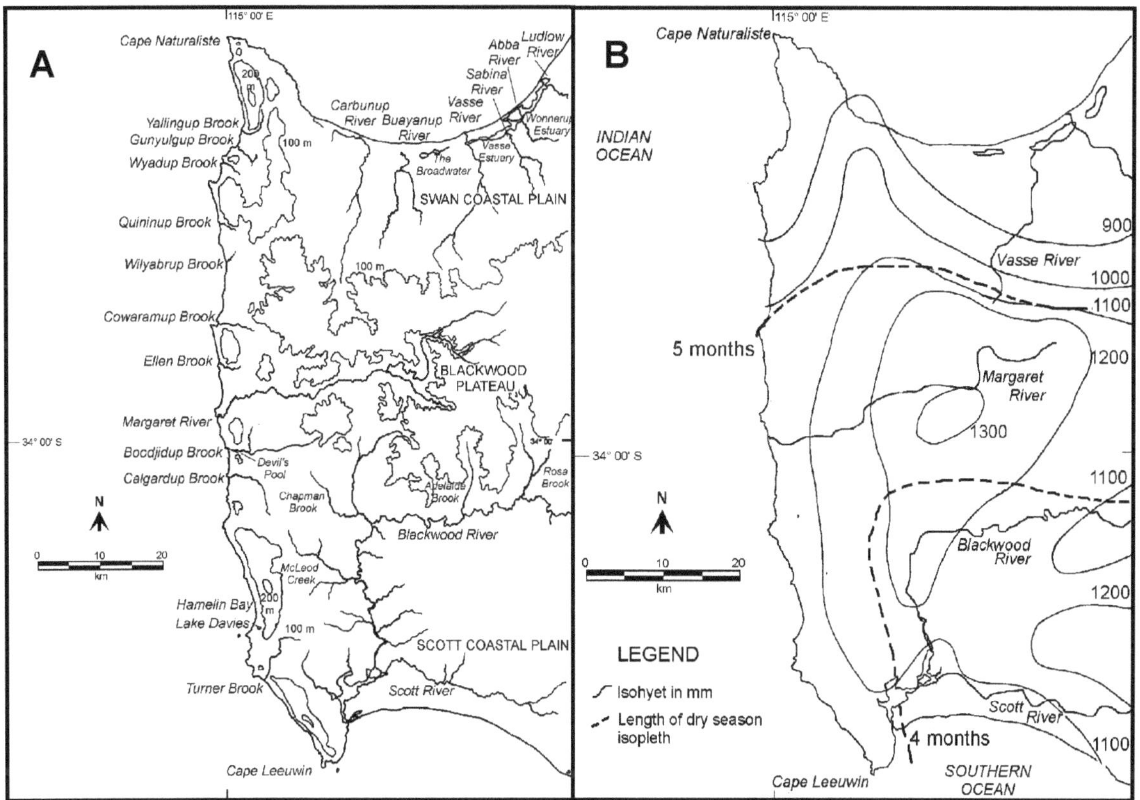

Figure 2.3 Drainage and topography (Figure 2.3a) and rainfall (Figure 2.3b), in the Leeuwin-Naturaliste Region (1:250,000 Mapsheet SH 50-5 and SH 50-9, Beard 1981: Figures 6, 9).

Table 2.4 Leeuwin-Naturaliste Region climatic data (from Bureau of Meteorology 1975).

Weather station	Position	Longitude	Latitude	Number of dry months[1]	Sum of mean monthly rainfalls (mm) Dry months	Wet months	Total	Summer water deficit (mm)[2]	No. rain days Dry months	Wet months	Total	Av. daily mean temperature (°C) Dry months	Wet months
Busselton	Coast, low	115° 20' E	33° 40' S	5	81	757	838	122	21	116	137	20.3	14.0
Cape Naturaliste	Coast, elevated	115° 0' E	33° 32' S	5	84	754	838	108	27	129	156	18.7	14.5
Margaret River	Inland, elevated	115° 4' E	34° 57' S	4	82	1110	1192	78	19	128	147	20.0	13.9
Cape Leeuwin	Coast, low	115° 8' E	34° 22' S	4	86	908	994	70	30	153	183	19.8	15.3

[1]Months where mean rainfall in mm is less than twice the mean daily temperature in °C, after the formula of Bagnouls and Gaussen (1957), cited in Beard (1981). At Busselton and Cape Naturaliste, the dry months are November to March inclusive, and at the southern stations, December to March inclusive.
[2]The column headed "summer water deficit" shows the north-south cline in summer evaporation rates, a strong influence on vegetation (cf Gentilli 1989).

Vegetation

The variations in soils, topography, seasonality, and rainfall across and along the Leeuwin Ridge result in an "intricate mosaic" of vegetation associations (Figure 2.4; Beard 1981: 194, F.G. Smith 1973). The seaward side of the Leeuwin Ridge features *Acacia*-dominated heath, scrub and thicket (Smith 1973). In sheltered areas, Peppermint (*Agonis flexuosa*), Banksia (*Banksia grandis*, *B. attenuata*, *B. ilicifolia*), or Jarrah (*Eucalyptus marginata*) form low woodland or low open-forest. Canopy cover and height increase towards the ridge crest, as onshore winds and sea-spray have less effect.

In the lee of the ridge crest, and where rainfall permits (generally in the southern part of the region), brown sands on calcarenite support Karri tall open-forest. Where the escarpment is abrupt, open heath shifts to Karri tall open-forest with no intervening zone of low woodland or open forest. Karri forms pure stands on red earths and mixes with Marri or Jarrah or both on brown sands.

On the Blackwood Plateau, Jarrah open forest grows on lateritic podzolised sands, with Marri a component on richer valley soils. Swamps between the Blackwood Plateau and the Leeuwin Ridge support sedges and *Melaleuca* species.

At the northern end of the Leeuwin Ridge, Jarrah-Banksia formations may include She-oak (*Allocasuarina fraseriana*) and Candle Banksia (*B. attenuata*). Marri open-forest is prominent on the leached sands extending over the sandy Swan Coastal Plain, and mixing with Peppermint and Yate (*E. patens*). Also prominent on the Swan Coastal Plain is Tuart (*E. gomphocephala*) tall open woodland. This tree grows only on yellow sand over calcarenite (Beard 1981). At Dunsborough and Busselton, Peppermint low woodland borders lagoons and wetlands of the estuary systems.

Along the Southern Ocean coastline east of Augusta, the poorly-drained, leached sands of the Scott Coastal Plain support Jarrah-*Banksia* low woodland, *Acacia* scrub, and large reed swamps.

Significance of vegetation for hunter-gatherer occupation

Vegetation in the Leeuwin-Naturaliste Region includes small areas of woodland, hemmed in by open-forest or tall open-forest on one side and coastal heath or scrub on the other. Where the open-forests border tall-open forests, the boundaries are complex and their respective taxa intermingle. If hunter-gatherers foraged in or occupied forests differentially, very much as their different conditions dictated, a zone consisting of a mosaic of forest types is a good place in which to try to detect evidence of such differential use. In such a fine mosaic, all occupation sites would be located close to examples of most other vegetation types, so it is the immediate surroundings of a site that may be most important for hunter-gatherers. Archaeological investigation can give an indication of the length and frequency of human occupation of a site, in relation to the surrounding vegetation, and thus permit one to infer the influence of that forest type. The site's proximity to water, accessibility, and attractiveness for occupation can also be estimated and taken into account in evaluating the attractiveness of surrounding vegetation.

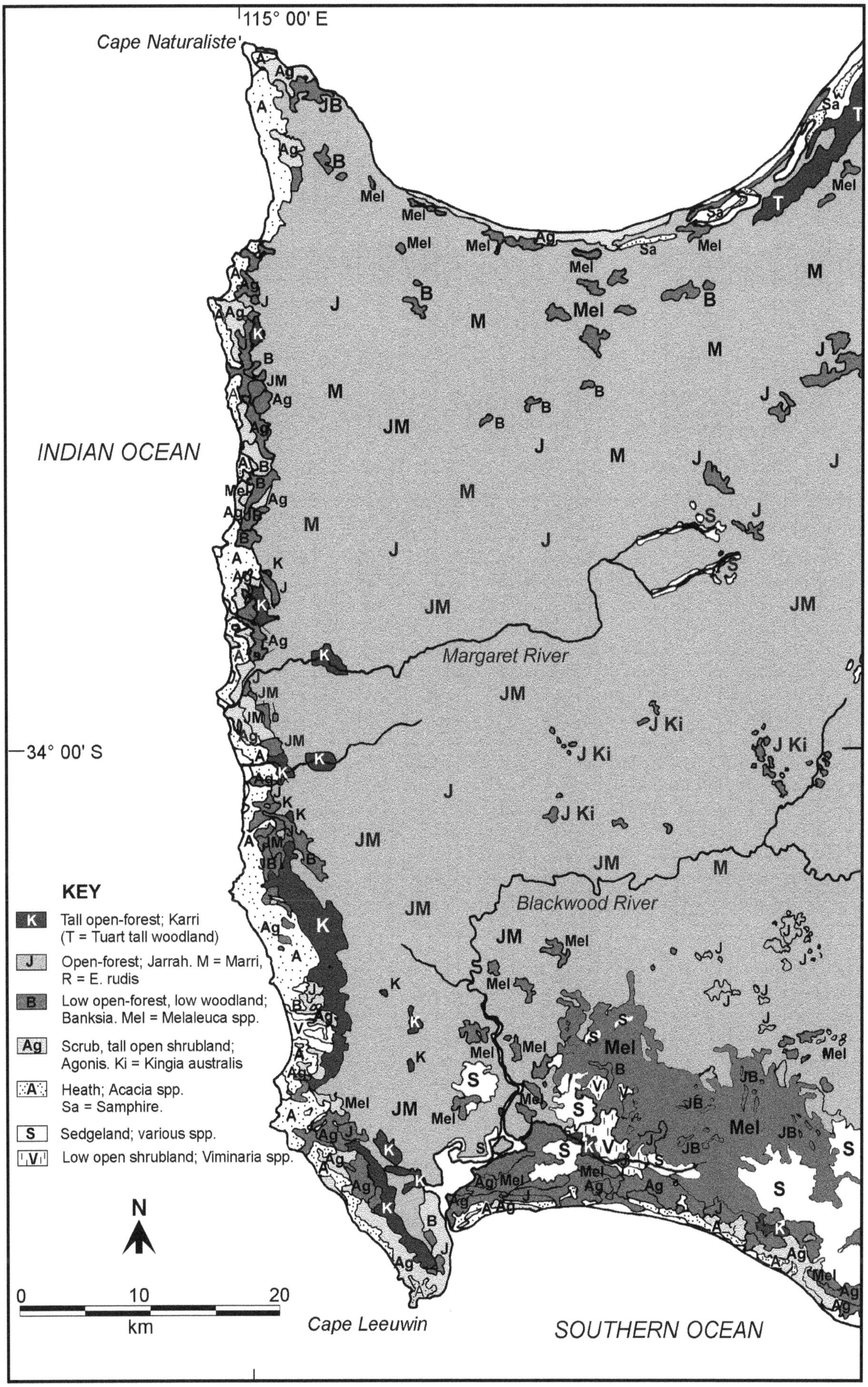

Figure 2.4 **Vegetation associations in the Leeuwin-Naturaliste Region, after F.G. Smith (1973)**

Table 2.5 Vegetation series and canopy heights for soils and rainfall, predicted for substrates on the leeward side of the Leeuwin Ridge (from Beard 1981: Table XIII and pp 97-98, 135-139; and F.G. Smith 1973)
For Leeuwin-Naturaliste Region rainfalls see Table 2.4.

<table>
<tr><th></th><th colspan="8">Minimum mean annual rainfall (mm)</th></tr>
<tr><th>Substrate</th><th>1,000</th><th>900</th><th>800</th><th>700</th><th>600</th><th>500</th><th>400</th><th>300</th></tr>
<tr><td>Podzolised sands with laterite gravel</td><td colspan="4">Jarrah-marri open-forest
25 m</td><td>*Banksia* low woodland
5-8 m</td><td colspan="2">Heterogenous scrub-heath
1-3 m</td><td>*Acacia - Allocasuarina* thicket
2 m</td></tr>
<tr><td>Deep sands</td><td>Karri tall open-forest and marri open-forest on brown sands on calcarenite
30-70 m</td><td>Jarrah-marri woodland
20 m</td><td colspan="3">*Banksia* low woodland on leached sands
6-8 m</td><td colspan="3">*Banksia- Xylomelum* scrub heath alliance on sandy yellow earths
1-5 m</td></tr>
<tr><td>Yellow sands on calcarenite</td><td colspan="2">Karri, Jarrah, and/or Marri low forest or woodland with Peppermint, *Banksia* understorey
30-70 m</td><td colspan="3">Tuart tall woodland with Peppermint, *Banksia* understorey
30-40 m</td><td colspan="3">*Banksia- Xylomelum* scrub heath alliance on sandy yellow earths
1-5 m</td></tr>
</table>

Another important aspect of Leeuwin-Naturaliste Region vegetation is that it demonstrates regionally-derived preferences among "vegetation series" for soils and rainfall as presented in the *Vegetation Survey of Western Australia* (Beard 1981). Beard defines "vegetation series" as the vegetation formations (structure) and plant taxa that are associated with a particular substrate in all climates. As mentioned, the entire south-western Australian climate is Mediterranean, and only internally differentiated on the basis of length of the dry season, the amount of summer evaporation or water deficit, and annual rainfall (Gentilli 1989). The first two factors vary inversely with rainfall (see Table 2.46), so rainfall is therefore a broad indicator of the different climates experienced in a vegetation series.

Table 2.5, derived from Beard's (1981) Table XIII and his discussion, shows how the Leeuwin-Naturaliste Region vegetation (on the left-hand, maximal rainfall side of the table) varies on the same substrates with decreasing rainfall (towards the right of the table). The three substrates presented here are the highly leached sands and gravels found on the Blackwood Plateau; the deep sands in the lee of the Leeuwin Ridge; and the leached yellow sands on calcarenite top of the ridge. An interesting feature of this table is that in different south-west regions, low annual rainfalls down to 300 mm do not produce vegetation different from that now found in many parts of the Leeuwin-Naturaliste Region (owing to its present structural and floristic diversity). Rainfalls lower than 300 mm are presently received only in the semi-arid "South-West Interzone" which has its southern boundary 600 km north-east of the Leeuwin-Naturaliste Region (Figure 2.1; Beard 1981: Figure 6).

With the rainfall data from Table 2.4, Table 2.5 implies that even if rainfall were as much as 70% lower during the last glacial, it need not have entailed that the vegetation in the Leeuwin-Naturaliste Region as a whole was very different. Whether these changes over time were nonetheless significant to south-western hunter-gatherer groups is another matter. My purpose is to deduce the influence that even small changes in floristic composition or structure may have had on people's behaviour - how long or often they camped in a place. How can changes in floristics or structure be known? The next section discusses what information may be gleaned from faunal remains in dated archaeological deposits.

Fauna

The remains of vertebrates, particularly mammals, are well preserved in the deposits investigated here. Interpretation of past environments from vertebrate remains is feasible if one assumes that their association or preferences for vegetation types has been uniform over long periods. This approach should not stand alone, and I propose to test my findings with direct botanical evidence for vegetation. However, I shall assume two aspects about prehistoric vertebrate ecology in south-western Australia:

Vertebrates' requirements for *shelter* and *habitat* can be summarised in terms of vegetation formation (e.g., open-forest, woodland, scrub), or *structure*. Thus one can infer from identified vertebrate remains the general structure of vegetation at the time of their deposition.

Vertebrates' *food* requirements would be satisfied in any south-western Australian vegetation type, as almost all of the herbivores are adapted to consume a range of plants throughout south-western Australia (Nichols and Muir 1989). Thus one can expect that the distribution of vertebrate foods varied little over time, unless rainfall was much lower than anywhere now experienced in this region.

Thus, changes in vegetation structure are the main palaeo-environmental inferences that one can draw from changes in the proportions of identified vertebrate taxa in dated archaeological horizons. Data that will be used to support these inferences are presented in tables below. The following discussion describes the fauna of potential interest for palaeo-environmental interpretation.

Partly because of the nutrient-poor environment, and in contrast to great diversity among south-western Australian invertebrates, vertebrates are moderately diverse, and mammals are the least diverse (Christensen 1992, Christensen *et al.* 1985, How *et al.* 1987). It is hard to assess the truth of historical reports of the great diversity of amphibians, reptiles, birds, and small mammals, because after recent land clearances many species would have become extinct or now have a very restricted habitat (How *et al.* 1987).

In assessing the distribution of ground-dwelling vertebrates in the coastal parts of the Warren Botanical Sub-District, How *et al.* (1987) record 11 species of frogs, one species of tortoise, 22 species of lizards, and 12 species of snakes. As a group, the lizards are abundant and diverse in areas of lower rainfall (Christensen 1992, How *et al.* 1987). Rainfall seems to be an important determinant of distribution. Owing to difficulties of identifying lizard remains to species level, no palaeo-climatic inferences from individual species can be made here (J. Mead, University of Arizona, pers. comm.), but the above findings suggest that the total number of lizard remains in archaeological deposits may indicate general aridity (Balme *et al.* 1978).

Christensen (1992) lists 145 south-western Australian birds, most of which have wide habitat preferences and would find suitable habitat in the varied Leeuwin-Naturaliste Region. Baird (1992) notes some bird species' specialised habitat preferences; his treatise on fossil bird remains from a Leeuwin-Naturaliste Region deposit is summarised in Chapter 3.

The mammal records made by How *et al.* (1987) include 15 marsupial species, five rodents, seven bats, and the dingo. Mammals are already of greatest interest to this investigation because their remains are abundant in Leeuwin-Naturaliste Region deposits (Merrilees 1984), and easier to identify than the other classes, like lizards. Christensen *et al.* (1985) calculate that 95% of the vertebrates in Karri forest also inhabit Jarrah and Wandoo forest and woodland to the north and east, suggesting that there is no distinct fauna for any botanical sub-district of south-western Australia. The main distinction is a northward trend from many mesic species (not strongly adapted to either wet or dry conditions), to many xeric species (adapted to dry conditions), which is shown strongly among lizards. Nonetheless, among mammals, one can identify distinct habitat preferences within their wide ranges.

Before assessing these preferences, I present Tables 2.6-2.8, which show the taxa one may expect to find in Leeuwin-Naturaliste Region deposits, divided according to the extent of knowledge about their presence in the region. From these tables, I select a list of taxa whose habitats I shall investigate further.

Table 2.6 lists mammals that exist now in the Leeuwin-Naturaliste Region, or recently existed there before 20th C land clearances. It includes rare or recently extinct species represented in collections of the Western Australian Museum.

Table 2.7 lists several mammals recorded only as fossils in the Leeuwin-Naturaliste Region (Merrilees 1984), that were extant elsewhere in Australia at the time of European settlement, and are therefore presumed to have been extinct in the Leeuwin-Naturaliste Region at that time. Those indicated as extant in Western Australia (most of them) are of interest to this study because their fossil remains show that their ranges once extended to the far south-west, indicating different environmental conditions at these times. The current status of *Sarcophilus harissii* is given as uncertain because of relatively recent ages (600-400 BP) for fossils found in the Leeuwin-Naturaliste Region (Archer and Baynes 1974).

A third list of mammals comprises mammals that are no longer found anywhere and whose fossil remains are found in the Leeuwin-Naturaliste Region (Table 2.8; Merrilees 1984). These fossils are of varying age, are generally rare, and the taxa will not be assessed for palaeo-climatic information here.

Table 2.6 Modern native mammals recorded in the Leeuwin-Naturaliste Region, after [1]Baynes *et al.* (1975) and [2]How *et al.* (1987), with updated species names after Strahan (1995).

Order / family	*Species*	*Common (or Nyoongar) names*
Chiroptera		
Vespertilionidae	*Chalinolobus gouldii*[2]	Gould's Wattled Bat
	Chalinolobus morio[1,2]	Chocolate-Wattled Bat
	Eptesicus regulus[2]	King River Eptesicus
	Falsistrellus tasmaniensis[1,2]	Tasmanian Pipistrelle
	Nyctophilus geoffroy[1,2]	Lesser Long-eared Bat
	Nyctophilus gouldii[2]	Gould's Long-eared Bat
	Nyctophilus timoriensis[2]	Greater Long-eared Bat
Rodentia		
Muridae	*Hydromys chrysogaste*[1,2]	Water Rat
	Pseudomys shortridgei[1]	Heath Rat
	Pseudomys praeconis[1]	Shark Bay Mouse
	Rattus fuscipes[1,2]	Southern Bush-Rat
	Rattus tunneyi[1]	Pale Field Rat
Polyprotodonta		
Dasyuridae	*Antechinus flavipes*[1,2]	Yellow-footed Antechinus (Mardo)
	Dasyurus geoffroi[1,2]	Native Cat (Chuditch)
	Phascogale tapoatafa[1,2]	Brush-tailed Phascogale (Tuan)
	Sminthopsis griseoventer[1,2]	Grey-bellied Dunnart
Peramelidae	*Isoodon obesulus*[1,2]	Southern Brown Bandicoot (Quenda)
Diprotodonta		
Burramyidae	*Cercatetus concinnus*[1,2]	Western Pygmy Possum
Tarsipedidae	*Tarsipes rostratus*[1]	Honey Possum
Pseudocheiridae	*Pseudocheirus peregrinus occidentalis*[1,2]	Western Ring-tailed Possum (Nworra)
Phalangeridae	*Trichosurus vulpecula vulpecula*[1,2]	Common Brush-tailed Possum (Gnuraren)
Potoroidae	*Potorous tridactylus*[1]	Long-nosed Potoroo
	Bettongia penicillata[1,2]	Brush-tailed Bettong (Woylie)
Macropodidae	*Setonix brachyurus*[1,2]	Quokka
	Macropus eugenii[1,2]	Tammar Wallaby
	Macropus irma[1,2]	Brush Wallaby
	Macropus fuliginosus[1,2]	Western Grey Kangaroo (Yongar)
Carnivora		
Canidae	*Canis familiaris dingo*[1]	Dingo

Table 2.7 Mammals recorded as fossils in the Leeuwin-Naturaliste Region ([1]Baynes *et al.* 1975, [2]Merrilees 1984), extant in modern times elsewhere in Australia, with updated species names after Strahan (1995). The right-most column indicates whether the animal was extant in modern times in Western Australia.

			Recently extant in WA
Chiroptera			
Megadermatidae	*Macroderma gigas*[1]	Ghost Bat	Yes
Rodentia			
Muridae	*Notomys mitchelli*[2]	Mitchell's Hopping-mouse	Yes
	Pseudomys albocineurus[1]	Ash-grey Mouse	Yes
	Pseudomys occidentalis[1]	Western Mouse	Yes
Monotremata			
Tachyglossidae	*Tachyglossus aculeatus*[2]	Short-beaked Echidna	Yes
Polyprotodonta			
Dasyuridae	*Sarcophilus harissii*[1]	Tasmanian Devil	Uncertain
Thylacinidae	*Thylacinus cyanocephalus*[1]	Thylacine or Tasmanian Tiger	No
Peramelidae	*Perameles bougainville*[2]	Western Barred Bandicoot	Yes
Diprotodonta			
Phascolarctidae	*Phascolarctos (cinereus ?)*[2]	Koala	No
Potoroidae	*Bettongia lesueur*[1]	Burrowing Bettong	Yes
Macropodidae	*Lagostrophus fasciatus*[2]	Banded Hare-wallaby	Yes
	Onychogalea (lunata ?)[2]	Crescent Nailtail Wallaby	Yes
	Petrogale lateralis[1]	Black-footed Rock-wallaby	Yes

Table 2.8 Mammals recorded as fossils in the Leeuwin-Naturaliste Region (Baynes *et al.* 1975, Merrilees 1984), extinct by modern times.

Monotremata		
Tachyglossidae	*Tachyglossus* sp.	large *Tachyglossus*
	Zaglossus hacketti	large echidna
Diprotodonta		
Diprotodontidae	*Zygomaturus trilobus*	diprotodontid
Vombatidae	*Vombatus hackettii*	wombat
Thylacoleonidae	*Thylacoloeo carnifex*	marsupial lion
Macropodidae	Protemnodon brehus	sthenurine
	Sthenurus brownei	sthenurine
	Sthenurus occidentalis	sthenurine

Here I select a list of species that I can identify in this project. There is no scope for identifying all species given in Tables 2.6-2.8, when collections of specimens number in the thousands and some species are difficult to identify or have limited value as palaeo-environmental indicators. Table 2.9 lists the species that I can identify, along with information about their ecology relevant to the analysis. I exclude from analysis mammals that:

1. appear to lack fossil remains dated to the terminal Pleistocene and Holocene, the age of deposits that are important to my analysis;
2. have few or no predators, and are thought to have ranged widely through many environments, suggesting that habitat and shelter are unimportant;
3. appear to have left few fossil remains anywhere; or
4. are so small that identifying to species requires lengthy examination under a microscope.

On the first point, I exclude the entire extinct fauna (all the fauna in Table 2.8, and *Phascolarctos*). On the second, I exclude *Dasyurus geoffroi*, *Sarcophilus*, *Thylacinus*, and *Canis*. On the third, I exclude *Tachyglossus*, *Lagostrophus*, *Onychogalea*, *Cercatetus*, and *Tarsipes*. On the fourth, I exclude all the Chiroptera, Rodentia, and the smaller Dasyuridae. These exclusions produce the selection of medium-size and large, extant herbivorous mammals shown in Table 2.9.

Table 2.9 also shows the habitat preferences and habits of each animal. The latter category gives an idea of animals' other requirements, such as tree-holes for nesting, thickets for shelter, etc. Some of these requirements are helpful to palaeo-environmental interpretation. All of the information about habitat and habit that relates to vegetation types is summarised in Table 2.10. This table indicates the number of habitats used by each of the marsupials analysed.

A potential objection to the principle of inferring habitats from the presence of mammal remains is that under strong pressure, most animals are sufficiently adaptable to survive in sub-optimal environments (Balme *et al.* 1978: 63). For example, Quokkas require shelter, shade, and access to water, but they also inhabit cleared and poorly watered terrain on Rottnest Island, 18 km off Perth, where they persist despite severe seasonal stresses (Kitchener 1995). However, it remains to be seen how viable are the Rottnest Island Quokkas in their present habitat, which has been greatly denuded. Another potential objection is that animals may use unfavourable environments as part of their normal range, but this observation should not be confused with their habitat requirements. For example, Tammar wallabies graze in open grassland at night, but their day-time sleeping places are permanently established in thickets with established runways; in fact, the thickets are colloquially known as Tammar thickets (Christensen 1980). In general, the presence of certain forests or other vegetation types in a locality may be indicated by a significant number of individual remains from certain species, as follows:

Significant presence or high numbers of *Perameles*, *Bettongia lesueur*, or *Petrogale* indicate the presence of open vegetation, such as woodland, low woodland, open scrub, heath, or grassland. *Petrogale* feeds mainly on grasses (Strahan 1995).

Significant presence or high numbers of *Pseudocheirus*, *Potorous*, *Setonix*, or *M. eugenii* indicate the presence of dense ground cover, thickets, and possibly tall open-forest. The former two may only require the cover provided by dense low shrubs; the latter two probably require equally dense but larger shrubs. As *Pseudocheirus* nests and forages in trees, it has less stringent requirements for shelter on the ground, except in open-forest where there is likely to be less canopy cover (Strahan 1995). *Potorous* and *Setonix* require access to water. *M. eugenii* is often associated with an infrequent fire regime relative to the surrounding forest; this regime is often experienced on the borders of Karri and other forests (Christensen 1980).

The presence of *Isoodon*, *Trichosurus*, *Bettongia penicillata*, *Macropus irma*, or *M. fuliginosus* suggest relatively little about the overall trend of environments as these animals are widespread throughout the main vegetation formations.

Table 2.9 Ecology of medium- to large-size herbivorous mammals likely to be found in Leeuwin-Naturaliste Region deposits. After Strahan (1995), Walton (1988).

Scientific name	Habitat	Habits
Isoodon obesulus	temperate, open-forest, woodland, open heath, prefers low shrubby ground cover.	nocturnal, terrestrial, predator, insectivore, digs shallow nest holes, excavates prey.
Perameles bougainville	temperate, woodland, open scrub, tall shrubland, open heath, low shrubland, sandplain.	nocturnal, terrestrial, omnivore, predator, digs shallow nest holes, excavates prey.
Pseudocheirus peregrinus	tropical, subtropical, temperate, closed forest, tall open forest, open forest with thick shrub understorey	nocturnal, arboreal, florivore, folivore, frugivore, shelters in tree holes.
Trichosurus vulpecula	tropical, subtropical, temperate, tall open-forest, open-forest, woodland.	nocturnal, arboreal, terrestrial, folivore, omnivore, nests in any tree hole, burrow, etc.
Potorous tridactylus	temperate, coastal, heath, tall open-forest with thick ground cover.	nocturnal, terrestrial, root-feeder, fungivore, insectivore, predator, solitary, sedentary, shelters in thickets.
Bettongia penicillata	tropical, subtropical, temperate, open-forest, open woodland, open scrub, tussock or shrubby understorey.	nocturnal, terrestrial, fungivore, omnivore, nest builder, solitary.
Bettongia lesueur	temperate, coastal, tussock grassland, hummock grassland, low shrubland (heath), open woodland, desert.	nocturnal, terrestrial, burrowing, granivore, fungivore, root-feeder, omnivore, builds warrens, gregarious.
Setonix brachyurus	temperate, coastal, tall open-forest, open-forest, low open forest, low closed forest, woodland, heath, swamp thickets.	nocturnal, crepuscular, terrestrial, folivore, territorial, shelters in thickets. requires access to water.
Petrogale lateralis	tropical, subtropical, temperate, coastal, mallee (low open woodland), open scrub, tussock grassland.	noctidiurnal, terrestrial, montane, folivore, shelters in rocky terrain near feeding areas, gregarious.
Macropus eugenii	temperate, open-forest, woodland, scrub.	nocturnal, terrestrial, folivore, territorial, shelters in thickets near feeding areas.
Macropus irma	temperate, open-forest, woodland.	nocturnal, crepuscular, terrestrial, folivore.
Macropus fuliginosus	temperate, coastal, open-forest, woodland, low woodland, low open woodland, open scrub, grassland.	nocturnal, terrestrial, crepuscular, folivore, gregarious.

Table 2.10 Marsupial preferences for vegetation units (based on communities with one significant layer, Beard 1981: Table V)

Ticks indicate that the animal prefers this formation.

Plant form: Tall trees > 30 m; Medium trees 10-30 m; Low trees < 10 m; Shrubs > 1 m; Dwarf shrubs < 1 m.

Canopy cover (pfc): Dense: 70-100%; Mid-dense: 30-70%; Incomplete: 10-30%; Sparse: < 10%; Very sparse: negligible

Vegetation formations:	**Tall open-forest**	**Open-forest**	**Woodland**	**Low woodland**	**Scrub**	**Thicket**	**Heath**	**Shrubland**	**Grassland**
Plant form, canopy cover	Tall trees, dense to mid dense	Medium trees, dense to mid dense	Tall to medium trees, incomplete to sparse	Low trees, incomplete to sparse	Shrubs and dwarf shrubs, incomplete to very sparse	Shrubs, dense to mid-dense	Dwarf shrubs, dense to mid-dense	Dwarf shrubs, incomplete to sparse	Bunch grasses and hummock grasses, dense to very sparse
Isoodon obesulus		✓	✓	✓	✓	✓	✓		
Perameles bougainville			✓	✓	✓			✓	
Pseudocheirus peregrinus	✓	✓	✓	✓		✓			
Trichosurus vulpecula	✓	✓	✓	✓					
Potorous tridactylus	✓	✓				✓ (for shelter)	✓		
Bettongia penicillata		✓	✓		✓		✓		
Bettongia lesueur			✓	✓	✓			✓	✓
Setonix brachyurus	✓	✓	✓			✓ (for shelter)	✓		
Petrogale lateralis				✓	✓				✓ (for food)
Macropus eugenii		✓	✓		✓	✓ (for shelter)	✓		
Macropus irma		✓	✓						
Macropus fuliginosus		✓	✓	✓	✓				✓

Summary

The Leeuwin-Naturaliste Region environment shows close associations between physiography, soils, climate, vegetation, and fauna. These associations help one infer palaeo-vegetation from limited sets of biotic remains. Tables 2.9 and 2.10 show the habitat preferences of the larger mammals and indicate that their prehistoric remains broadly indicate past vegetation formations as defined by the dominant (canopy) layer. The discussion of the distributions and ecology of the major vegetation formations suggest that the dominant trees have particular rainfall requirements on certain soils. Hence, different rainfall regimes inferred for past climates could have encouraged shifts in vegetation boundaries, significant within small localities.

In this study I will also analyse remains of dominant trees that directly indicate changes in local palaeo-vegetational floristics. Archaeological, faunal, and botanical evidence can be compared with regional palaeo-environmental records to infer whether local changes in vegetation are more consistent with known palaeo-environmental parameters or with changes in human occupation patterns and a hunter-gatherer fire regime.

It is important to recall the reasons for assuming such a fire regime existed, and to summarise its potential effects. People may have frequently set fires in forests to improve foods for herbivorous animals and to manage forest understoreys and animal habitats. Very frequent artificial firing in marginal rainfall areas may convert tall open-forest to open-forest. Under high rainfall such firing depletes nutrients and potentially converts tall open-forest to dense scrub. Dense vegetation would exclude people, but eventually revert to a more varied structure due to the natural, lightning-induced fire regime, perhaps involving moderately frequent fires, as would be expected in the region's seasonally stressed, regularly dry Mediterranean climate. Moderately frequent fires, induced by lightning or managed by hunter-gatherers, would probably create a drifting but generally stable vegetation mosaic experiencing fires of varying intensity and extent, and occasional intense wild-fires in tall open-forest stands where dense understorey predominated would continue.

In this scenario, fire regime helps determine vegetation structure and floristic composition in canopy and understorey, within the constraints of climates and soils. Hunter-gatherers could have wielded most influence on vegetation in regions with rainfall marginal for that vegetation; conversely, the development of optimal rainfall would have limited the effect of artificial firing. However, these propositions cannot be verified from present-day knowledge, since observed fire regimes often involve intense fires and combine with severe effects such as logging and agricultural clearances. The question as to how much influence on vegetation is wielded by a regularly firing, small hunter-gatherer population can only be addressed in the final chapter of this study.

The next chapter assesses the documented palaeo-environmental record for south-western Australia and Australia as the regional context for local environmental change in the Leeuwin-Naturaliste Region.

Chapter 3 Palaeo-environmental change in south-western Australia

This chapter assesses the evidence for environmental change in south-western Australia since the time of the last glacial maximum (LGM), 18-22,000 BP. Two records reflect long-term environmental change in detail: the faunal and charcoal remains from the archaeological deposit at Devil's Lair in the Leeuwin-Naturaliste Region (Balme et al. 1978, Burke 1997), and the pollen sequence from lakes in suburban Perth (Pickett 1997). I discuss the potential effects of climatic changes inferred from these records on Leeuwin-Naturaliste Region vegetation, considering the ecological information given in Chapter 2. I conclude by assessing the possible effects of vegetational changes on hunter-gatherer occupation.

Climatic changes from the LGM to the late Holocene

Throughout the Quaternary, orbitally-induced changes in solar radiation have induced successive glacial and interglacial periods (Kutzbach and Webb 1993, Sturman and Tapper 1996, Williams *et al.* 1993). The most recent glaciation, and one of the most severe, was the Last Glacial Maximum (LGM), in the period 18-22,000 BP. During this period, temperatures fell, the polar ice-caps expanded, rainfall decreased, and ice sheets covered 30% of the earth's land surface. The world's oceans, comprising less water, covered less of the continental shelves.

At about 15,000 BP, world temperatures began to rise again, the ice-sheets began to retreat, and sea-level rose. The pace of change quickened after 12,000 BP, and by 9,000 BP present day temperatures and rainfalls prevailed, with the world's coastlines more or less at their present positions (Chappell and Shackleton 1986). In Australia, some researchers suggest that temperature and precipitation continued to rise, by 6,000 BP achieving levels warmer and wetter than those of the present-day, and remaining so until 4,000 BP, in a phase termed the "mid-Holocene optimum" (Harrison and Dodson 1993, Kershaw 1995, Wasson and Donnelly 1991). Afterwards, climates became slightly cooler and drier, to attain present conditions. However, some regions, including south-western Australia, provide little evidence for a mid-Holocene optimum (Newsome and Pickett 1993). Certainly, there is limited evidence anywhere for glacio-eustatic sea-level changes after 6,000 BP (Chappell 1993: 44).

Australian evidence

In Australia, Quaternary glaciations caused aridity, de-vegetation, temperature falls, and shifts in vegetation boundaries. These changes are documented by lake levels (Bowler 1976, Bowler and Wasson 1984, Harrison and Dodson 1993); dune morphology and sedimentology (Wasson 1989); and vegetational shifts inferred from pollen cores (e.g., Colhoun 1991; Dodson 1987; Hope 1978; Kershaw 1976, 1986; Macphail 1979, 1984; MacPhail and Colhoun 1985; Singh *et al.* 1981). These sources also indicate post-glacial trends, as do remnant glacial and periglacial features (Hope 1989); coastal landforms (Head 1988); and aquatic fossils such as ostracods and diatoms (O'Connor 1986, Yassini and Kendrick 1986). Sea surface and atmospheric temperatures are inferred from deep ocean cores of foraminifera deposits (Bé and Duplessy 1976, Connolly 1967, Shackleton and Opdyke 1973, Wells and Wells 1990), Antarctic ice cores (Jouzel *et al.* 1987), and the oxygen isotope record from dated coral reef terraces on the Huon Peninsula, Papua New Guinea (Chappell and Shackleton 1986). From the rate of tectonic rise at the latter site, the record of past sea-levels for Australasia has been calculated. At Lake Eyre, amino-acid racemization analysis of fossil eggshell indicates long-term regional temperature change (Miller *et al.* 1996).

Limitations with these records do not seriously challenge the general trends inferred from decades of research. The limitations are due partly to the records' restricted distribution and incompleteness, and partly to the limited responses to climatic change from the plants and animals whose remains provide many of the records, and which tolerate severe aridity relatively well (Kershaw 1995). In Australia, mass migrations of flora into newly deglaciated regions, as took place across the Northern Hemisphere, occurred only at very high altitudes (Hope 1989). Many past vegetation associations have no modern analogue, hindering the assessment of climatic and vegetational dynamics. However, the palaeo-climatic evidence that has accumulated is generally consistent across Australia.

Scarce palaeo-environmental evidence from the arid zone and the western half of the continent is not only consistent with the relatively detailed evidence from the east, but also with computer-generated "general circulation models" based on oceanic data (Harrison and Dodson 1993). These circulation models simulate changing climatic conditions, predict broad-scale effects in poorly documented areas, and suggest the causes of climatic change (Wright 1993). The Community Climate Model (CCM) predicts a poleward shift in the westerlies at the LGM, due to a southward expansion of the subtropical high pressure belt. This model is consistent with glacial period dune orientation and southern Australian aridity (Harrison and Dodson 1993). This shift would have brought dry summer winds to the southernmost parts of the continent, intensified anti-cyclonic flow in the interior, and displaced the moist westerlies southward. All these effects could have been expressed in the Leeuwin-Naturaliste Region (Chapter 2). In any mid-Holocene optimum, these trends would have been reversed, with an equatorward shift of the subtropical high pressure belt.

Sequence of climatic change in Australia, from 40,000 BP to present

During a pre-glacial stage, 40-25,000 BP, temperatures were perhaps slightly lower than present (Sturman and Tapper 1996, Aharon and Chappell 1986). Dated former lake shorelines indicate that lake levels were close to present-day levels but began to lower from about 25,000 BP with the onset of glacial period aridity (Wasson and Donnelly 1991). Dune building in the arid zone intensified from about 22,000 BP, suggesting higher wind speeds (+20%) and lower rainfall (-40%) by the LGM (Wasson 1989).

Under full glacial conditions, 18-15,000 BP, Australian sea-levels fell 130 m, exposing the continental shelves linking Australia, New Guinea, and Tasmania, and forming the land mass termed Sahul (Chappell and Shackleton 1986). Glaciers advanced in the mountains of New Guinea and Tasmania, lowering snowlines and vegetation belts by 1,000-1,500 m (Hope 1989). In eastern Australia, low temperatures and low and fluctuating rainfalls caused scrub and shrub-steppe to expand, and *Eucalyptus* and *Casuarina* forest and woodland to contract to valleys and coasts (Hope 1989). In the southern semi-arid zones, including the south-western interior, low lake levels and the formation of lunette dunes suggest deteriorating climate and vegetation cover, and increased windspeeds (Bowler 1976). Expanded coastlines reduced inland rainfall: vegetation belts on the Nullarbor Plain shifted south as the coastline retreated seawards 160 km from its present position (Martin 1973).

Timing of initial, post-glacial warming varied from 15,000 to 10,500 BP (Colhoun and Peterson 1986, Kershaw 1976). The period 12-9,000 BP saw the most extensive vegetational changes. As plant growth increased, many pollen sequences across the continent only began in this phase (Kershaw 1995). *Eucalyptus* woodland and forest expanded rapidly across south-eastern Australia and alpine regions, *Asteracae* steppe and heathland retreated, and rainforest appeared in valleys in south-west Tasmania and along the continental divide. Lake shorelines rose in mountainous and arid regions throughout eastern Australia (Harrison and Dodson 1993).

In the early Holocene, 9-6,000 BP, climatic warming ended, achieving in some areas higher temperatures and rainfalls and denser vegetation than today. In the arid zone, dune building, linked to faster winds and less vegetation, ceased at 6,000 BP, and resumed after 4,000 BP (Wasson and Donnelly 1991). South-eastern and south-western Australian lakes were full 7-4,000 BP (Bowler 1976). In Tasmania, high pollen counts of the rainforest taxon *Nothofagus* and the wet sclerophyll indicator *Pomaderis* centre on the date 6,000 BP (Macphail 1984). For the period 6-4,500 BP in south-western Australia, Churchill (1968) inferred higher rainfall from high Karri (*Eucalyptus diversicolor*) values, though recent studies at his main site, Boggy Lake, find the evidence equivocal (Newsome and Pickett 1993). Throughout Australia, sea-level stabilisation led to the formation of large coastal dunes and in-filling of estuaries (Head 1988, Woodroffe *et al.* 1985).

Late Holocene climate approximated modern conditions by 3,000 BP (Harrison and Dodson 1993). Intensified human occupation and firing may explain evidence for frequent burning, increases in sclerophyll forest and heath, and re-activation of desert dunes from 4,000 BP (Dodson *et al.* 1992, Wasson 1989). An alternative explanation is that climatic instability and a slight decrease in rainfall, inferred from hydrological data, caused frequent vegetational shifts and increased erosion (Kershaw 1995).

Late Pleistocene palaeo-environmental records from south-western Australia

Foraminifera

Foraminifera identified in deep sea cores off Western Australia, in latitudes 15°-35° S, suggest that LGM summer sea surface temperatures were 4-6° cooler than present day (Bé and Duplessy 1975, Connolly 1968, Wells and Wells 1990). The effects are more marked in low latitudes where the Arafura Sea is now fed by warm water travelling through the Torres Strait from the Pacific (Gentilli 1991), which would have been cut off during the LGM when sea-levels lowered and Torres Strait ceased to exist (Wells and Wells 1994). In higher latitudes, sea surface temperatures are likely to have been colder if the offshore Western Australian current, emanating from the Southern Ocean, became colder or stronger (McCorkle *et al.* 1994, Wyrwoll 1979). A colder ocean surface would have formed fewer clouds and brought less rain to the coast; meanwhile, low sea-levels would have retracted the humid influence of the maritime climate to the LGM coastline 30-40 km to the west (cf Gentilli 1991).

Lunette dunes

Bowler (1976) dated a dune-building episode around four inland salt lakes 400-600 km east of Perth, to 20-15,000 BP. These dates suggest a drying phase coincident with that documented from playas in south-east Australia. Despite Bowler's promising preliminary results, no further work has been done at Pleistocene lunette dunes.

Devil's Lair faunal record

Research on this deep, well-stratified cave deposit forms a major palaeo-environmental record for south-western Australia (Baird 1992; Balme *et al.* 1978; Baynes *et al.* 1975; Lundelius 1960, 1983; Merrilees 1979a, 1979b, 1984; Tyler 1985). Thousands of mammal remains from this site and other caves, chiefly analysed by Western Australian Museum palaeontologists under the direction of D. Merrilees, indicate shifts in the proportions of open woodland and more closed forest around the locality. The inference is based on the same type of assumptions given in Chapter 2. Many of the remains, deposited by non-human predators before 31,000 BP and Aborigines and other predators from 31,000 BP, are from species that the researchers associated with either closed vegetation ("forest mammals") or woodland and scrub-heath habitats ("non-forest mammals"). These mammals are listed in Table 3.1.

Table 3.1: Forest, non-forest, and other animals at Devil's Lair, identified by Balme et al. (1978)

Forest mammals		Non-forest mammals	Habitat unknown
Rattus fuscipes	*Trichosurus vulpecula*	*Macroderma gigas*	*Vombatus hacketti*
Hydromys chrysogaster	*Phascolarctos cinereus*	*Pseudomys* spp.	*Protemnodon brehus*
undifferentiated bats	*Potorous tridactylus**	*Notomys mitchelli*	*Sthenurus browneii*
Cercatetus concinnus	*Bettongia penicillata**	*Tarsipes spencerae*	*Sthenurus occidentalis*
Phascogale tapoatafa	*Macropus eugenii*	*Perameles bougainville**	*Zygomaturus ?*
Antechinus flavipes	*Macropus irma*	*Bettongia lesueur*	
*Sminthopsis murina**	*Macropus fuliginosus**	*Petrogale lateralis**	
*Dasyurus geoffroii**	*Sarcophilus harrisii*	*Lagorchestes ?*	
Isoodon obesulus	*Thylacinus cynocephalus*		
*Pseudocheirus peregrinus**	*Setonix brachyurus*	*remains are abundant in the deposit	

Mammals from "forest" or "non-forest" are represented throughout the deposit, suggesting that open and closed vegetation formations existed close to the site at all times. The mammals mostly have wide tolerances, so continuous records for these species are consistent with moderate fluctuations in climate or vegetation cover. By the same argument, "notable absences" of some species that are abundant in other parts of the deposit suggest strong vegetational change. Moreover, the proportion of "forest" mammals in the entire vertebrate assemblage increases after the LGM, as the proportion of "non-forest" mammals declines, suggesting that some vegetation units around the cave have expanded or become more closed.

Balme *et al.* (1978) argue that one boundary between vegetation units would not have shifted, regardless of the vegetation on either side. This is the present boundary between coastal heath and forest or woodland, at the crest of the Leeuwin Ridge and 2 km west of Devil's Lair. The reason is that prevailing winds and the soil substrate are key determinants of the vegetation west and east of the ridge crest. These are unlikely to have changed much since the late Pleistocene and the most likely vegetational changes are likely to have been in structure. For example, the high proportion of "non-forest" relative to "forest" animals at the glacial maximum could be attributed to a scrub or woodland formation west of the ridge and Jarrah or Marri open-forest east of it. By the early Holocene, these formations could have shifted to the present scrub and Jarrah woodland on the west and Karri tall open-forest on the east.

The sparse late Pleistocene record of *Potorous*, *Setonix*, and *Hydromys*, implies the survival of thickly vegetated water courses, and that forest contraction at the LGM was insufficient to extinguish these forest mammals. Terminal Pleistocene increases in *Potorous* and *Setonix* and possibly *Antechinus*, *Cercatetus*, and *Rattus* suggest increasing understorey density, temperature, and rainfall. Remains of *Macropus fuliginosus*, *M. irma*, *Petrogale*, *Bettongia penicillata*, and *B. lesueur*, all species that prefer open formations (see Chapter 2), are found from c.30,000 BP to c.12,000 BP. These macropods then appear to decline, with *Petrogale* and *B. lesueur* rare or absent in nearby cave deposits that are younger than 8,000 BP (J. Dortch 1996, Porter 1979). Balme *et al.* (1978) treat *M. fuliginosus* and *M. irma* as forest mammals, but they are rare in tall open-forest (Christensen and Kimber 1975), and are perhaps only common in the Leeuwin-Naturaliste Region today because of land clearances (Baynes *et al.* 1975). Thus from 12,000 to 8,000 BP some "non-forest" species became locally extinct, and others declined. The Devil's Lair faunal record ends with sealing of the cave entrance some time after 8,000 and before 6,500 BP (Balme *et al.* 1978, Lundelius 1960).

Late Quaternary migrations of the coast probably did not cause the local extinctions at 8,000 BP, since marine transgression probably slowed around 10,000 BP, and later contractions in these animals' potential range were probably insufficient to produce any stress (Balme *et al.* 1978). Moreover, lizards are animals whose small ranges would not be seriously affected by the encroachment of the coastline, and confirm the inferences of climatic change. Lizard numbers, corresponding with aridity, are high from 19-12,000 BP, and peak 15-14,000 BP. Difficulties with identifying frog bones preclude many species identifications, and the total number of frogs throughout the deposit merely indicates the continuation of suitable frog habitat throughout the late Quaternary (Tyler 1985). Climate was probably the main factor in determining habitats around Devil's Lair, but it need not have eliminated any habitat.

Balme *et al.* (1978) conclude that the forest in the lee of the Leeuwin Ridge, where Devil's Lair and Tunnel Cave are located, probably shifted from Jarrah or Marri open forest to Karri tall open-forest by the early Holocene. Assuming no alteration of human hunting or firing practices, nor any change of animal predator habits, the lizard maximum and the abundance of non-forest mammals suggests an arid phase that continued almost to the terminal Pleistocene, 12,000 BP: "...while the climate need not have differed greatly from the present regime, such differences as there were must have been in the direction of lower or more markedly seasonal rainfall during late Pleistocene time than now" (Balme *et al.* 1978: 63). Extinctions of mammal species, as the most tolerant of all vertebrates considered by Balme *et al.*, suggest a marked climatic change. The authors note that more detailed and conclusive evidence would come from studies of plant fossils, lower vertebrates, or invertebrates.

Devil's Lair avifauna

Baird (1992) examined the Pleistocene avifauna from Devil's Lair and found specimens of the Mulga Parrot (*Psephotus varius*) in layers dated from 12,000 to 20,000 yrs BP, and others of the Noisy Scrub-bird (*Atrichornis clamosus*) and Bristle-bird (*Dasyornis longirostris*), only in layers younger than 12,000 yrs BP or older than 32,000 yrs BP. Today, *P. varius* requires an open shrub understorey and five or more dry months per year, and the latter two species inhabit thick vegetation in high rainfall areas (Serventy and Whittell 1967). The presence of *P. varius* in the older layers indicates a drier climate and wider areas of open understorey during the LGM, and the presence of *Atrichornis* and *Dasyornis* in the younger layers suggest that vegetation became denser and rainfall increased after 12,000 BP. The inferences drawn from bird remains about glacial and post-glacial vegetation and climate are consistent with those drawn from mammal remains (cf. Balme *et al.* 1978).

Devil's Lair charcoal

Burke (1997) provides a palaeo-vegetational record for Devil's Lair based on charcoal fragments, that suggests that Karri trees grew at the site after the blocking of the cave's former entrance some time after 6,500 BP. Karri charcoal is common in a surface layer of humic sediment (layer A) deposited when the present entrance formed, and dated 320 BP[1]. Layer A has no remains of the "non-forest" mammals, such as *Petrogale*, *Bettongia lesueur*, and *Perameles*. Conversely, Karri charcoal is absent from parts of the cave's deposit that date to the period when the cave's former entrance was open, from the terminal Pleistocene to the early Holocene.

The charcoal assemblage that pre-dates 6500 BP includes Jarrah, Marri, Tuart, and species of *Banksia*, which, as Burke (1997) infers, probably derive from open-forest or woodland. These plants' remains are stratigraphically associated with those of the "non-forest" mammals, up to the time that the former entrance became blocked and deposition ceased, only to resume with the opening of the present entrance.

[1] This charcoal derives from undisturbed parts of layer A located away from the main excavation (Burke 1997).

Burke's (1997) analysis agrees well with that of Balme *et al.* (1978). An important difference is that he finds no evidence for Karri in the Devil's Lair locality prior to 6,500-8,000 BP, or when the former cave entrance collapsed and temporarily halted deposition of biotic remains. He suggests that the changes in fauna 12-8,000 BP, when non-forest mammals and reptiles declined, and forest mammals increased, are attributable to changes in vegetation structure. This statement is supported by evidence that the diameter of vessels in Jarrah charcoal increased significantly from 18,000 to the period 12-8,000 BP. In some eucalypts, vessel diameter is proportional to tree height, and Burke notes that large Jarrah trees are today associated with wetter climates and dense understoreys.

Swan Coastal Plain pollen records

Pickett (1997) assesses pollen sequences extending back 40,000 years, from Lake Bangenup, Bibra Lake, and North Lake, which are located on the Swan Coastal Plain in the southern suburbs of Perth about 5 km from the coastline. Before 28,000 BP, and after 11,000 BP, eucalypt woodland was the dominant regional vegetation. From 23,000 BP, a *Banksia-Casuarina-Eucalyptus* association similar to open woodland belts now bordering the Darling Scarp, 20-30 km inland, appears to have expanded westward to the lakes. This woodland phase included an arid episode at 15,000 BP, when *Eucalyptus* declines and *Casuarina* is prominent. From 11,000 BP, the *Eucalyptus-Casuarina* association, similar to modern vegetation at the site, returned. There are asynchronous changes in the mid-Holocene vegetation records at each lake, suggesting local variations in ground-water, soil development, and fire regime.

These findings reveal no major vegetational shifts on the Swan Coastal Plain as a whole. Rather, the LGM and post-LGM floristic shifts that are expressed in the lakes' pollen records represent east-west movements of vegetational belts lying parallel to the coast (Beard 1981). Global changes in atmospheric and oceanic circulation, mentioned above, met with the vegetation's natural resilience. Major influences on vegetation were marine regression and transgression, which controlled the position of the coastal rain belt; the presence or absence of the Leeuwin Current (cf. Fahrner 1993); and variations in the height of the coastal plain's shallow, unconfined groundwater system.

Abundant charcoal fragments in the cores suggest that vegetation was always fire-prone. Charcoal became more abundant, and hence fires could have been more extensive or frequent, in the Holocene. Since the Holocene was a wet period, the increase in charcoal is probably due to more abundant vegetation, rather than on the whole drier conditions. The summer drought of the modern period (in what is still a wet climate) had probably persisted and hence permitted seasonal fires throughout the Holocene.

Computer-generated climate models

One interpretation of south-western Australian climates conflicts directly with the emerging detailed local records. According to this interpretation, based on a computer-generated climatic model, the LGM in south-western Australia was wetter, not drier, than present.

Hubbard (1995a) discusses the first version of the Community Climate Model (CCM0), which predicts that the southern boundary of the southern hemisphere high pressure belt moves poleward in glacial conditions. Modern weather patterns suggest that a long-term southern expansion of the sub-tropical high pressure belt (as posited for the LGM) would keep rain-bearing westerlies south of the continent and account for glacial aridity as evinced by most regional Pleistocene records. Hubbard believes that some Australian data (especially in the south-west and the south-east) contradict this model. She concludes that the model is flawed and needs testing against improved regional records. Subsequently Hubbard (1995b) notes that CCM0 never allowed for likely LGM expansions of Antarctic sea-ice, which today (she notes) are correlated with equatorward movements of the Antarctic high pressure belt. If this high pressure belt moved northward at the LGM, then the westerlies would have been forced northward also. Assuming increased evaporation (Hubbard's estimate) over the ocean from increased windspeeds (Wasson 1989), then an intensified westerly influence would bring wetter conditions to southern Australia. Hubbard revises the predictions of CCM0 and proposes that the LGM climate in south-western Australia was wetter than present. She concludes that this prediction is consistent with the south-western palaeo-environmental data as she interprets it.

A computer model that includes sea-ice effects may or may not indicate a LGM change in this direction. However, the models that Hubbard cites have deficiencies in the southern hemisphere. CCM0 and the improved version, CCM1, are inconsistent with southern data such as palaeo-lake levels, possibly because southern hemisphere ocean currents are more complex than suspected when the models were designed (Harrison and Dodson 1993, Qin *et al.* 1998). Moreover, Harrison and Dodson (1993) note that the LGM extension of Antarctic sea-ice would *not*, as Hubbard suggests, have forced the westerlies equatorward, as probably happened in the Northern Hemisphere with Arctic sea-ice expansion. The Northern Hemisphere ice-sheet acted as an orographic barrier, but no such barrier formed between Antarctica and the other Southern Hemisphere continents. Various data from southern Australia and South America indicate that the Southern Hemisphere westerlies contracted *poleward*, and must have been constricted by expansion of the subtropical and Antarctic high pressure belts on either side (Harrison and Dodson 1993).

A problem with Hubbard's assessment of the evidence for her model is revealed by her statement on the Devil's Lair faunal record: "after ... 25,000-22,000 yr BP non-forest species decline relative to forest species until about c. 14-16,000 yr BP" (1995a: 154). On the contrary, Balme *et al.* (1978: Figure 12) illustrate the reverse in their graph. Non-forest species declined after 25,000 BP, then increased from 20,000 BP, peaked at 14,000 BP, and declined from 12,000 BP. Contrary to Hubbard's statement (1995a:154), Balme *et al.* (1978:60) infer no late Pleistocene expansion of forest, apart a possible spread of open-forest or woodland associations relative to tall open-forest. Moreover, the post-12,000 BP changes in non-forest and forest species are much greater than any variation in the ratio of these two groups in the period 20-12,000 BP. In the period 20-12,000 BP, non-forest species were prominent and forest species were scarce. As far as fossil vertebrate records can agree with a climatic model, Devil's Lair faunal trends agree with the original CCM0, not with Hubbard's revised version.

Hubbard also cites a LGM high water level at Storey's Lake, one of the above-mentioned south-western inland sites, but inspection of the original data indicates that Bowler (1976: Table 4) infers active dune growth (i.e., a *dry* lake-bed) 19-17,000 BP. Hubbard's detailed assessment of other Australian regions is not reviewed here, but this re-assessment of her south-western Australian model has implications for other views of late Pleistocene human occupation patterns (O'Connor *et al.* 1993, cf. Dortch and Smith 2000).

Holocene palaeo-environmental records

Evidence for Holocene climatic changes in south-western Australia is conflicting, and does not document a mid-Holocene optimum as reported for south-eastern Australia. Holocene palaeontological sites are as sparsely scattered as the Pleistocene ones.

The first reported pollen sequences from Boggy Lake, Weld River Swamp, and Flinders Bay Swamp (all coastal sites between Augusta and Walpole), suggest a wetter phase favouring Karri forest, 6-5,000 BP (Churchill 1968). In contrast, Kendrick (1977) and Yassini and Kendrick (1988) infer low or sporadic mid-Holocene rainfall from analysing fossil mussel beds in the Swan River estuary. Semeniuk (1986) argues that the extent of calcrete in coastal sands of the southern Perth Basin also suggests a mid-Holocene arid phase, and O'Connor (1986) reports a mid-Holocene dry phase in a diatom sequence from Hertha Road Swamp near Perth. In contrast, Newsome and Pickett (1993), who compare all these Holocene records to their pollen sequences at Boggy Lake and Loch McNess (at Perth), find limited evidence for any significant regional climatic variation.

Boggy Lake and other pollen sites

Churchill (1968) argues that the moisture requirements of Karri and Marri forests suggest changes in rainfall according to palynological evidence of their past distributions. His sequence suggests a wetter phase, favouring Karri, 6-4500 BP, followed by a dry phase favouring Marri, 4500-2500 BP, and then later and minor fluctuations. Newsome and Pickett (1993) identify a similar vegetation sequence to Churchill's at Boggy Lake, but argue that several factors undermine Churchill's conclusions: a large percentage of *Eucalyptus* pollen is always unidentifiable; the ratio of two species, rather than a range of taxa, is subject to erratic variation because of variable representation of Marri pollen; ecological controls on Karri are complex, and soils, unconsidered by Churchill, restrict its present distribution, rather than rainfall alone (Beard 1981). It seems unlikely that Karri was more widely distributed than present in the high-rainfall Boggy Lake locality, since Karri forest already occupies most substrates that are suitable for Karri there. Newsome and Pickett (1993) suggest that changes in pollen counts could result from changing climatic conditions, if these climatic changes favoured pure stands of one species. However, it is also possible that local sand movements, unrelated to rainfall, destabilised vegetation. Thus Boggy Lake offers equivocal evidence for a wetter mid-Holocene.

Devil's Pool

McNicol (1999) analyses the upper part of an 8.0 m deep pollen core from Devil's Pool, which is a lake on Boodjidup Brook (Figure 2.3a), and c.200 m from an archaeological site investigated here (Witchcliffe Rock Shelter; see Chapter 5, J. Dortch 1996). The uppermost two metres of core (the remainder is under analysis: B. Smith, University of Cape Town, pers. comm.) records changes in pollen floristics since 2,000 BP. The Devil's Pool core, as presently known, broadly supports the continuation of similar vegetation from 2,000 BP to the present day (notwithstanding changes associated with land clearance by settlers at c. 150 BP). More definite conclusions await completion of the analysis of the core's lowermost six metres. However, floristic changes at 800 BP suggest a slight loss of moisture in the locality (e.g., a lower water table) or an alteration in fire regime (McNicol 1999). As suggested by the nearby archaeological site, people were probably active in the locality from at least this date, and human firing perhaps could have influenced vegetation more than they had before, if there had been a decrease in rainfall or lowering of water table at 800 BP.

Skull Cave vertebrate remains

Skull Cave, located approximately 12 km north of Cape Leeuwin and 15 km south of Devil's Lair, has a fossil vertebrate record dated 7,000 BP to less than 3,000 BP, that confirms the Holocene extinction or contraction of fauna identified at Devil's Lair (Baird 1991, Porter 1979). Mammals (*Bettongia lesueur*, *Petrogale*, and *Notomys*) requiring woodland or heath disappeared from the deposit by 7,000 BP. By 3,000 BP, species with similar habitat requirements, *Perameles* and *Pseudomys albocineurus*, had also disappeared. Mammal and bird remains support the existence of forest close to the site throughout the period of deposition, and the persistence of heath at presumably some greater distance.

Swan Coastal Plain aquatic fauna

At Guildford on the Upper Swan Estuary, Kendrick (1977) reports a date of 6,700 BP for a fossil bed containing 31 mollusc species that today have marine affinities. These fossil molluscs are located at a modern fresh-water confluence, where they would not have tolerated modern levels of seasonal freshwater discharge, even allowing for mid-Holocene transgression that extended the river's oceanic connection and enhanced marine conditions in the estuary. The Guildford fossils are stratigraphically associated with a soft calcareous siltstone, perhaps deposited in a marine environment (Kendrick pers. comm.), unlike the poorly sorted quartz sand deposited by winter floods today. Kendrick infers that mid-Holocene discharge, and therefore rainfall, was lower, sporadic, and aseasonal, relative to present.

Investigations at Point Waylen in the Middle Swan Estuary reveal a sequence of foraminifera and ostracods, including two phases of estuarine-lagoonal fauna dated 6-4,000 BP (Yassini and Kendrick 1988). As at Guildford, these fauna indicate a transgression, with low or sporadic freshwater discharge in the estuary. The fauna of a third phase, starting 4,000 BP, suggests regression of c. 1 m and seasonal, regular discharge; i.e., a shift to the modern sea-level, hydrological regime, and climate. The disappearance of similar mollusc fauna at Peel Inlet, 60 km south of the Swan River (Brown 1983), at 4,000 BP, suggests a regional phenomenon of regression.

Continuing work by Kendrick (pers. comm.) shows that these fossil shell beds on the Swan-Canning Estuary include molluscs now living in warm waters off Geraldton (500 km north) and cool waters off Albany (400 km south). This diverse, enriched fauna disappeared from 4,000 BP with the above-mentioned regression, but regression is not easily admitted as the sole cause. The deep, saline mid-Holocene Swan Estuary has no modern structural analogue in Western Australia, and temperature differences between then and now cannot be yet inferred. All that one can say is that the evidence strongly indicates a mid-Holocene hydrological regime in the Swan Estuary very different from that of the present day.

O'Connor (1986) reports a diatom sequence from a core at Hertha Road Swamp, near Perth, dated 8100 BP to present. A 170 cm section of the core, dated 6800-4400 BP, and containing no diatoms, indicates a dry phase in the swamp's history. Only shallow water flora lived there at other times, and higher water level than present would support a wider range of these unicellular plants rather than none at all.

Swan Coastal Plain calcrete

Semeniuk (1986) identifies calcrete horizons in the profiles of four east-west transects on the Swan Coastal Plain, and relates the extent of calcrete formation in dune sands to rainfall, among other factors. Southern transects, with extensive calcrete beds forming above the modern water table, have limited mid-Holocene beds; northern transects have rare occurrences of calcrete, none of which are dated to the mid-Holocene. Semeniuk suggests that these investigations support an arid phase on the lower west coast lasting 7-2,000 BP, considerably longer than that proposed by Kendrick (1977) and Yassini and Kendrick (1988). The findings of the calcrete formation analysis have little support at present; possibly the relationship between water table height and rainfall, sea-level, and vegetation cover is still inadequately understood (Pickett 1997).

Summary of regional palaeo-environmental change

During the LGM, low rainfalls, high wind speeds, and marked seasonality reduced vegetation cover and destabilised the landscape throughout Australia. At the Pleistocene-Holocene transition, when climate changed most rapidly, forest and woodland expanded throughout southern Australia. In some Australian regions, from 8-4,000 BP, temperature, rainfall, and vegetation cover peaked, afterwards decreasing to present day levels.

The limited Pleistocene records in south-western Australia indicate aridity during glacial periods, consistent with evidence from around Australia and the world. As with other parts of southern Australia, conditions appear to have been cool and dry 18-12,000 BP, with vegetation and fauna most affected about 18-15,000 BP. However, the south-west data, at present, do not suggest that change was as extreme as shown in parts of south-eastern Australia or Tasmania (Harrison and Dodson 1993, Kershaw 1995).

The Devil's Lair and North Lake sequences suggest that woodland replaced forest, and forest replaced tall open forest, probably with subtle changes in floristic composition. There is very little data to show the regional extent of glacial period changes. At the end of the Pleistocene, c. 12,000-10,000 BP, rapid shifts towards modern conditions are suggested by interpretations of deep-sea cores, pollen records, lunette dune radiocarbon dates, and Devil's Lair faunal remains. Hubbard's 1995 assessments of a computer-generated climatic model are the only investigations that contradict these interpretations, but the climatic model is obsolete and the regional evidence cited in support of it is incorrectly interpreted.

If the Boggy Lake cores fail to show a mid-Holocene optimum in south-western Australia, they also show no sign of the enhanced aridity inferred from Swan Coastal Plain or estuary sites. Newsome and Pickett (1993) add that pollen cores at Loch McNess, also located on the Swan Coastal Plain, suggest no arid trend of the extent visualised by Semeniuk (1986).

Moreover, Pickett (1997) identifies limited evidence for changes after 11,000 BP, and argues that if there were any, they could not have been towards greater aridity. In fact, small Holocene changes in the coastal lakes' arboreal and aquatic pollen records may be sufficiently explained by sea-level, coastal rainfall, and water table fluctuations rather than climatic variations. These factors may be significant in the other, coastal Holocene south-western Australian sequences. Variations in water table height may be significant at Devil's Pool, for example (McNicol 1999).

The terminal Pleistocene faunal sequence from the Leeuwin-Naturaliste Region and the Swan Coastal Plain pollen record suggest the following: Holocene climates tended towards present-day, warm, humid conditions; and there may have been considerable variability between different localities. The small number of sites, their restricted distribution, and the inconsistencies observed at Holocene sites mean that while south-western Australian climates throughout the Quaternary are broadly understood, vegetation distributions can be assumed at very few localities. For constructing histories of site use and vegetation in the Leeuwin-Naturaliste Region, additional local records are needed. The available evidence does suggest that important changes would have happened at the local level.

Local vegetational change in the Leeuwin-Naturaliste Region

Deducing past vegetational responses to climatic changes depends largely on the assumption that on a given soil, rainfall is a key determinant of vegetation. This assumption is derived partly from vegetation surveys showing the existence of vegetation series, or sequences of vegetation responses on a limited set of soils to variations in the key climatic factor, rainfall (see Chapter 2; Beard 1981, 1983). It is also based on interpretations of meteorological records kept across south-western Australia throughout the last century (Beard 1981, Gentilli 1989). These records suggest that evaporation, the radiation budget, and length of summer drought vary approximately together and inversely to total rainfall. The likely contractions in rainfall or rainy season experienced in the Leeuwin-Naturaliste Region in the LGM may have been a major influence on vegetation, which may have changed within each *vegetation series* as defined and tabulated in the previous chapter (see also Table 2.5).

An important corollary of this table is that climatic changes would not have introduced vegetation that is new to the Leeuwin-Naturaliste Region unless annual rainfall was 70% less than present (see Chapter 2). Such a change is not indicated by the existing evidence, particularly that presented by Pickett (1997) and Balme *et al.* (1978). All of the biotic remains discussed by these authors derive from species whose modern distribution lies within the Darling Botanical District or even within several kilometres of the palaeontological sites. The preferred vegetation formations for the fauna identified at Devil's Lair and elsewhere are still present within all the botanical sub-districts and in many cases, parts of the Leeuwin-Naturaliste Region (see also Baird 1991, 1992).

However, two tree species are at the extremes of their range in the Leeuwin-Naturaliste Region and immediately to the north. Tuart (*Eucalyptus gomphocephala*) appears to be limited to annual rainfall zones of c. 600-900 mm, and it extends no further south than Busselton (Figure 2.4). Karri (*E. diversicolor*) requires at least 750 mm of annual rainfall and normally does not tolerate a summer drought longer than four months. It grows in isolated patches in the northern Leeuwin-Naturaliste Region, where there is a four-to-five month summer drought (Figure 2.3b), and where, presumably, micro-variations in climate, topography, and position relative to the zone of coastal rainfall allow it to persist in this marginal zone (Beard 1981, F.G. Smith 1973).

These observations suggest that small contractions in rainfall, perhaps, as visualised by Pickett (1997), reflecting shifts in the coastal rainfall zone, would have allowed Tuart to colonise suitable limestone substrates in locations south of its present distribution. The same rainfall contractions, in both seasonal and geographical senses, would have excluded Karri from parts of the northern and central Leeuwin-Naturaliste Region, although of course they did not remove the few patches that persist today (Figure 2.4; Beard 1981).

Given that there was probably relatively slight climatic change throughout the Holocene, it is difficult to envisage Karri forest expanding across the Leeuwin-Naturaliste Region in the period after the LGM to these widely-spaced locations and then contracting to mere outliers by the present day. Karri outliers are more likely to be relics of wetter climates experienced long before the LGM. Hence the changes experienced regionally in the LGM could not have been so extreme as to totally eradicate the Karri outliers.

As will be shown in Chapter 9, climatic changes at the LGM did extend the range of Tuart and reduced that of Karri. The changes may have been small, and almost certainly not large enough to have radically altered regional vegetation. But local vegetational changes may have been significant at hunter-gatherer occupation sites, perhaps more so if they were rapid or resisted mediation by human burning. Variations in vegetation structure and composition over distances of a few hundred metres can be significant to many human food species. Accessibility of sites to humans through tall open-forest or thickets may also have been significant.

Humans today are capable of altering local and even regional vegetation drastically. While one cannot infer prehistoric human fire-control of vegetation from charcoal horizons in fire-prone environments (Head 1988, Pickett 1997), one can infer human control of forests from human occupational records in archaeological sites. If human occupation at sites continued through times when changes in environmental parameters favoured new forest types at those sites, there are implications for the extent of hunter-gatherer control of forests.

Chapter 4 Hunter-gatherers in tall open-forests

This chapter reviews archaeological and historical evidence for hunter-gatherer occupation of tall open-forests. In Chapter 1, I mention claims that hunter-gatherers failed to occupy the least favourable vegetation types in a region. In south-western Australia this claim has been made for Karri tall open-forest (Ferguson 1985, Hallam 1975). Ecological and limited ethnographic research suggests that this vegetation formation contains few food resources and tends to have a dense understorey that is difficult for people to move through or control by fire (see Chapter 2 and below). Yet there is some evidence for a hunter-gatherer presence in Karri tall open-forest, nevertheless.

The first piece of evidence is from Tunnel Cave, an archaeological site located in a small stand of Karri tall open-forest in the southern part of the Leeuwin-Naturaliste Region. In the uppermost part of the sandy deposit in the cave's entrance area, there is a hearth (called Feature 1) made by Aboriginal people, dated *c.*1,400 BP (J. Dortch 1996). Because no rapid changes in climatic conditions can be envisaged in the last fourteen centuries, that would have caused a sudden incursion of Karri forest, the hearth was conceivably built at a time when Karri trees grew at the site.

The second piece of evidence is from Dombakup 24, which is a forest "compartment" of the Western Australian Department of Conservation and Land Management (CALM), located near Northcliffe, 150 km south-east of the Leeuwin-Naturaliste Region. Dombakup 24 includes a large stand of pure Karri tall open-forest. J. Dortch *et al.* (1998) report a test-pit survey in this stand that found stone artefacts a few centimetres below the surface. In one test-pit, a radiocarbon-dated charcoal sample identified as Karri, taken from below artefacts, suggests an Aboriginal presence in the centre of the present stand after *c.*550 BP (J. Dortch *et al.* 1998). Because of the longevity of Karri trees, the investigators conclude that this date and the estimated thousands of near-surface artefacts in Dombakup 24 indicate recent hunter-gatherer occupation of this particular Karri tall open-forest. It is particularly significant that the stand now contains extremely dense understorey, which the field workers had to either smash down or crawl through. To camp in the area defined by the present compartment, Aboriginal people would have probably cleared the understorey with fire. If they fired the understorey in Karri tall open-forest here, might they have also fired and occupied other stands of pure Karri?

The Dombakup 24 evidence, the uppermost hearth from the Tunnel Cave sequence, and a few historical documents (discussed below) constitute limited evidence for hunter-gatherer use of Karri tall open-forest. In the following chapters I present detailed evidence, in the form of my analysis of Leeuwin-Naturaliste Region archaeological sites, that helps to answer the question of hunter-gatherer occupation in tall open-forest. Before I turn to this detailed evidence I summarise the evidence for hunter-gatherer use of all forest types, and of tall open-forest in eastern Australia and in Tasmania. The archaeological and ethnographic records from these forested regions suggest the potential for hunter-gatherer occupation in the south-western forests.

In this chapter I also assess the full archaeological and historical records that pertain to hunter-gatherer occupation of south-western Australian forests, as some of this data is also consistent with Aboriginal use of tall open-forest. These records and the ecological constraints discussed in Chapter 2 are then summarised to indicate the potential for hunter-gatherer occupation in, or exclusion from, tall open-forest.

Hunter-gatherers in forests

Hunter-gatherers in all regions have close relationships with vegetation. They all have developed techniques for using vegetation types that contain few resources or are dense and hard to move through. These techniques include the following:

1. controlling vegetation by firing it. Firing is a worldwide practice; examples are found in North and Central America (Chapman *et al.* 1982, Lewis 1973, Patterson and Sassaman 1988, Piperno 1994, Piperno *et al.* 1990), Europe (Clark 1975, Dimbleby 1961, Mellars 1976, Mellars and Reinhardt 1978, Simmons *et al.* 1981, A.G. Smith *et al.* 1989), Australia and Tasmania (Gould 1971; R. Jones 1969; Lourandos 1983a, 1983b; Thomas 1994).

2. submitting otherwise inedible foods to intensive processing or specialised techniques. In Australian forests such techniques might include leaching and grinding of cycad fruit (Beaton 1977, Cosgrove 1996, Harris 1989, Horsfall and Hall 1990); an analogous technique is the grinding of seeds in deserts (Gould 1977a; M.A. Smith 1986, 1989).

3. improving the number of potential foods by foraging in many different vegetation types in both inland and coastal or riverine zones. Broad-base or diverse foraging is documented ethnographically in island south-eastern Asia (Endicott and Bellwood 1991), historically on the New South Wales coast (Belshaw 1978; Bowdler 1983a, 1983b; Byrne 1983; McBryde 1977) and in south-western Australia (Anderson 1984, Hallam 1987, Meagher 1974), and archaeologically in south-western Europe (Straus 1990).

Around the world at the end of the last glacial period, about 10,000 years ago, vegetation changed rapidly, and groups that had not developed some of these techniques are proposed to have abandoned the most inhospitable regions. These regions were only recolonised relatively recently, as environments improved, population pressure increased, or with technological or economic innovations (Bailey *et al.* 1989, Binford 1968, Cosgrove *et al.* 1990, Driver *et al.* 1996, Lourandos 1983a, Ferguson 1985, B.D. Jones 1994). Most of these regions are resource-poor, difficult environments: high latitude boreal forests or equatorial rain-forests. Among these studies, only Ferguson (1985) proposes a prehistoric abandonment of a temperate region forest.

Contrasting with these arguments, that environment determines settlement and subsistence patterns, an alternative idea is that social organisation and historical trajectories are most influential in changing settlement and subsistence patterns; the main role of environment is to set outer limits (cf. Bender 1978, de Bie 1997, Ingold 1984, Jochim 1996, Lourandos 1985).

Standing apart from these competing environmental and social approaches to explaining cultural change is the concept that environments are artefacts of human activity. Environmental modification has been argued for many Australian regions (cf Head 1988, 1989, R. Jones 1969, Lourandos 1980, Merrilees 1968). In south-western Australia, different forest types respond differently to fire. Within any part of the forest mosaic, the abandonment of a site, and the cessation of Aboriginal firing in its immediate vicinity, could have allowed one vegetation type to supplant another. This transition is known to have happened in the period of European settlement, which saw the end of the pre-existing fire regime (Rayner 1992, Ward 1998, Ward and van Didden 1997, for other parts of Australia see Gell *et al.* 1993, Head 1989, 1994).

The above-mentioned research on hunter-gatherers in forests worldwide suggests there are as many techniques for occupying forests as there may be forest types. The present study has only scope to address the potential for human occupation of one type. Tall open-forests, a peculiarly Australian formation (Ashton and Attiwill 1994), have led to a range of archaeological perceptions, from "difficult to occupy" to "capable of manipulation". They therefore have substantial interest to the question of human responses to environment.

Hunter-gatherer occupation in tall open-forests

How did hunter-gatherers occupy tall open-forests, if at all, in Australia? And how were hunter-gatherers affected by vegetation change there? This section examines records from Tasmania, south-eastern Australia, and south-western Australia. Many records for eastern Australian regions refer to rain-forest rather than tall open-forest. These are quite different formations but they are discussed here because on the eastern seaboard of the continent, tall open-forest is often associated with rain-forest. The extent of tall open-forest is to some degree dependent on the same factors that control the distribution of rain-forest: high rainfall and suitable soils. Moreover, the dense understorey in both tall open-forest and rain-forest may have meant that these formations posed similar problems to human occupation.

Tasmania

Archaeologists and palaeo-ecologists have been investigating prehistoric hunter-gatherer interactions with plant formations in Tasmania since the 1960s. Although there is relatively little literature on hunter-gatherers in tall open-forest, there is much on their relationship with rain-forest. This formation is widespread in the western half of the island and as in other parts of eastern Australia, interfingers with tall open-forest. In central and eastern parts of Tasmania, tall open-forest interfingers with open-forest. Because tall open-forest is viewed as intermediate between rain-forest and open-forest in difficulty of occupation and susceptibility to firing (MacPhail 1980, Lourandos 1983a), and because archaeological models of hunter-gatherer use of tall open-forest are lacking in Tasmania, I discuss concepts about occupation of rain-forest (e.g., Cosgrove *et al.* 1990, Thomas 1994) under the assumption that they apply to some degree to tall open-forest also.

An ecologist, Jackson (1965, 1968), was the first to implicate Aboriginal firing, climates, and soils as contributing to the large extent of Tasmanian grasslands, sedgelands, and grassy understoreys in woodlands. Later palynological research confirmed that distributions of other vegetation formations were related chiefly to climate and to some extent, fire regimes imposed by people (e.g., MacPhail 1979, 1980; Mount 1979).

R. Jones (1968, 1971) laid the framework for archaeological research in Tasmania and argued that its environment always restricted population, especially during and after Holocene sea-level rise and forest expansion. Jones's (1971, 1977) comments on north-western Tasmania were that the Rocky Cape sites exemplifed the north-western region's coastal economies. *Nothofagus* rain-forest fringing the coastline here gave few resources and blocked inland travel. For much of the time, meat and presumably, plant food, came from the littoral (e.g., seals and, after *c.*8,000 BP, shellfish). Throughout the late Quaternary the dense rain-forest of the hinterland in western Tasmania presented a barrier rarely or never traversed by hunter-gatherers. This interpretation was complemented by fieldwork in eastern Tasmania, where open-forests and woodlands were probably favourable environments for Aboriginal groups (Lourandos 1977). The pre-European economies based on the western littoral and the eastern half of the island were probably "regionally co-ordinated" (R. Jones 1977) and based on the two-fold vegetational division of coastal heaths and sclerophyll forests. This generalization, maintained in recent research, implied that dense rain-forests excluded people, who therefore visited and controlled little more than their fringes (Cosgrove *et al.* 1990, R. Jones 1990).

Bowdler (1974, 1979, 1984) demonstrated Pleistocene human occupation at Cave Bay Cave on Hunter Island, which was formerly part of Tasmania's north-western tip before rising sea-level cut it off 6,000 BP. Human visits began before 23,000 BP and ceased 18,000 BP, then resumed from 6,000 BP, at first intermittently, and from 2,000 BP more frequently. Faunal and botanical evidence suggests that the abundance of food resources and the openness of local vegetation determined whether people camped at Cave Bay Cave. Before 12,000 BP, dry grasslands with scattered eucalypts provided prey animals, but afterwards and up to 7,000 BP, regional rainfall increased, promoting woodland and scrub, and animals used by people disappeared. Bowdler (1984) suggested that in the early Holocene, dense rain-forest on the mainland restricted access to any embarkation point to Hunter Island. After 2,000 BP, greater capability to control rain-forest with fire perhaps allowed people to travel to Cave Bay Cave and other west coast sites.

In the late 1970s and early 1980s the discovery of Pleistocene-age artefacts at Beginner's Luck Cave, in the Florentine Valley (Goede and Murray 1977, Goede *et al.* 1978), and Kutikina, in the Franklin River valley (Kiernan *et al.* 1983) indicated that people had once inhabited the south-western river valleys, now thickly forested and therefore, for the same reasons as proposed by R. Jones (1977), uninhabited. Beginner's Luck Cave provided relatively few artefacts but the numerous artefacts and hearths at Kutikina suggested intensive site use and hunting of Red-necked Wallaby (*Macropus rufogriseus*) in grasslands on the edges of Pleistocene glaciers. This way of life would have been terminated by terminal Pleistocene rain-forest advance. These theoretical possibilities have been enlarged by research at other south-western sites under the aegis of the Southern Forests Archaeological Project (SFAP; see below, Allen 1996, and references therein).

Lourandos (1983a) viewed environmental change as providing opportunities for Holocene cultural and demographic change, but not determining it. From 12,000 BP, rising sea-levels and increasingly dense vegetation had restricted the range of Tasmanian hunter-gatherers and limited their possible subsistence strategies. But after 6,000 BP, climate became cooler and drier and hunter-gatherers could fire vegetation more often. Warragarra rock shelter provides some evidence for this. It is located in a stand of tall open-forest which is part of a mosaic complex with rain-forest. Large numbers of artefacts in the upper part of the shelter's archaeological deposit suggested to Lourandos more intense human occupation there after 3,500 BP, as if people more easily controlled these vegetation formations only from that time onwards. These views of the development of human firing in relation to vegetational changes were mirrored in analyses of southern Tasmanian pollen records. MacPhail (1979, 1980, 1984, and MacPhail and Colhoun 1985) argued that when late Quaternary climates allowed, firing helped create dynamic relations between plant formations, on the model of Jackson (1968). Fire regimes *complemented* climatic conditions in the late Pleistocene and in the late Holocene. Fire regimes *opposed* climatic conditions in the early Holocene, when rain-forest and tall open-forest expanded. In the early Holocene, the apparent effects of climates and soils on vegetation distributions suggest the model of Mount (1979), which discounts fire as a determining factor. The co-dominance of tall eucalypt and rain-forest tree species in early Holocene records shows that the former, fire-promoting trees were not the only canopy trees in tall open-forests. This canopy mixture implies that tall open-forests were an ecotone between rain-forest and open-forest in which fires were infrequent (MacPhail 1984). Since climate was becoming increasingly wetter at this time it seems likely that fires had progressively less effect until *c.* 7,000 BP.

From this time onwards, the trend reverses so sharply, with rain-forest species declining and eucalypts increasing, that new or intensified Aboriginal firing is either implicated or at least feasible. People probably fired vegetation during the last glacial, but from *c.*12,000 BP, human firing, if there was any, did not stop the encroachment of rain-forest and tall open-forest in central and south-eastern Tasmania. From the mid-Holocene onwards, human firing was once more a factor that probably limited the extent of rain-forest and tall open-forest.

MacPhail and Colhoun (1985) recognized further that climatic conditions still permitted many different vegetation types in one area. At Ooze Lake, in southern Tasmania, open-forest, grassland, and sedgeland, which contain many ethnographically recorded Aboriginal food plants and animals, persisted throughout the phases of rain-forest expansion. But in the late Pleistocene and late Holocene, "it is difficult to avoid concluding that the development of grassland and sedgeland-heath complex was party due to fire and that Aborigines... were partly responsible for these fires" (MacPhail and Colhoun 1985: 47).

Palynologists and ecologists continued to argue for the importance of Aboriginal firing where suitable climatic conditions prevail (Bowman and Brown 1986, Thomas 1993). There is ethnographic evidence dating from the 17th C to the late 19th C that Aboriginal people fired all Tasmanian plant formations, with the possible exception of rain-forest (Thomas 1994). Their likely influence in the past implies that there were subtle and dynamic reactions from vegetation, resulting in much variation in local vegetation distributions. However, archaeologists at this time continued to make broad regional generalisations about the influences of vegetation, appropriate to long-term records.

The SFAP, mentioned above, located at least seven other Pleistocene archaeological deposits in western Tasmania (Allen and Cosgrove 1996: Figure 1). A key factor determining the direction of SFAP research was the density of vegetation, which makes the open sites difficult or impossible to find. In any case, many open sites had scanty deposits unsuitable for dating. Nevertheless, open sites in a previously densely forested landscape in the King River valley, now exposed by nearby industrial pollution, were dated to the late Holocene (Freslov 1993). Equally densely forested valleys further inland could have been visited. However, in general, the very few artefacts on the surfaces of carbonate-capped cave deposits that contain great quantities of Pleistocene archaeological material suggested Holocene visits to caves were much less frequent than in the Pleistocene. It is possible that cave occupation ended only in preference to open-air camping in the same valleys (Thomas 1993), but to the SFAP researchers, the impenetrability of rain-forest understoreys was strong reason to believe that Aboriginal people had infrequently visited the south-western valley sites since their invasion by rain-forest.

The Pleistocene cave deposits suggested a human ecology very different from any inferred for late Holocene Tasmania (Cosgrove *et al.* 1990). The predominance of smashed *Macropus rufogriseus* and *Vombatus* bones at all these sites confirmed the indications from Kutikina that glacial-period hunter-gatherers targeted these animals and extracted marrow from their bones. In periods of low biomass and extreme cold, animal fat would have been a prized resource. This subsistence strategy, along with evidence for seasonal cave use and broad palaeo-vegetational records, was probably part of a specialised and seasonal inland economy based on hunting wallabies and other animals in sub-Alpine herbfields at the edge of then-substantial glaciers. But from 15,000 BP, climate became wetter, and rain-forests, unsuitable for the prey animals, began to expand into the valleys.

Western Tasmanian sites used after 13,000 BP were those at the edges of the rain-forest vegetated karst zone: to the north, Mackintosh, Warragarra, Parmerpar Meethaner; to the east, ORS7. But even these sites, and those on Bass Strait islands that had been hills on the now-submerged Bassian Plain, were used infrequently after the glacial period, 18-13,000 BP (Brown 1993, Cosgrove 1995). Terminal Pleistocene changes in rainfall and sea-level drastically altered Tasmanian landscapes. Cosgrove (1995) argues that as forests expanded and lowlands were submerged, occupation of caves ended, along with tightly-focused subsistence strategies requiring intensive use of these sites.

In all their various interpretations, archaeologists have explicitly searched for very broad influences of environment on human populations. The problem is that these interpretations address the interaction between hunter-gatherers and vegetation on a scale bigger than can be experienced by individuals or communities (Thomas 1993). Interactions between plants and people happen within their lifetimes, not over millennia. Colinvaux (1987) argues that one can assume no history of millennia for any association or community of plants seen in the modern world. Plant species grow according to conditions that suit them, not their community of the moment. The argument, based on the "Gleasonian" view in ecology (Colinvaux 1987: 4), implies that humans interact with vegetation communities that are too short-lived to be identified in many archaeological sites. Archaeologists interested in human-vegetation interactions may be forced to study interactions with individual plant species. To date, archaeologists working in Tasmania have not identified the movements of plant species in sufficient detail, in time or space, to infer ways in which people might have responded to them. Moreover, the geographical range of archaeologists' interpretations is much larger than the range of many plant species in question (e.g., all of Tasmania and emergent parts of Bass Strait are the focus of Cosgrove, 1995). The required fine resolution is lacking from many Tasmanian archaeological sites partly because of the slow rate of deposition, as little as 2 cm per millennium (Allen 1996, Cosgrove *et al.* 1990), and partly because there are few environmental records from the sites or their immediate localities (Thomas 1993, 1995; but see Dolby 1995, for preliminary analysis of archaeobotanical remains from the south-western sites).

Archaeologists argue that Tasmanian hunter-gatherers were limited by vegetation, and ecologists assume that they have for a long time manipulated vegetation with fire. These are not conflicting statements. The major event that gave rise to the ecologists' assumption is the marked vegetational change in the 19th century that occurred after the violent usurpation of Aboriginal occupation by British settlers. The rapid development of dense scrub and understoreys is recorded in historical documents and in pollen cores (Bowman and Brown 1986, Jackson 1968, Thomas 1993). Ecologists found further support in current practices of Aboriginal people in northern Australia, who burn carefully, deliberately, and in accordance with to the distribution of plant formations. For example, people burn woodland understoreys to the edges of rain-forest thickets, a practice that preserves the thickets from wildfires and maintains diverse habitat for food animals and plants (Bowman 1998, Bowman and Panton 1993). Such carefully applied combinations of regular fire and fire-exclusion suggest that hunter-gatherers have the potential to influence all parts of their range, including vegetation formations that are infrequently visited.

The disparity between the views of archaeologists and ecologists is likely to derive from different regional scales and of the distribution of the archaeological evidence. While the archaeological evidence suggests that human control in the south-western valleys was either absent or ineffective at times of major climatic change, the question of whether vegetational change was deleterious to people occupying all of south-western Tasmania is still unresolved because of the limited numbers of sites outside the valleys. The view that rain-forest and tall open-forest present great challenges to hunter-gatherer manipulation is most plausible in the extreme conditions of Tasmania, but even here, a particularly exhaustive long-term archaeological research programme may have revealed highly localised rather than regional responses.

South-eastern Australia

Hunter-gatherer interactions with vegetation in mainland south-eastern Australian forests have not been as much analysed as in Tasmania, but the region provides much data on human occupation in different forest types nonetheless. Important studies incorporating archaeological and palaeontological investigations and historical observations are Bowdler (1983a, 1983b), Byrne (1983), Head (1983, 1988, 1989), Lourandos (1983b) and McBryde (1977, 1978).

South-eastern Australian forests are botanically diverse and form mosaics with great contrast. Along the entire eastern seaboard, the rugged topography of the Dividing Range and the equable climate (rain for most of the year) promote a vegetation mosaic that includes rain-forest, tall open-forest, and open-forest. In coastal Queensland, rain-forest occupies a relatively larger proportion of the terrain and (eucalypt) tall open-forest exists mainly in small pockets. In southern and eastern Victoria, tall open-forest reaches its greatest height, with some individual *Eucalyptus regnans* (Mountain ash) exceeding 100 m (Ashton and Attiwill 1994).

Bowdler (1983a) commented that a north-south gradient in the extent and food species diversity of rain-forest had implications for hunter-gatherer distributions. In the northern, more extensive and more diverse rain-forests, people were adapted to rain-forest. They lived more or less entirely within it and had a unique set of techniques for living in it. At the other extreme, people avoided southern, cold rain-forests, extremely lacking in resources (of which the extreme form was in Tasmania). In Bowdler's words, rain-forest formed a continuum from north to south, from the "colonised" to the "coloniser". The continuum is necessarily a broad concept and Bowdler emphasised that there would be departures from it, especially in its centre in eastern New South Wales. The following discussion suggests that the rain-forest geographical continuum has a parallel with the southern and northern tall open-forests of the eastern seaboard. It also points out some of the variations and how they may relate to the extent of human occupation in tall open-forest.

Ethnographic and archaeological evidence suggests Aboriginal hunter-gatherers in eastern Australia favoured regions with the maximum diversity of resources. In the Five Forests region, on the southern New South Wales coast, Byrne (1983) found archaeological sites in open-forest, tall open-forest, and rain-forest formations. According to ethnographic evidence, most hinterland forests were used seasonally by most groups to relieve stress on coastal, riverine, or estuarine resources. The greatest number of sites were on the coast, but in areas with no quick access to the coast (more than 30 km away), sites were concentrated in stream valleys containing rain-forest and usually one or two other types of forest. Recorded plant and animal foods were distributed in the same way as sites, with the greatest number of foods recorded in valleys with the greatest number of plant formations. Areas with only one formation, especially formations that were dominated by a single species, had few foods and few sites. These formations were *E. pilularis* tall open-forest, and *E. considineana* and *E. botryoides* open-forests.

Archaeological and historical research in other parts of New South Wales confirm that no formation presents a particularly great obstacle to human occupation. Sites in the Sydney region and on the northern NSW coast are located in all formations (Attenbrow 1982, Belshaw 1978, Campbell 1978, McBryde 1977, Pierce 1978, Prentis 1984, Sullivan 1978). Particularly important for food animals are junctures of different formations where each one was of limited extent. In all forests, yams and other tubers were important vegetable foods. In northern NSW and south-eastern Queensland, historical-period Aboriginal groups hunted small marsupials and reptiles in rain-forest, tall open-forest, and dense scrub that bordered estuaries and swamps (McBryde 1977, Pierce 1978, Prentis 1984, Sullivan 1978). Although people camped less often in dense formations (Belshaw 1978), plant foods existed in them, and they were sometimes cleared, presumably by firing, to facilitate access (Campbell 1978).

Bowdler (1983b), reviewing Aboriginal sites in NSW forests, notes that they are often simply small artefact scatters, suggesting small or short occupations. The nature of resources, especially in tall open-forest, suggests that occupations in forest were often short and apart from ceremonial sites, involved small groups of people. Few Aboriginal foods are found solely in the tall open-forests. Yet occupation of tall open-forest and open-forest probably relieved stress on other resource zones, and they could be exploited relatively easily in the varied environment of the eastern seaboard.

In Victoria, there is less evidence of hunter-gatherer occupation in forests. As evidence for economic intensification and associated population redistribution, Lourandos (1983b) cited late Holocene establishment of the Seal Point midden on Cape Otway, on the south-western Victorian coastline. From the site, hunter-gatherers could have reached many productive areas. Its location on the edge of tall open-forest, and the remains of forest animals in the midden, suggests that by the late Holocene, people exploited the tall open-forest and had developed forest-firing techniques.

In the interior of Victoria, evidence for hunter-gatherer occupation of tall open-forest is scarce. In western Victoria, the main dated evidence for past hunter-gatherer presence in forests is from open-forests in Grampians Ranges rock-shelters (Coutts and Witter 1977). In eastern Victoria, Head (1989) cites the low fire frequency in tall open-forest and an unpublished report by Gell and Stuart (1987, in Head, 1989), to argue for infrequent Aboriginal occupation in the large tall open-forest of Gippsland.

Yet in Gippsland, Howitt (1890) inferred that the exclusion of Aboriginal people and cessation of regular firing after European settlement in 1840 had caused dense scrub and young forest to cover many areas previously covered by open-forest and large trees. Howitt noted that Aboriginal people told him that forests were more open before settlement and that clearing of dense scrub often revealed stone artefacts indicating that Aboriginal people had visited the forests. He surmised that Aboriginal people had probably prepared the open vegetation so attractive to the first settlers "by their *annual* burnings" (1890: 111, my emphasis). This is a very frequent rate of burning, and would almost certainly promote grassy understoreys. Howitt (1890: 110, 111) mentions several eucalypts as associated with dense scrub, and previously forming more mature and more open forest with grassy understoreys. Among these are eucalypts typical of tall open-forest and mixed open-forest: *E. sieberi* (Howitt's *E. sieberiana*) and *E. viminalis* (Ashton and Attiwill 1994: Table 6.1).

Gippsland palynological records suggest that there must have been some variation in the importance of Aboriginal firing. Gell *et al.* (1993) argue for relatively little Aboriginal influence on the plant communities that contributed to the Tea Tree Swamp pollen site, which is located 900 m ASL. The communities are rain-forest, tall open-forest, and montane woodland. Grass pollen and charcoal particles, indicating frequent firing, increased significantly only after an inferred date of 1885, about the time of European mining and first settlement of the locality. The inference of infrequent pre-European burning in this moderately high altitude part of East Gippsland is supported by a scarcity of archaeological sites and historical records (Howitt 1904), suggesting sparse Aboriginal occupation.

All of the above evidence suggests considerable variation in the influence of Aboriginal people on tall open-forests. Long pollen records from north Queensland and Tasmania suggest that Aboriginal firing could have been important in the creation of forest forms, provided that environmental conditions were conducive, *i.e.*, drier (Kershaw 1986; MacPhail 1979, 1980, 1984). Such conditions existed in the Last Glacial Maximum and in the late Holocene. In the Tasmanian pollen sites and Kershaw's long Quaternary record from Lynch's Crater, firing complemented and accelerated expansion of eucalypt forest at the expense of rain-forest in the late Pleistocene and again in the late Holocene. In a similar manner, increased human impact in the late Holocene is argued for woodlands and swamps in the Discovery Bay region of south-western Victoria (Head 1983, 1988, 1989).

In contrast, early Holocene advances of rain-forest and tall open-forest were not halted by Aboriginal or natural firing, whether in the north or the south. In MacPhail's Tasmanian pollen sites, and at locations in the Daintree rain-forest in Queensland (Hopkins *et al.* 1993), firing that opposed climatic influences on vegetation in the early Holocene, when climates became wetter, had relatively little effect.

There is no delineation of hunter-gatherer distributions in relation to tall open-forest in eastern Australia, but like the rain-forest continuum, there may have been more resources in the northern tall open-forests, where the mosaic is finer. There appear to be many sites in eastern New South Wales, where stands of tall open-forest cover small areas. In contrast, the Gippsland tall open-forests cover large areas and high altitude parts may have been less often visited. Broad-scale views of vegetation affecting human populations may be appropriate in cold temperate climates, as in Pleistocene Tasmania and montane parts of Victoria, but at lower latitudes and altitudes, it is probably more important to analyse changing extent of tall open-forests on a smaller scale.

South-western Australia

South-western Australia is a large low-lying region and in most areas, its vegetation forms a fine-scale mosaic. However, hunter-gatherer occupation of the forests here has been modelled in rather similar terms to the Tasmanian, broad-scale approach (Hallam 1975, Ferguson 1985). The actual evidence for occupation of different forest types is slim, but the following discussion of known hunter-gatherer adaptations in this region, which are well-documented, shows whether the model for occupation is appropriate. The evidence for hunter-gatherer adaptations in south-western Australian forests is divided into two parts, ethnohistorical and archaeological.

Ethnohistorical evidence

Ethnohistoric evidence comprises both primary and secondary historical sources documenting traditional life of the Nyoongar people in south-western Australia. Before the 1880s, European settlement was relatively restricted, enabling Nyoongar people to continue their traditional hunting and gathering activities throughout much of the lower south-west (Hammond 1933: 9). Records kept by the first European settlers, explorers and navigators provide detailed evidence of many subsistence activities.

The major study of ethnohistoric documents on forest adaptations is Hallam's (1975) *Fire and Hearth*, a comprehensive review of historical records of Aboriginal firing. Reviews of Aboriginal foods in different plant formations, discussed in Chapter 2, are provided by Meagher (1974) and Meagher and Ride (1979) and "salvage" ethnographic research into Aboriginal foods is reported by Bird and Beeck (1988).

Meagher (1974) and Hallam (1975, 1987) use ethnohistoric sources to infer land use and subsistence patterns before European arrival in the lower south-west. To quote from one regional study that summarises their inferences:

In broadest terms these [patterns] seem to be that the [regional hunter-gatherer] economy was based on the systematic and to a very large extent seasonal exploitation of a very wide variety of food resources of coastal and adjacent inland districts; that economic activities often required co-ordinated group effort within a system of group land tenure; and that regular burning, used in vegetation management and animal drives, was a significant controlling device.
(Dortch and Gardner 1976: 263)

Information about subsistence practices show that coastal woodland and open-forests contained many foods. Meagher (1974) and Hallam (1987) suggest that food resources included a wide variety of terrestrial and aquatic animal and plant foods, obtained by methods given below.

Techniques for gathering these resources included digging for roots and burrowing reptiles and mammals; grinding acacia seeds; removal of toxins from cycad (*Macrozamia*) nuts and some frogs; hand-capture or gathering of plant foods, insect larvae and eggs, crustaceans, amphibians, tortoises, reptiles, and eggs and young of birds; bringing down birds with the boomerang; trapping and spearing marine fish in stone and brush-wood fish-traps built across estuaries and inlets; spearing reef fish; opportunistic scavenging of stranded marine mammals; cutting footholds in possum-trees with the *kodj*, a stone axe; spearing medium to large size game (macropods and emus); the use of fire, snares, dogs, and pits for trapping terrestrial game of all sizes. Virtually the only edible foods discussed by neither Meagher (1974) nor Hallam (1975), were molluscs, e.g., abalones and fresh-water mussels. However, C.E. Dortch *et al.* (1984) have shown that marine molluscs were sometimes eaten in small amounts, and fresh-water mussels have been recovered from archaeological sites (C.E. Dortch 1979, J. Dortch 1996). Thus the only resources known to the Nyoongar, but not eaten, were supposedly poisonous fish and possibly, sharks and sting-rays, which at King George's Sound were hunted for sport but not eaten (Nind 1831: 33). Also, fish that could not be speared or trapped close to shore were inaccessible, since boats, nets, fish-hooks and lines were not known to the Nyoongar.

This diversity of resources and the subsistence economy meant that food was available throughout the year, if people visited coasts, swamps, or wetlands (Meagher and Ride 1979: 77). Anderson (1984) and Hallam (1987) show that people moved in accordance with seasonal abundances and occasional shortages of food and water, usually using coastal and estuarine resources in summer and moving inland in winter.

Nind (1831) summarises these movements for the people living around King George's Sound:

I believe that in winter (when the sea coast tribes go into the interior) they are in small parties, and much scattered, living upon opossums, bandicoots, and kangaroos, &c. They begin to return to the coast about September or October, and at this season they chiefly subsist on roots. In calm weather, however, they procure a few fish. As the season advances, they procure young birds and eggs, and their numbers increase. About Christmas they commence firing the country for game, and the families who through the winter have been dispersed over the country, reassemble. The greatest assemblages, however, are in the autumn (pourner), when fish are to be procured in greatest abundance. Towards the end of autumn, also, they kill kangaroos, by surrounding them.
(Nind 1831: 36)

Information on the distribution of populations in relation to plant formations is limited. Hallam (1975) assesses documents pertaining to the Swan Coastal Plain, the northern Jarrah forest, and the early settlement at King George's Sound (Albany). However, she speaks only in broadest terms about human activity in tall open-forests, and she believes that occupation of the Karri and southern Jarrah (Menzies Sub-district) was limited or non-existent. In contrasts there is abundant evidence for Aboriginal firing of the Jarrah and Marri open-forests and woodlands on the coastal plain, and east of the coastal plain in the

York region (Figure 2.1). This evidence is corroborated in more recent historical research by Ward (1998). The Nyoongar fired open-forests annually or biennially throughout the summer, and especially at the end of summer when fuel was driest. Most historical observers saw a direct connection between frequent Aboriginal firing and the results: vigorous regrowth of understorey herbs and grasses, restricted areas of shrubby understoreys, large populations of game animals such as kangaroos, and open, park-like stands of trees very attractive to European farmers. I have already discussed, in Chapter 2, studies of fire scars in tree trunks that support the inference of frequent cool fires in various forests, dating from at least the 17th C (Burrows *et al.* 1989, Rayner 1992, Ward and van Didden 1997). Ward (1998) mentions the frequent historical descriptions of native grasses, which are now rare or extinct in south-western Australia.

Hallam (1975) contrasts these characteristics of Aboriginal firing in Jarrah forest with Karri forest, which, she states, has few foods, a dense understorey, and is too wet to burn. The first two factors are probably true for many Karri stands, but the third is more complex than she suggests. As suggested in Chapter 2, the climate throughout south-western Australia actually promotes summer fires. Although the Karri forest experiences a shorter summer drought and its soils remain damp for a greater part of the year, inhibiting fire (Burrows 1987), so it would have experienced fires less frequently, people and lightning-strikes would have still set fires in the whole Karri forest region, perhaps every year.

As mentioned also in Chapter 2, there are great contrasts in the number of historical observations from the coast and from the interior. Historical records of Aboriginal people occupying coastal open-forests and woodlands are so frequent, and well-described in Hallam (1975), that I merely list some of them here. They include descriptions of campsites and huts in coastal woodlands on the Swan River estuary (de Vlamingh 1696-97 [Major 1859]; cf. Figure 2.1), Geographe Bay (Marchant 1982), and King George's Sound (Vancouver 1801); numerous observations of Aboriginal people in coastal woodlands and open-forests by British officials and settlers (e.g., Bussell 1833, A.P. Bussell n.d. [1834], Grey 1841, Moore 1884, Nind 1831, Roth 1903); and reminiscences of coastal groups' traditional movements through open-forest and woodland (A.J. Bussell n.d. [1941], Hammond 1933, n.d.).

In contrast, there are few observations of Aboriginal people in forests in the interior. Captain Thomas Bannister (1833), Captain Collet Barker (Mulvaney and Green 1992), the Reverend John Bussell (1833), William H. Graham (Stephens 1954), Sir George Grey (1841), and Dr Thomas Braidwood Wilson (1835) number among the British colonists who passed through the southern tall open-forests between Albany and Augusta. None of these early writers refer to Aborigines occupying or using pure Karri forest, and Bannister, Bussell, Grey, and Wilson note the density of tall open-forest understoreys compared to the open-forests and woodlands.

However, Hammond (1933: 17, see also Ferguson 1985) mentions that Aboriginal people maintained trackways through forests "like cattle pads [paths]" across the south-western interior. In separate trips in 1830, Wilson (1835) and Barker (letter to Dr Wilson in Wilson 1835: 267-271) followed Aboriginal trackways maintained through Karri forest. They also walked across "fine open land" (Wilson 1835: 250), amongst tall trees of "enormous circumference and altitude" (*ibid.*: 256), through "narrow strips of finely timbered forest land" (*ibid.*: 264), and often saw "fine blue gum [Karri[1]]" (*ibid.*: 268). Wilson and Barker frequently met "impenetrable underwood" (*ibid.*: 268), but their writings suggest that they travelled through forests as much as around them, and that the understorey was not uniformly dense. In 1861, W.H. Graham observed Nyoongar firing bush around "groves of blue gum" near Broke Inlet (Stephens 1954). It would have been feasible for the Nyoongar, as adroit users of fire, to have occupied areas near Karri forest, and perhaps areas within it.

In the remaining accounts cited above, Sir George Grey and Captain Bannister lost the Aboriginal trackways they had been following, so their impression of dense forest with few human inhabitants might only be accurate for the compass bearings that they unhappily followed in order to extricate themselves from the forest. The Rev. John Bussell was also lost, but eventually found recently burnt-over forest where travel was easier. At points along their respective journeys, most of the explorers met Aboriginal people. However, there are scant details about which forests they were in. Only Graham and Wilson say they met Aboriginal people near Karri, and Grey is the only person to state unequivocally that he saw no one, in his journey through dense understorey in tall Jarrah forest.

These early observations suggest that dense understoreys existed in patches in the tall open-forests and in the taller, southern Jarrah forest also. Such understoreys would not have been occupied by people. However, the observations of cleared or fired vegetation, the occasional encounters between historical observers and Aboriginal people, and Hammond's description of Aboriginal trackways, suggest that people could have at least travelled through open-forest and tall open-forest. There is little or no historical evidence to say whether they occupied tall open-forest or not. The small number of observers (Wilson, Barker, Bussell, Graham) who saw Aboriginal people an

[1] Long-term residents of Manjimup, Northcliffe, and Pemberton, all in Karri areas, have confirmed this early name for Karri to me (G. Gardner, G. Kelly, J. Sebire, pers. comms).

unspecified number of times in or near the wetter, southern eucalypt forests does not imply that Aboriginal people were rarely there. As mentioned, only one observer who reported from the southern forests states that he failed to meet Aboriginal people. Hallam's (1975: 25-27, 55, 107, 110) statements that the southern forests were little frequented by Aboriginal people are not in fact testable with the published historical evidence, despite her plausible argument.

Archaeological evidence

Aboriginal people have occupied the forested and wooded landscape of south-western Australia for more than 47,000 years (see Chapter 6, and Turney *et al.* 2000), but few archaeological sites have been recorded in Karri tall open-forest. Conventional transect surveys are ineffective in tall open-forests as dense understoreys and deep litter layers make it difficult to walk through the forest and see the ground surface (Bowdler 1983b; Byrne 1983, 1987; Cosgrove 1990; Dortch and Gardner 1976). Road surveys and test-pit surveys are often used to surmount the visibility problem (Krakker *et al.* 1983, Lightfoot 1989, Lovis 1976, Nance and Ball 1986) but road surveys cover relatively little area, and only one test-pit survey, the one mentioned at Dombakup 24, has been carried out in south-western Australian forests (J. Dortch *et al.* 1998).

Archaeological surveys confirm ethnographic evidence that people frequently travelled in open-forest, at least in coastal or coastal hinterland areas. Perhaps the most extensive archaeological surveys in south-western forests are Pearce's 1970s and early 1980s surveys in Jarrah open-forest near Collie, made in a series of management studies commissioned by bauxite miners (Pearce 1982). Pearce's investigations revealed 300 sites (mostly artefact scatters) in an area of 280 km². This density of sites was similar to that found on the Swan Coastal Plain, a region of abundant and varied resources in woodland, although the Jarrah forest artefact scatters were somewhat smaller (Anderson 1984, Hallam 1987).

Anderson (1984) proposes that prehistoric open-forest use followed the pattern of ethnohistoric seasonal movements. Groups occupying the Swan River region as their core territory visited Jarrah open-forest in the Darling Range, a hilly region rising steeply above the coastal plain, in winter. Locations in the Jarrah forest too far from permanent water in summer could have been attractive campsites in winter, since no location in the Jarrah forest is more than 1 km from an ephemeral water source (Ward and van Didden 1997). Groups' seasonal movements would have reduced over-use of coastal plain resources. Although there is as yet no archaeological evidence of the season of site use in the Darling Range, Anderson found predominantly small sites in the open-forest there, suggesting the small or short occupations of the ethnographic model for the inland forest.

Pearce (1989) points out that the northern Jarrah forest (Figure 2.1), where the surface geology is hard rock, is more likely than the sandy coastal plains to contain rock outcrop suitable for stone artefact manufacture. While rock such as chert had a restricted supply on the coast, outcrops of granite, dolerite, and quartz in the interior could have made forests comparatively attractive. These rocks were probably quarried at outcrops throughout open-forests and perhaps tall open-forests. The Northcliffe Silcrete Quarry, located next to the Karri forest at Dombakup 24, is the only recorded quarry near tall open-forest.

Lilley (1993) surveyed open-forests and coastal woodland along the Margaret River valley, in the Leeuwin-Naturaliste Region. Finding no sites in the open-forests, he proposed that poor resources in forests kept populations low always, effectively restricting Aboriginal hunter-gatherers to woodlands and wetlands on the coastal plains of south-western Australia. Lilley's remarks support Hallam's (1975) and other researcher's arguments (see below).

Given the above-mentioned findings from Tunnel Cave and Dombakup 24, and the existence of archaeological records in eastern Australian tall open-forests, technical problems with surveys and the limited number of surveys so far carried out in tall open-forests are reasons to doubt the apparent lack of sites in tall open-forests. However, the contrast with archaeological records from open-forests, and Hallam's (1975) arguments for different occupation patterns in Karri and Jarrah forest, encouraged Ferguson's (1985) research programme.

In his doctoral thesis, Ferguson (1985) argues that the apparent lack of sites in Karri tall open-forest is genuine, because Karri forest is naturally impenetrable, resource-poor, and difficult for hunter-gatherers to manage or limit by firing (Ferguson 1985: 49-50). Ferguson's thesis has some acceptance (Dodson *et al.* 1992, Lourandos 1997, O'Connor *et al.* 1993), and aspects of it are almost certainly based on real differences between forest formations. However, the thesis is probably too simplistic to be used further, for reasons as follows:

The argument is based on rather little ecological evidence. Ferguson states that either cool fires killed the thick understorey without removing it (citing Rule 1967: 14) or hot fires went out of control, burning very large areas (Christensen and Kimber 1975: 104-5). Limited by the available ecological research at the time of his writing, Ferguson was unaware of the apparently higher fire frequency (and probably cooler fires) in Karri forest in the pre-European period (Rayner 1992), and the potential for grassy understoreys and a mixed canopy to be created within Karri stands by regular burning (Bradshaw 1985, McCaw 1986). But he does not discuss vegetation mapping published in the 1970s and early 1980s, which shows that Karri stands have limited extent because of the intricacy of the soil mosaic and limited extent of suitable soils (Beard 1981, F.G. Smith 1973). Moreover, the comment by Rule (1967) is one retired forester's recollection of thickets in Karri forest, not an observation repeated generally in ecological literature. These oversights are not discussed further because I have already discussed more conventionally accepted explanations of fire regimes in Karri tall open-forest (Chapter 2), and I wish to address problems in Ferguson's archaeological analysis (see also Dortch and Smith 2001).

Ferguson (1985) analyses occupational records from 12 sites, including some that he excavated. He proposes that the likely expansion of Karri forest since the end of the last glacial period into areas previously vegetated by more open formations caused people to abandon those areas, most of all in the period 6,000-4,000 BP. He suggests that this hypothesis is testable at several south-western Australian sites where he argues that the number of artefacts deposited per millennium represents populations using those sites. He argues that across south-western Australia, from about 6,000 BP, artefact numbers declined, and people gradually abandoned all sites. At the same time, according to pollen records amassed by Churchill (1968), Karri forest replaced more open formations. The mid-Holocene Karri expansion, Churchill infers, was caused by increases in rainfall. According to Ferguson (1985), the same archaeological sequences show that people only returned to the south-west after Karri forest contracted to its present extent, about 4,000 BP.

Against these arguments, Dortch and Smith (2001) have questioned the validity of inferring population sizes from stone artefact distributions in poorly stratified and often unconsolidated sand dune sites. Apart from this and other objections mustered by Dortch and Smith, M.V. Smith (1993) shows that Ferguson (1985) incorrectly interpreted occupational sequences at several sites; in fact, occupation at three sites continued throughout the Holocene. These sites are the Northcliffe Silcrete Quarry, occupied from 7,000 BP to after 3,000 BP (Dortch and Gardner 1976); Dunsborough, 7,000 to 4,000 BP (Ferguson 1980a); and Walyunga, c.30 km north-east of Perth, 8,000-1,300 BP (Pearce 1978). Dunsborough and Walyunga are located in the Drummond Botanical Sub-District, some hundreds of kilometres from the main Karri forest belt, and perhaps were unaffected by Karri forest expansion. However, the Northcliffe Silcrete Quarry, located in the high-rainfall, far southern part of the south-western forests, and presently next to Karri tall open-forest, belies Ferguson's thesis – as suggested by the title of Ferguson's seminar, given at the ANU in 1981: "Or did they all go to Northcliffe?" (Ferguson 1985: iii).

Ferguson argues that most artefacts at Northcliffe are younger than 4,000 BP, and hence this age indicates the end of the "de-population phase". However, the 4,000 BP age estimate for the majority of artefacts is not a radiocarbon date, but Ferguson's assumed age for the lowermost microlith (backed blade) in the deposit. Contrary to Ferguson's statement (Ferguson 1985: 467, 475), it is now believed that microliths appeared in various parts of Australia before 4,000 BP (Hiscock and Attenbrow 1998). It may be that the microlith manufacturing technique produced more flakes, hence whatever date they began to be made at Northcliffe would explain the sudden increase in flake numbers midway through the sequence. This is the argument made about the sequence at Kalgan Hall, near Albany, a site that Ferguson excavated and had used in support of his thesis (Gardos 1997). Gardos shows that variations in the numbers of artefacts, which Ferguson takes to signify population levels, are accounted for by adoption of microlith manufacturing technology, and that there is no need to posit changes in populations using the site.

Furthermore, Dortch and Smith (2001) point out that Ferguson's argument is partly based on deposits which cannot be used as Holocene master-sequences: the Arumvale site, in the Leeuwin-Naturaliste Region, has a badly mixed deposit (Dortch and McArthur 1985), and the Devil's Lair cave site was sealed in the proposed depopulation period (C.E. Dortch 1979).

The total number of occupations for any millennium is so small that they are probably not regionally significant, and Smith (1993) suggests that local Aboriginal groups' slight adjustments to periodically scarce resources probably would account for them. A few new discoveries could drastically alter the evidence. To illustrate these points, all occupation layers in south-western Australia are listed in their respective sites in Table 4.1 (based on Smith 1993: 42-43, 64-68). Figure 4.1 graphs the number of dated occupation layers per millennium. It suggests a possible population increase towards the present day, or alternatively, a bias of preservation, but no regionally significant, mid-Holocene changes in occupation.

Firstly, one expects younger archaeological material to be better preserved. Secondly, occupational records from the interior are under-represented (Figure 4.2). Archaeological research in south-western Australia has had a coastal bias because it was directed at shell middens (sites 1-6, 22, 23, 26, 32, 37, 38, 40, 47), cave sites in calcarenite (a coastal dune formation; sites 7, 24, 25, 27, 28, 35, 36, 39, 50), sites containing artefacts deriving from nearby submerged off-shore sources (8, 20, 22), or sites unearthed by urban development on the Swan Coastal Plain (11, 12, 19, 21). Onset of El Niño Southern Oscillation (ENSO), *c.*5,000 BP, which may have improved marine productivity in Australia (Sandweiss *et al.* 1996), perhaps provoked late Holocene population increases Australia-wide (Rowland 1999). In south-western Australia, however, marine or littoral resources were probably never as important as ones on the coastal plains (Hallam 1987).

Table 4.1 **Number of dated occupation layers per millennium in south-western Australia (continued next page)**

Sites are arranged by increasing latitude on the west coast (Green Head to Arumvale) and increasing longitude on the south coast (Lake Bolghinup to Barndi Cave). Because only one site has layers dating to the millennia between the 41st and 50th, and 31st and 40th, these columns are combined.

Millennium before present:																																		
No.:	**Site:**	41st to 50th	31st to 40th	30th	29th	28th	27th	26th	25th	24th	23rd	22nd	21st	20th	19th	18th	17th	16th	15th	14th	13th	12th	11th	10th	9th	8th	7th	6th	5th	4th	3rd	2nd	1st	**Reference**
1	**Green Head**																												1					C.E. Dortch *et al.* (1984)
2	**Middle Head**																											1			1			C.E. Dortch *et al.* (1984)
3	**Sandy Point**																																1	C.E. Dortch *et al.* (1984)
4	**Sandy Point North**																												1					C.E. Dortch *et al.* (1984)
5	**Sandland Island**																												1					C.E. Dortch *et al.* (1984)
6	**Moore River**																												1					C.E. Dortch *et al.* (1984)
7	**Orchestra Shell Cave**																										1			1		1		Hallam (1975)
8	**Walyunga**																									1	1		1	1		1		Pearce (1978)
9	**Brigadoon**																														1	1		Schwede (1990)
10	**Frieze Cave**																													1		1	1	Hallam (1975)
11	**Upper Swan**																																	Pearce & Barbetti (1981)
12	**Helena River**			1																				1	1					1		1	1	Schwede (1990)
13	**Minim Cove**																							1										Clarke & Dortch (1977)
14	**North Lake**																														1			Pearce (1979)
15	**Soldier's Road**																															1		Pearce (1979)
16	**North Dandalup**																															1		Anderson (1984)
17	**Boddington Cave**																													1				Anderson (1984)
18	**Collie Refinery**																											1						Pearce (1982)
19	**Armitage Road Burial**																																1	AAD Site File SO 2877
20	**Dunsborough 1**																									1			1					Ferguson (1980a)
21	**Dunsborough 2**																					1	1											J. Dortch (1995)
22	**Quininup Brook**														1								1											Ferguson (1980b)
23	**Ellen Brook**																												1				1	Bindon & Dortch (1982), Lilley (1993)
24	**Rainbow Cave**																																1	Lilley (1993), this volume
25	**Witchcliffe Rock Shelter**																																1	J. Dortch (1996), this volume
26	**Calgardup Brook**																													1				C.E. Dortch *et al.* (1984)

Table 4.1 Number of dated occupation layers per millennium in south-western Australia (continued from previous page)

Millennium before present:																																		
No.:	**Site:**	**41st to 50th**	**31st to 40th**	**30th**	**29th**	**28th**	**27th**	**26th**	**25th**	**24th**	**23rd**	**22nd**	**21st**	**20th**	**19th**	**18th**	**17th**	**16th**	**15th**	**14th**	**13th**	**12th**	**11th**	**10th**	**9th**	**8th**	**7th**	**6th**	**5th**	**4th**	**3rd**	**2nd**	**1st**	**Reference**
27	**Tunnel Cave**										1	1		1		1	1				1			1	1							1		J. Dortch (1996), this volume
28	**Devil's Lair**	5	4				1	1	1	1		1		1		1	1			1	1						1							Balme *et al.* (1978), C.E. Dortch (1979a), Turney *et al.* (in prep.), this volume
29	**Arumvale**														1							1		1										Dortch & McArthur (1985), C.E. Dortch (1986b)
30	**Lake Bolghinup**																							1										C.E. Dortch, unpublished data
31	**Lake Jasper**																													1				C.E. Dortch (1997)
32	**Malimup**																																1	C.E. Dortch (1985)
33	**Dombakup 24**																																1	J. Dortch *et al.* (1998)
34	**Northcliffe**																										1			1				Dortch & Gardner (1976)
35	**Nookanellup Rock Shelter**																														1	1	1	Dortch & Kelly (1996, 1997)
36	**Yongar Bogal Rock Shelter**																												1					Dortch & Kelly (1997)
37	**Conspicuous Cliffs**																																1	C.E. Dortch *et al.* (1984)
38	**Lights Beach**																																1	C.E. Dortch *et al.* (1984)
39	**Katelysia Rock Shelter**																														1	1		Dortch (1997, 1999)
40	**Herald Point**																																1	C.E. Dortch *et al.* (1984)
41	**Moorillup**																									1					1			Ferguson (1985)
42	**Kalgan Hall**														1									1				1		1		1	1	Ferguson (1985)
43	**Moingup Spring**																																1	Ferguson (1985)
44	**Kambellup Pool**																												1			1	1	Ferguson (1985)
45	**Waychinicup River**																									1					1			Ferguson (1985)
46	**Bennett Lake**																																1	Bird (1985)
47	**Cheyne Bay**																																1	C.E. Dortch *et al.* (1984)
48	**Whalebone Point**																																1	C.E. Dortch *et al.* (1984)
49	**Cape le Grand**																															1		M.V. Smith (1993)
50	**Cheetup**																			1	1	1	1		1						1	1	1	M.V. Smith (1982, 1993)
51	**Barndi Cave**																														1	1	1	M.V. Smith (1993)
	Total number of dated layers	5	4	1	0	0	1	1	1	1	1	2	0	2	3	2	2	0	0	2	3	3	3	6	3	4	4	3	9	9	9	15	21	

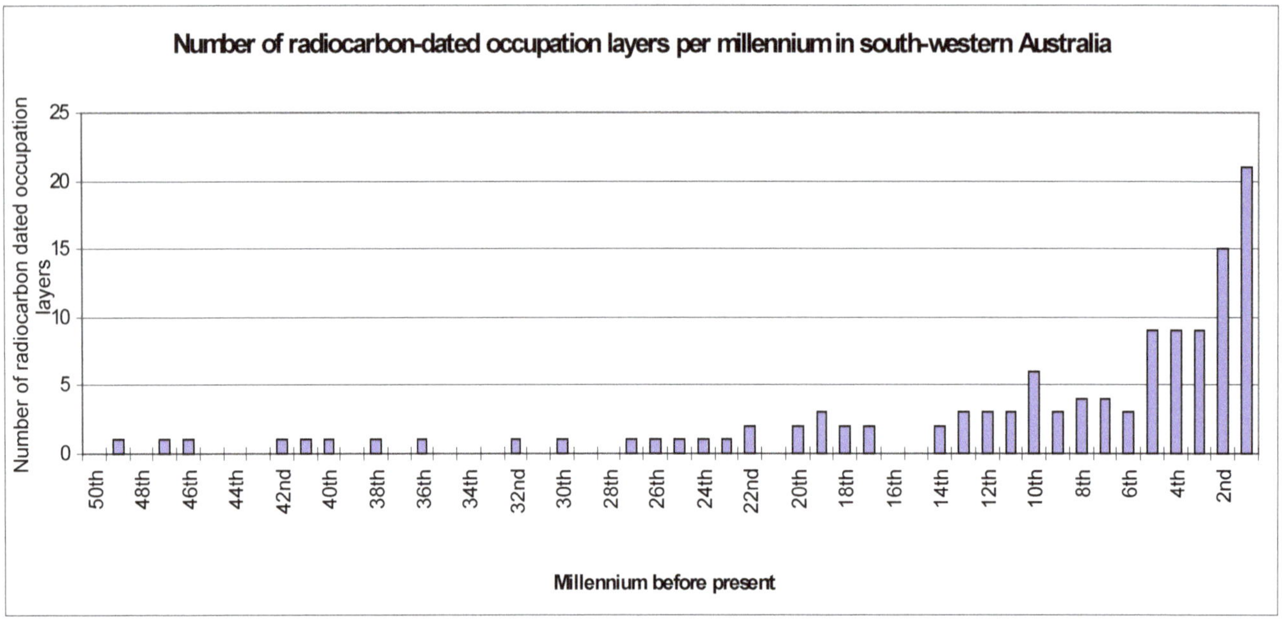

Figure 4.1 **Number of radiocarbon-dated occupation layers per millennium in south-western Australia**

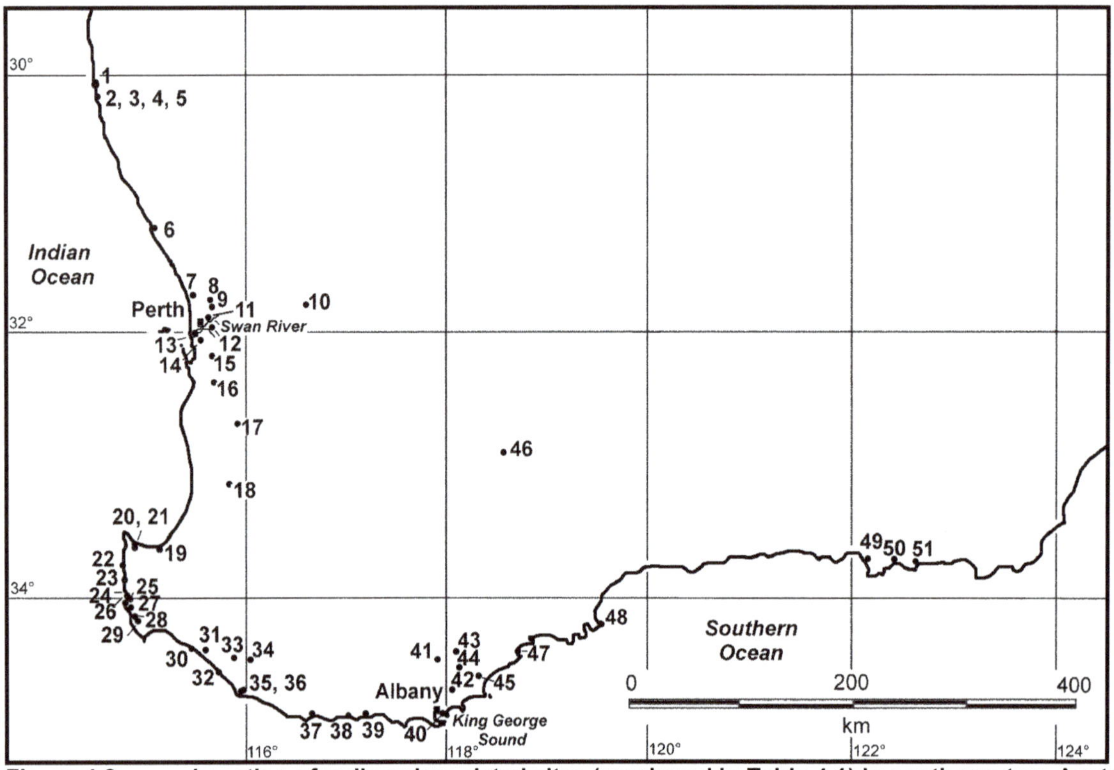

Figure 4.2 **Location of radiocarbon-dated sites (numbered in Table 4.1) in south-western Australia**

Rather than regional population growth, coastal sites could reflect local responses to continuing estuary basin sedimentation and consequent changes in ecological productivity (C.E. Dortch 1999). Since estuary development could have preceded sea-level stabilisation *c.*6,000 BP, many sites occupied as bases for exploiting estuarine resources may now be submerged. Coastal site numbers may be biased indicators of population increase.

Thirdly, around the 6-4,000 BP period critical to Ferguson's thesis, four sites were occupied in the seventh millennium BP (7,000-6,000 BP), three in the sixth millennium BP, and nine in the fifth, fourth, and third. These are not figures that can support any thesis about mid-Holocene population changes.

One can also surmise, from comparing the extent of the modern Karri forest (200,000 ha) with that of the modern Jarrah forest (3 to 5 million ha), that a mid-Holocene spread of Karri forest at the expense of Jarrah forest must have been extraordinary, to have caused a regional depopulation across the lower south-west (area estimates from Christensen 1992, Dell *et al.* 1989, cf. Figure 2.1). The final criticism of Ferguson's thesis is that evidence for mid-Holocene climatic change is conflicting and probably relevant only to small localities (see Chapter 3). Notably, Churchill's (1968) pollen record from Boggy Lake, the basis for the proposed mid-Holocene expansion of Karri forest, is not replicated by recent research there (Newsome and Pickett 1993).

Ferguson's depopulation hypothesis was for several years the only attempt to identify hunter-gatherer interactions with south-western forests. For this reason, it was given cautious support by Lilley (1993), and assumed to be correct in syntheses by Dodson *et al.* (1992), Lourandos (1997), and O'Connor *et al.* (1993). The first two syntheses are not discussed here because they do not deal with any new evidence other than Ferguson's. The third synthesis, O'Connor *et al.* (1993) is based partly on Hubbard's (1995a, b) environmental interpretations, shown to be incorrect in Chapter 3.

The synthesis of O'Connor et al. (1993) is also flawed in this case for following Ferguson's assumption that artefact and site numbers are proportionate to regional populations, a problem discussed already (M.V. Smith 1993, Dortch and Smith 2001). There are still too few sites across south-western Australia to draw any inference about regional depopulations on the scale suggested by Ferguson (Dortch and Smith 2001). Research in Tasmania offers an instructive example why these inferences are dangerous.

Tasmania is a region covering a similar area to the Darling Botanical District of south-western Australia, but with much more archaeological evidence of occupation and abandonment, and a great range of environments. These factors seem to give credence to inferences that people reacted to vegetational change (they abandoned the Tasmanian south-west), yet some archaeological evidence and the probable existence of artificial fire regimes in the past implies a human presence even in putatively abandoned regions. Even in this island of extreme environments, archaeologists should be dubious of theories of hunter-gatherer exclusion from regions that contain unfavourable plant formations.

A plausible argument by Horton (1982) provides another example of the problems with generalisation at the continental or super-regional scale. Horton (1982) argues that hunter-gatherers have little influence on environment. His concept *fire potential* embraces important aspects such as fuel load (standing vegetation and litter accumulation) and fuel quality (dryness, litter decomposition rate). These aspects are all controlled by climate. In central and northern Australia, vegetation burns frequently at low intensity; on the continental periphery, and to the south, wet forests burn infrequently at high intensity. Fire potentials are distributed according to climate, soils, and topography, apparently belying the fire-stick farming hypothesis of R. Jones (1969), and the proposition of Aboriginal modification of landscapes by fire (Howitt 1890, Merrilees 1968, Tindale 1959).

Lightning-strikes are sufficient source of fire ignition even with heavy rain. High lightning strike rates recorded for northern Australia account for frequent fire there (see also Hopkins *et al.* 1993). Correspondingly, in many parts of southern Australia, low lightning strike rates correlate with low fire-potential woodlands. The exceptional southern regions experiencing frequent lightning-strikes are the south-westerly coastal forests, where the topography destabilises fast onshore westerly winds. These coastal forests are those of Tasmania, southern Victoria, and south-western Australia. Horton argues that the south-west coastal, high fire-potential forests provide no evidence that they were altered from the state in which natural fire frequencies would have left them. But this point can be rejected for Tasmania, at least, on the basis of palynological and historical observations of the history of sedgelands and grasslands, given above (Bowman and Brown 1986).

Horton's conclusion is that Aborigines used fire, thus providing an alternative or additional source of ignitions, but they did not alter the existing fire regime and could not have shifted vegetation patterns by themselves. There may, as a result of Aboriginal firing, be minor local differences from the pattern that would be left by a natural fire frequency. But it is precisely these "minor" differences that may be important to people (Bowman and Brown 1986, Head 1988, 1989). Bowman and Panton (1993) and Head (1994) show that northern Australian Aboriginal people take advantage of the natural fire regime to adjust and maintain the distribution of plant formations in a way that favours ecological communities and hence human food resources. In southern Australia, there can be no ecological studies of "fire-stick farming", which is no longer practised, but one might expect that it too, while failing to determine regional vegetation structure or floristic composition, could have controlled local vegetation.

All of this implies that the relatively few sites in south-western Australia are nonetheless useful in identifying hunter-gatherer occupation of forests, provided that their archaeological and palaeo-environmental records are interpreted in detail and on a smaller scale. Detailed archaeological and palaeontological records derive from cave and rock shelter sites in coastal limestone or calcarenite (Balme 1980; Balme *et al.* 1978; C.E. Dortch 1979a; J. Dortch 1996; C.E. Dortch 1997; Dortch and Dortch 1996, 1997; Dortch and Kelly 1996, 1997; Lilley 1993). Evidence from these sites provides most of the archaeological evidence for animal foods that forests and woodlands provided, and the economies that they supported. The sites of greatest interest here are Devil's Lair, Tunnel Cave, Witchcliffe Rock Shelter, and Rainbow Cave, as these are located in the Leeuwin-Naturaliste Region. These sites are discussed in detail in subsequent chapters. Preliminary observations are given below.

The other limestone shelter sites, Katelysia Rock Shelter and Nookanellup Rock Shelter, are of relatively less interest to the forest question as the first is located directly on an estuary shore and the other on a heath-covered coastal plain (J. Dortch 1997; Dortch and Kelly 1996, 1997). Both sites indicate recent Aboriginal exploitation of marine and terrestrial food animals. They are only discussed below where their records are relevant to the Leeuwin-Naturaliste Region sites.

Devil's Lair is the longest-occupied limestone cave site known in Australian forests (Turney *et al.* in press, cf. Cosgrove 1995). It contains abundant biotic and artefactual remains and was visited by Aboriginal people from *c.*47,000 BP to 12,000 BP (Balme *et al.* 1978, C.E. Dortch 1979a, Turney *et al.* in press; see also Chapter 6). The cave is now surrounded by Karri tall open-forest, but because the site was sealed by roof-collapse *c.*6,500-12,000 BP, before Karri tall open-forest surrounded Devil's Lair (Burke 1997), the site can provide no evidence to indicate whether people occupied Karri tall open-forest. When the site was occupied by people, the surrounding forest or woodland included the eucalypts Jarrah, Marri, and Tuart (Balme *et al.* 1978, Burke 1997). The record from Tunnel Cave is mentioned above, and in a published report (J. Dortch 1996). The site was mostly occupied in the Pleistocene, when Karri forests are inferred to have been of restricted distribution. The site probably remained accessible, but was not occupied, throughout the Holocene, when Karri forest probably expanded. The site appears to support the concept that people avoided Karri tall open-forest: the evidence to test this concept is presented in detailed analysis in later chapters.

Witchcliffe Rock Shelter and Rainbow Cave both contain short records dated to the last millennium and they are located in coastal woodland and scrub (J. Dortch 1996, Lilley 1993). They are analysed below because they demonstrate occupation patterns in vegetation that can be assumed to have differed only slightly from the present-day, known vegetation.

The emerging archaeological record for south-western Australian tall open-forests is still very restricted. Anderson and Pearce worked entirely in the northern and central Jarrah forest. Field surveys in the southern forests by MacDonald Hales and Associates (1994) had inconclusive results, and owing to poor ground visibility, these authors do not reject the possibility of an archaeological record in tall open-forest. Lilley (1993) does not report surveying tall open-forest. In any case, his surveys along the Margaret River valley cannot have included large areas of tall open-forest (cf. Figure 2.4). Even after Holocene tall open-forest expansion, Tunnel Cave is still located close to the edge of the Karri stand it is located in. Dombakup 24 remains the sole evidence for Aboriginal occupation of tall open-forest, and the limited biotic remains from its deposit (known to include charcoal) have yet to be investigated for their potential as palaeo-environmental records (J. Dortch *et al.* 1998).

The abundant evidence from the Leeuwin-Naturaliste Region cave and shelter sites relates largely to the site localities (Balme *et al.* 1978, Burke 1997). It is detailed enough to show vegetational changes at the sites, and whether they were significant for human occupation there. These factors make the Leeuwin-Naturaliste Region cave sites the primary choice for investigation of small-scale hunter-gatherer interactions with vegetation (Dortch and Dortch 1997).

Summary: Aboriginal use of south-western Australian forests

In most of south-western Australia, the stable Mediterranean climate has probably influenced human settlement patterns for a long time. For the Nyoongar groups recorded by early European settlers at Perth and Albany, summer and autumn were attractive times for large numbers of people to gather near the coast, in areas that received occasional rain and had many wetlands providing food and water (Anderson 1984, Hallam 1975). However, coastal resources could not sustain large groups all year round. The onset of the rainy season permitted people to disperse into the drier, forested interior, though small groups or individuals may have returned to the coast in winter. Since food resources were scarce inland, the large gatherings were generally broken into family-sized groups, and no forest site was occupied for long.

Dispersed, opportunistic winter-time foraging was necessary. Having an alternative resource base meant that coastal foods were not over-used. In the Warren Botanical Sub-District, heavy rains are experienced for a large part of the year. As Dortch and Gardner (1976) observe, the seasonal climatic limits that more northerly localities experienced may not have been as strict, and social organisation may have been correspondingly loose-knit. This concept has had some support from recent fieldwork in the high rainfall Northcliffe district (Dortch and Kelly 1996, 1997).

If the forests in well-watered districts could be exploited for a greater part of the year than other areas, then settlement pattern and social organisation may relate to vegetation types as much as the seasonality of rainfall. For hunter-gatherers, factors in the exploitation and occupation of different forests would be the geographical, seasonal, and inter-annual distribution of food resources and water and the ease of movement through the understorey. These factors vary significantly between Karri and Karri-Marri tall open-forest and Jarrah or Jarrah-Marri open-forest (see Chapter 2).

Pure Karri tall open-forest has a high canopy, dense understorey, a limited number of species, and fewer resources than other forest types. Jarrah-Marri open-forest has a moderately high canopy whose animal inhabitants, such as possums and birds, are still accessible to people; more convoluted tree-trunks providing hollows for animal dens and nests; a heterogenous composition (implying more diverse resources); and an open understorey that is easier to travel through. Mixtures of Karri and Jarrah or Marri are mid-way on the continuum between these extremes.

Forest composition and structure is at least partly a product of fire regime, i.e., the frequency, intensity, and season of fires. Silviculturists' experiments show that fires would be required every three to five years, to kill regenerating Karri and understorey shrubs before they produce viable seed (Christensen and Abbott 1989, Christensen and Annels 1985, McCaw 1986). Frequent, "cool" fires, favouring fire-resistant lignotuberous species over fire-sensitive leguminous seed species, can convert mixed Karri forest to Jarrah forest, especially where soils are marginal (Bradshaw 1985). Ferguson (1985) claims that cool fires kill understorey without consuming it, leaving an impenetrable, long-lasting thicket (cf. Rule 1967), but these thickets are patchy, and would not cover the entire forest floor (Christensen and Kimber 1975). The existence of open and closed patches benefits the greatest number of animals, many of which inhabit one vegetation unit and forage in another. Faunal diversity is therefore partly a result of the fire regime.

With infrequent fires, Karri forest tends to develop into pure even-aged stands, as the occasional fires are wildfires, clearing the understorey and the younger trees and encouraging rapid regrowth. Because Karri saplings and some leguminous understorey plants grow faster than others, the stand maintains its pure composition, age structure, and hence fire regime. But given frequent fire, some Karri stands have the potential to develop a mosaic structure of mixed species of varying ages and responses to fire.

Achieving this latter pattern, probably desirable for hunter-gatherers seeking the greatest diversity of resources and ease of access, would depend on how frequently stands can be visited and burnt. Prehistoric firing of tall open-forest by hunter-gatherers would therefore depend on their settlement pattern and population density (MacPhail 1980, Ward and van Didden 1997, Ward 1998).

Vegetational changes identified in many temperate regions, such as the early Holocene spread of rain-forest and tall open-forest in south-eastern Australia and Tasmania, seem not to have been altered by hunter-gatherers, even if they had wished to limit this type of vegetation. At the same time hunter-gatherers were not displaced from regions because closed vegetation was nowhere sufficiently widespread to eliminate all of their resources or impede their movement across the whole region. Even major vegetational changes cannot imply abandonment of regions. But vegetational changes of the order seen in Tasmania probably caused adjustments in the small-scale distribution of hunter-gatherers. In following chapters I analyse the evidence for vegetational change in the Leeuwin-Naturaliste Region and attempt to identify changes in site occupation patterns, as an indication of population distribution.

Chapter 5 Location and test-excavation of archaeological deposits

This chapter describes a search for and identification of human occupational records in Leeuwin-Naturaliste Region limestone cave and rock shelter floor deposits. The aim of this search was to identify potential occupation sites in caves and rock shelters by inspecting a large number of sites, and to confirm the presence of occupational records by excavating a sample of four sites. Two of the four deposits excavated were found to contain occupational records comprising relatively brief episodes of occupation separated by centuries or millennia. At these sites, and two others previously excavated, occupational remains such as artefacts and hearths were stratigraphically associated with abundant biotic remains, evidence that would enable inferences about human occupation intensity and vegetation characteristics.

This chapter reviews other researchers' surveys for archaeological remains in Leeuwin-Naturaliste Region caves and rock shelters. It describes the first systematic search for potential occupation sites in the region's limestone caves and rock shelters (J. Dortch 1996), and several attempts to find archaeological material by augering and test-excavating deposits. Augering gave inconclusive results but test-excavations revealed occupation at Tunnel Cave and Witchcliffe Rock Shelter. Each of these sites contained many hearths, which provide strong evidence for episodes of human occupation. Adding to existing records from Devil's Lair (Balme *et al.* 1978, C.E. Dortch 1979a, Turney *et al.* in prep.) and Rainbow Cave (Lilley 1993), discussed in the next chapter, they make an intermittent, long-term record of human occupation in the southern part of the Leeuwin-Naturaliste Region.

Previous cave surveys

By the 1880s, the locations of many significant caves and sinkholes, almost certainly known to Aboriginal groups occupying the Leeuwin-Naturaliste Region, had been either communicated to, or discovered by, the first generation of British settlers (Bastian 1986; Battye Library Research Note 396/2: letter from R. Murray). The subsequent development of caves as tourist sites led to the discovery of more caves and of rich fossil remains at Mammoth Cave.

Palaeontological excavations in this cave's floor deposit by E.A. Le Souef began in 1904 and were resumed in 1909 by Ludwig Glauert of the Western Australian Museum (Dortch and Dortch 1997; Glauert 1910, 1948; Woodward 1909, 1910). The Mammoth Cave excavations revealed tremendous numbers of animal bone fragments, among them bones of extinct animals definable as "megafauna", but none of the investigators referred to evidence of Aboriginal occupation. They may have believed that none would be found, given contemporary European assertions that Aboriginal people avoided caves for spiritual reasons (Hallam 1975: 82-97; Battye Library Research Note 396/1: letter from T. Connelly, referring specifically to Witchcliffe Cave).

After these early excavations, cave investigations virtually ceased, owing to the collapse of cave tourism and the overgrowth of tracks and cave entrances. They resumed in the 1950s with the formation of the Western Australian Speleological Group (WASG; Bastian 1986), and in 1954, E. Lundelius' pioneering palaeontological excavations at Devil's Lair, which provided the first radiocarbon dates for the region (Lundelius 1960, 1966). D. Merrilees, Curator of Palaeontology at the WA Museum, instigated further palaeontological and later archaeological research at Devil's Lair and other sites (Archer *et al.* 1980; Balme *et al.* 1978; Baynes *et al.* 1975; Davies 1968; Merrilees 1968, 1975, 1979, 1984). Meanwhile, many new caves were found, thanks to the efforts of B. Loveday and other cavers who from the 1970s to the present day have mapped caves and plotted their geographical positions and extent on a 1:5,000 scale map of the entire Leeuwin Ridge (B. Loveday, no date).

The most recent Australian Karst Index lists 247 limestone features in the Leeuwin-Naturaliste Region (Matthews 1985). Features listed are pothole entrance caves, "walk-in" entrance caves, cliffs (that may include overhangs or very shallow shelters), and subterranean stream channels. These 247 features and at least thirty more, documented since that index was published, have been planned at 1:500 scale and located on Loveday's 1:5,000 scale map (B. Loveday, pers. comm.).

Present survey

After investigating these historical and current records, and consulting Loveday and other cavers about the "walk-in" entrance caves and the cliff overhangs mentioned in the Karst Index and in Loveday's notes, and examining Loveday's 1:500 scale plans and sections of every cave, I selected for inspection 90 caves and rock shelters, whose plans indicated they had some potential for human occupation (Table 5.1; cf. Attenbrow 1982). These karst features, numbering about 180, had what I considered were minimum requirements for human occupation and for archaeological excavation: a "walk-in" entrance, shelter for one person, and a sandy floor deposit, necessary for excavation. Conceivably, prehistoric people in this region used ropes to gain access to pothole-entrance caves, but they are unlikely to have regularly occupied such places. Excavations in two pothole-entrance cave deposits, which contain no traces of human occupation (Porter 1979, Prideaux *et al.* 1997), confirm this impression.

I made the survey for potential archaeological sites in caves in March 1993. Thick understoreys cover most cave areas, which are mostly in the Leeuwin-Naturaliste National Park, so Loveday's map was essential. The network of roads and tracks in the Park gave me access by two-wheel drive vehicle to within a few minutes' walk of most caves.

At each cave I made a rough plan, estimated the size of the cave, and the area of sandy deposit. I searched for surface artefacts and scanned cave walls for paintings, stencils, or engravings, as are found on some south-western Australian rock outcrops and caves (cf. Clark 1980; Hallam 1974, 1975: 86-87; Morse 1984). In some places, people or animals had dug holes in the floor deposit, or parts of the deposit had slumped, enabling me to examine vertical sections for artefacts or hearths. I evaluated caves and shelters in terms of how easily a person could enter them, whether the entrance let in much light, how dry the cave was (conversely, whether the roof dripped, even in late summer, the time of survey), and how much sandy floor deposit was under shelter. I assumed that a sandy floor deposit indicated potential for excavation and for occupation, since people would probably prefer to camp on soft sand rather than a hard rock surface. I also considered that any cave or rock shelter that satisfied the above criteria, and was close to a perennial spring, stream, or lake, would attract human occupation.

Under the conditions of my cave-visiting permit given by the Caves Management and Access Committee (CMAC), I do not give the precise locations of the caves that I visited. Figure 5.1 shows their approximate locations; Table 5.1 their names and comments.

Present survey – results

In the cave survey I failed to find six caves recorded by Loveday. These are either located in dense scrub or understorey vegetation, or on private property to which I was refused access. I can make no judgement about the human occupation or archaeological excavation potential of these caves. Of the other 84 locations, at least 12 caves or rock-shelters would probably have prime importance in any excavation programme aiming to assess Aboriginal occupation patterns in Leeuwin-Naturaliste Region limestone caves and rock-shelters (Table 5.2). These 12 locations are either known sites, or they have easy access, a large area under shelter that is still open to natural light, and a large sandy deposit, suggesting that they would be attractive as occupation sites (see Appendix 2). Another 24 caves or rock shelters are potentially useable as occupation sites, being easily accessible to people, and having some sandy deposit in the cave entrance where light penetrates. The other 46 locations that I inspected are unlikely to have abundant archaeological remains - these places have difficult access, admit little or no natural light, or they have almost no sandy deposit suitable for people to camp on or where long chrono-stratigraphic records might have accumulated.

At the time of the survey, only three of the twelve locations with most archaeological excavation potential (Devil's Lair, Rainbow Cave, and Old Kudardup Cave) had been shown to contain artefacts. Of the others, the large deposits at Wi-38 (Mammoth Cave) and Wi-49 (Calgardup Cave) had been excavated or test-excavated and no artefacts found (Glauert 1910; Lowry and Lowry 1968). The remaining seven locations with good potential had never been test-excavated: Wi-6 (Goanna Cave), Wi-16 (no name), Wi-37 (Orchid Cave), Wi-97 (east entrance of Tunnel Cave), MR-1 "B" (Witchcliffe Rock Shelter), Co-1 (Quininup Lake Cave), and Ya-40 (Barbilla Cave).

It seems likely that some caves would contain no artefacts, even allowing for modern disturbances. Visitors and excavators might have inadvertently or deliberately obscured or removed surface and sub-surface artefacts, but with the exception of Mammoth Cave (Glauert 1910), they affected only a small part of each deposit. The impression that a few apparently attractive locations are not sites may well be correct. Nevertheless, considering the small proportions of caves and rock shelter floor deposits that have been excavated, the large number of potential archaeological sites, the small size and scarcity of stone artefacts in some of the known sites, and the failure on the part of some amateur excavators to use fine-mesh sieves, only a small proportion of cave deposits can be regarded as well-tested.

Therefore, I selected two deposits for immediate excavation: Orchid Cave and the east entrance of Tunnel Cave. The following year I augered deposits in four limestone caves and rock shelters, as an additional means of locating archaeological deposits.

Augering

The search for cave sites was hampered by the probable disturbances caused by recent human visitors, such as trampling of surface artefacts and hearths into the floor deposits. Augering seemed to be an efficient means of finding occupational remains below the surface. Although augering might miss small stone artefacts, it would have less chance of missing laterally extending hearths.

I used a sand auger with a 50 cm bit and 10 cm bore to auger five cores at Quininup Lake Cave (Co-1), one at an overhang in the Mt Duckworth sinkhole (Ya-6), three at Wi-16, and one at Orchid Cave (Wi-37), this last being an attempt to confirm the results of my test-excavation there. Augered sediments from 50 cm spits were sieved through 3 and 5 mm screens, scrutinised on site, and retained for careful laboratory examination.

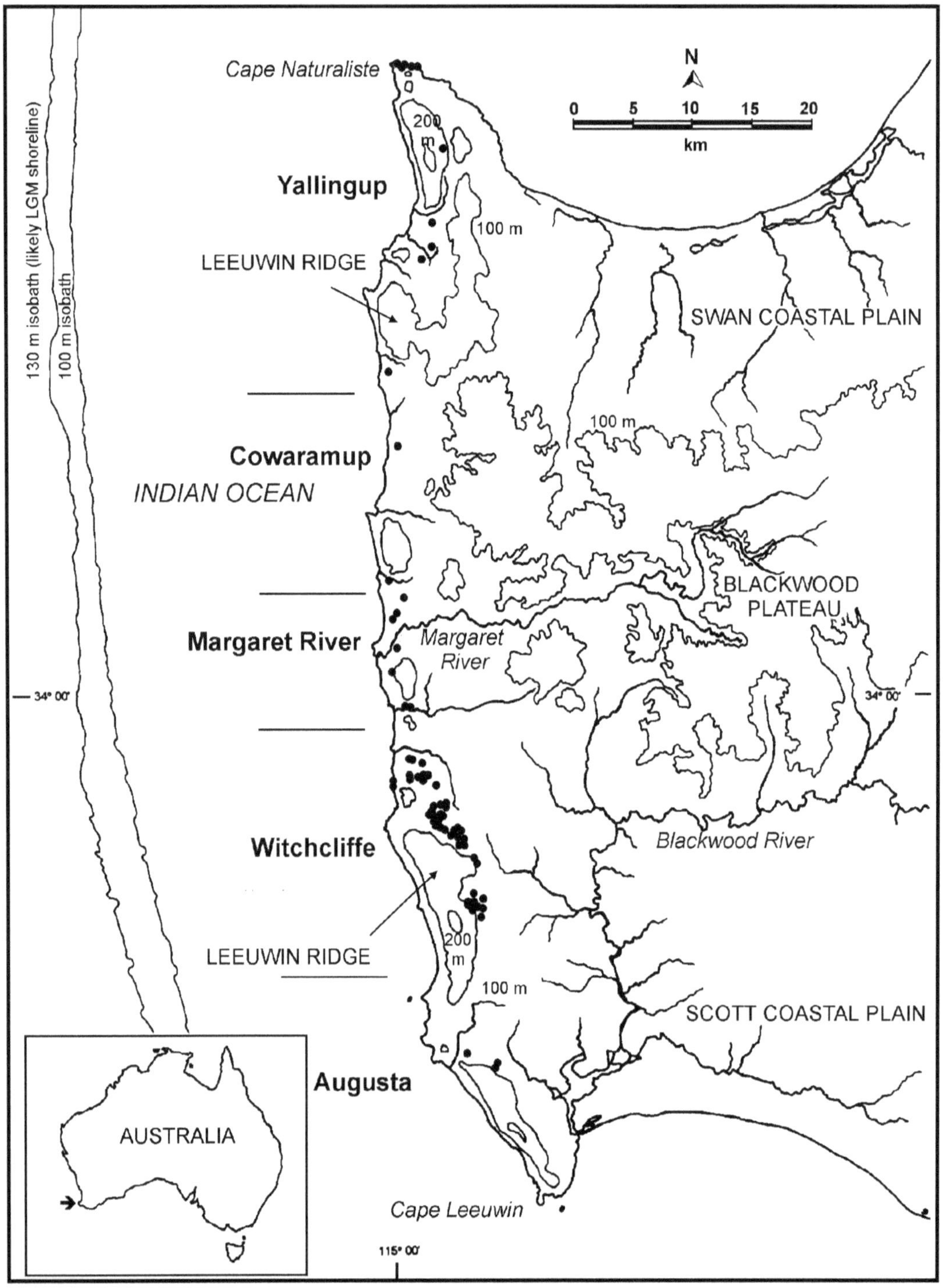

Figure 5.1 Approximate locations of limestone caves and rock shelters (black dots) inspected as part of a survey for potential archaeological sites.

Place names in bold print indicate the cave districts as they appear in the Australian Karst Index (Matthews 1985). Dots showing cave locations are too large to infer their exact position, as requested by CMAC.

Table 5.1 Leeuwin-Naturaliste Region caves and rock shelters selected for survey

The Australian Karst Index gives each limestone cave entrance, shelter, pothole, or cliff a number based on a cave district (cf. Figure 5.1).

Karst Index number and name		*Comments*
Augusta	**No. inspected caves: 3**	
Au- 1	Deepdene Cave	Excavated, no artefacts found (Lowry & Lowry 1968)
Au- 9	Old Kudardup Cave	Archaeological site, artefacts collected (Morse 1984)
Au- 10		
Au- 17		Not found in present survey
Au- 22-25	Deepdene Gorge	Access refused by landowner; no artefacts found in previous excavations (Archer & Baynes 1974).
Witchcliffe	**No. inspected caves: 50**	
Wi- 1		
Wi- 2	Green Cave	
Wi- 5	Midgie Hole	
Wi- 6	Goanna cave	
Wi- 7	Skittle Cave	
Wi- 8		
Wi- 10		
Wi- 11		
Wi- 13	Golgotha Cave	
Wi- 16		Augered deposit
Wi- 17	Mordang Dar	
Wi- 21A	Giants Cave	
Wi- 21B	Giants Cave, eastern cavern	
Wi- 22		
Wi- 25		
Wi- 28		
Wi- 29		
Wi- 33	Tunnel Cave, west entrance	
Wi- 35		
Wi- 37	Orchid Cave	Excavated, no artefacts found, augered deposit
Wi- 38	Mammoth Cave	Excavated (Glauert 1910), bone artefacts claimed (Archer *et al.* (980).
Wi- 39	Mammoth Cave, west entrance	
Wi- 42	Terry Cave	
Wi- 43		
Wi- 45		
Wi- 46		
Wi- 49	Calgardup Cave	Excavated, no artefacts found (Lowry and Lowry 1968)
Wi- 51	Ruddocks Cave	Artefact collected by caver, P. Bridge (WA Register of Aboriginal Sites, no date)
Wi- 53	Zamia Nut Cave	
Wi- 55		
Wi- 61	Devil's Lair	Excavated, archaeological site (C.E. Dortch 1979a)
Wi- 62	Crystal Cave	
Wi- 67	Acoustic Pot	
Wi- 69		
Wi- 70		
Wi- 84		
Wi- 95	Arnor Cave	
Wi- 97	Tunnel Cave, east entrance	Excavated, archaeological site (J. Dortch 1996)
Wi- 99		
Wi- 100		
Wi- 107	Fisherman's Cave	
Wi- 108		
Wi- 110		
Wi- 116		
Wi- 122	Pentorafice	
Wi- 123		
Wi- 134		
Wi- 148		
Margaret R.	**No. inspected caves: 7**	
MR- 1A	Witchcliffe Cave	Excavated, no artefacts found (I. Lilley pers. comm., J. Dortch 1996)
MR- 1B	Witchcliffe Rock Shelter	Excavated, archaeological site (J. Dortch 1996)
MR- 3	Rainbow Cave	Excavated, archaeological site (Lilley 1993)
MR- 4	Wallcliffe Cave	
MR- 9		
MR- 10	Foxhole Cave	
MR- 11		Not found in present survey
MR- 17	"Rabbit's Lair"	Excavated, no artefacts found (I. Lilley pers. comm.)

Table 5.1 Leeuwin-Naturaliste Region caves and rock shelters selected for survey (continued from previous page)

Karst Index number	*Name (if any)*	*Comments*
Cowaramup	**No. inspected caves: 3**	
Co- 1	Quininup Lake Cave	Augered deposit, no artefacts found
Co- 7	Cowaramup Cave	
Co- 8	Meekadarabee Cave	
Co- 9		Not found in present survey
Yallingup	**No. inspected caves: 21**	
Ya- 6?	"Mt Duckworth Doline"	Augered deposit, no artefacts found
Ya- 3	Seven Sisters Cave	
Ya- 11		
Ya- 13	Warrigal Cave	
Ya- 14-22	sea caves, Cape Naturaliste	
Ya- 40	Barbilla Cave	
Ya- 42 - 48	sea caves, Cape Naturaliste	

Table 5.2 Archaeological excavation potential of the 84 inspected Leeuwin-Naturaliste Region caves and rock shelters (cf. Appendix 2)

Locations by Karst Index number	No. of caves and rock shelters with poor archaeological excavation potential	No. of caves and rock shelters with moderate archaeological excavation potential	No. of caves and rock shelters with good archaeological excavation potential
Augusta	1 (Au-1)	1 (Au-10)	1 (Au-9, a known site)
Witchcliffe	26 (Wi-1, 2, 8, 10, 11, 17, 22, 25, 28, 29, 35, 39, 43, 46, 53, 55, 60, 62, 70, 84, 95, 108, 110, 116, 123, 148)	17 (Wi-5, 7, 13, 21A, 21B, 33, 42, 45, 51, 67, 69, 93, 99, 100, 107, 122, 134)	7 (Wi-6, 16, 37, 38, 49, 61, 97; 2 are known sites)
Margaret River	3 (MR-4, 9, 10)	2 (MR-1A, 17)	2 (MR-1B, 3; both are known sites)
Cowaramup	1 (Co-8)	1 (Co-7)	1 (Co-1)
Yallingup	17 (Ya-3, 14-22, 42-48)	3 (Ya-6, 11, 13)	1 (Ya-40)
TOTAL	48	24	12

Augering results

No auger-sample contained artefacts or hearth sediments. The auger samples did contain quantities of charcoal, bone fragments, snailshell fragments, and plant remains. The absence of occupational remains in some auger samples is inconclusive, because several auger holes had to be abandoned at relatively shallow depth (1 m or less). Sand-augers are built of thin steel and easily stopped by small stones, common enough common in limestone cave and rock shelter floor deposits.

However, it seems likely that I would have intersected archaeological layers if I had augered a deposit comparable to those at Devil's Lair, Tunnel Cave, Witchcliffe Rock Shelter, or Rainbow Cave, where remains of Aboriginal occupation include wide features such as hearths and layers containing hundreds of burnt bone fragments. The lack of these types of remains in the augering sites suggests that they contain few or no archaeological remains, at least at the depths reached in the auger holes (maximum depth 2.8 m). Depth of deposit may be a critical factor at some sites, e.g. Wi-16 and the Mt Duckworth doline, where the position of the deposit at the base of a doline suggests it could have trapped large quantities of sand in relatively short time. At sites that would not have trapped so much sediment, notably Orchid Cave and Quininup Lake Cave, augering may have established an absence of hearths. The latter site remains attractive for excavation due to its proximity to water, but, at the request of CMAC, I did not investigate it further, because a bat colony (*Chalinolobus morio*) resident there was subsequently thought to be vulnerable to disturbance from activities such as excavation. I therefore test-excavated at two other locations, Witchcliffe Rock Shelter and the adjacent Witchcliffe Cave. These two excavations represented my final attempt to find additional sites in Leeuwin-Naturaliste Region caves and rock shelters.

Test-excavations

The sites test-excavated were, in chronological order of excavations: Orchid Cave, Tunnel Cave, Witchcliffe Rock Shelter, and Witchcliffe Cave. Throughout the four months of excavations (in 1993 and 1995), I obtained help from up to three volunteers at a time. About 30 people offered their help over the two field seasons.

Section 16 of the WA *Aboriginal Heritage Act 1972* restricts the area of test-excavations to 2 m² at the base of the excavation, or 10% of the surface of the site, whichever is the smaller. I excavated 1 m² of surface deposit at Orchid Cave and Witchcliffe Cave, 1.5 m² at Witchcliffe Rock Shelter, and at Tunnel Cave, I excavated a test-pit measuring 4.5 m² at the surface and 1 m² at the base.

Excavation involved scraping and lifting of sediment with small trowels; identification of layers in the field; mapping, photographing, and recording in three dimensions the extent and position of archaeological features such as hearths and artefacts *in situ*; sieving excavated sediment through 3 and 5 mm mesh screens; and collection of artefacts, faunal remains, other biotic material, and the unexamined parts of the sieve residues. All collected materials, *in situ* finds, and sieve residues, whether scrutinised in the field or not, were returned to UWA for careful examination in better lighting.

Preliminary laboratory sorting

A preliminary aim of analysis was to identify archaeological and faunal remains from all layers for analyses described below (Chapters 7 and 8). The analyses required scrutinising hundreds of kilograms of 3 and 5 mm sieve residues and counting and weighing items in many categories of archaeological remains.

Volunteer assistants and I identified stone artefacts, emu eggshell, mollusc shell, bone fragments, and charcoal (except for very small fragments in the 5 mm residue, and all the fragments in the 3 mm residue). Table 5.3 shows the preliminary classifications and the reasons for making them. All items except charcoal were counted and all classes of material were weighed. Vertebrate taxa were identified from teeth and jaw fragments, parts that are well preserved and have good diagnostic features. Counts and weights of all categories were entered into a computer spreadsheet, with each spit as a new entry. Weights and counts from spits were then amalgamated into their respective layers.

At the time of writing, all of the remains collected, including unexamined material from squares not analysed, and sieve residues already examined, are stored at the Western Australian Museum, Perth.

I identified no plant remains apart from charcoal and roots; the latter I discarded or left in sieve residues. Most uncarbonised plant material is poorly preserved in sandy deposits (Beck *et al.* 1989), and in the scope of this project, it proved unfeasible to identify phytoliths, leaf cuticles, or other microscopic plant fossils. Charcoal samples chosen for radiocarbon dating were submitted to the Waikato University Radiocarbon Laboratory, New Zealand.

Table 5.3 Preliminary sorting classes for material recovered from Leeuwin-Naturaliste Region limestone cave and rock shelter sites

Sorting class	Description
Artefactual material	
Stone artefacts	Chert, quartz, and silcrete artefacts, carried to the shelter by people; calcrete ones possibly flaked from rock at the site.
Ochre	Probably carried to the site by people. The source is unknown.
Bone artefacts	Points (polished, shaped pieces) and miscellaneous pieces incised by stone tools.
Emu eggshell	Almost certainly left by people, as emu eggs are recorded as Aboriginal food, emus are not known to lay eggs in caves or rock shelters, and no other animal is likely to have carried quantities of eggs or eggshells to these sites. Divided into burnt (blackened) and unburnt (brown) eggshell.
Other eggshell	Probably artefactual but less common, perhaps due to poor preservation (more fragile). Not attributable to species.
Aquatic mollusc shell, fish and crustacean	As emu eggshell, and all documented as an Aboriginal food items. None of the three classes are common. Identified to species where possible.
Bone, not necessarily artefactual	
Teeth & jaws	Used for identifying vertebrates.
Burnt bone	Relates to hearth-building. Divided into burnt white and burnt black, as an indication of the intensity of burning.
Unburnt bone	Bone fragments, yellow or brown in colour.
Potentially identifiable (diagnostic) bone	Bone fragments that are identifiable to at least anatomical part. This bone is assumed to been less fragmented (comminuted) by the actions of animals or people.
Unidentifiable (non-diagnostic) bone	Totally unidentifiable bone fragments. A high proportion of unidentifiable bone in the total bone assemblage suggests a high degree of comminution by scavengers or people (Grayson 1984).
Other material	
Snailshell	Derived from landsnail, mostly *Bothriembryon* spp., a possible environmental indicator.
Charcoal	Carbonised wood. Very common in these deposits.
Seeds	In lower levels, only burnt seeds were found, suggesting that carbonisation is required for preservation in these deposits.
Coprolite	Nodules of hard, yellowish sediment containing bone fragments, and often retaining the form of carnivore scats. Most likely derived from Tasmanian Devil (*Sarcophilus harissii*; examined and identified by B. Marshall, La Trobe University).
Earth nodules	Miscellaneous lumps of earth, sometimes resembling red ochre, but more granular, and do not leave a stain.
Unburnt plant remains	Wood, leaves, seeds, etc, common only in uppermost spits.
Limestone nodules	Left as sieve residue along with fine charcoal and plant remains.

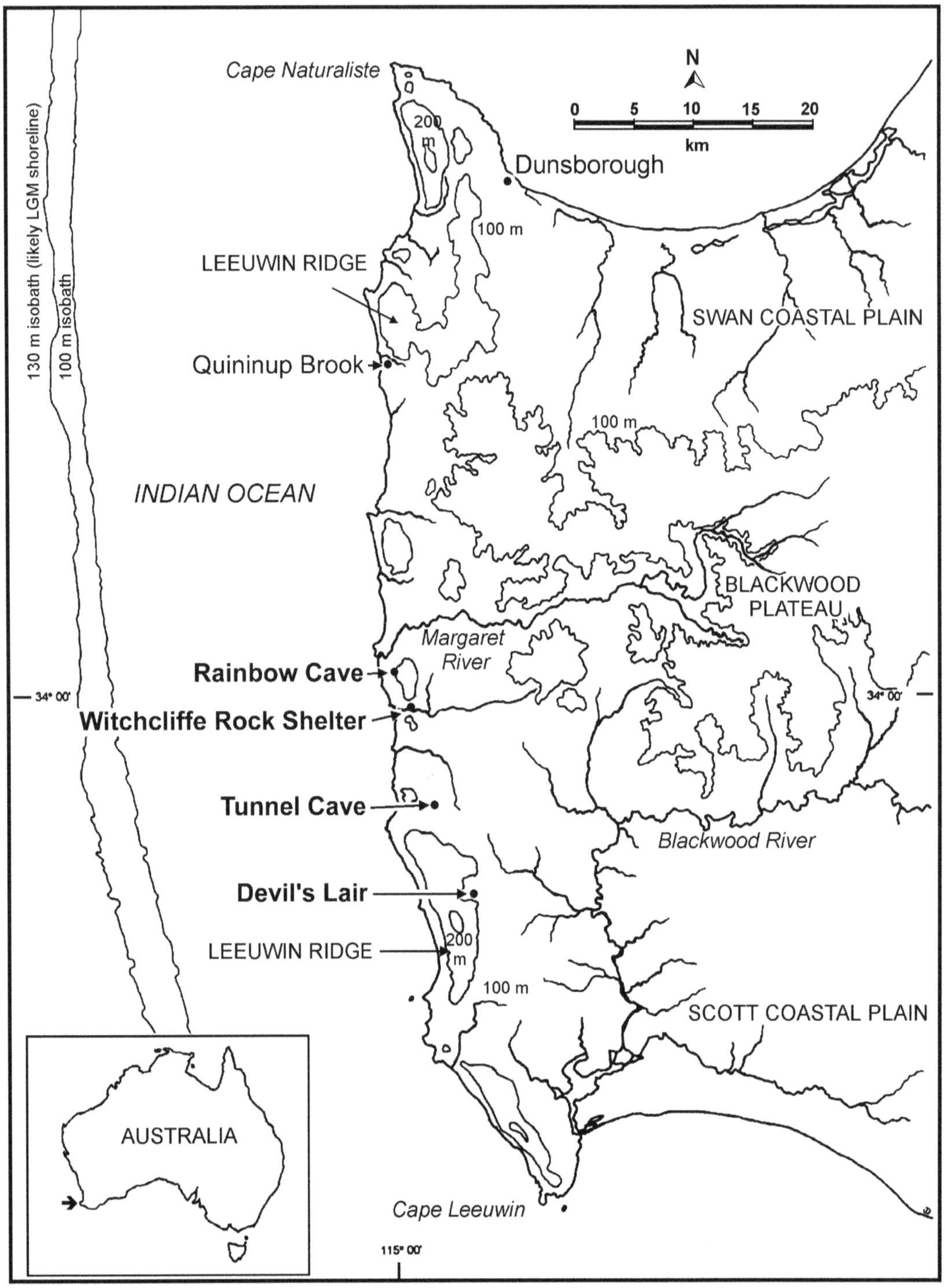

Figure 5.2 Locations of four Leeuwin-Naturaliste Region limestone cave and rock shelter sites (bold print) and other places mentioned in the text.

Summary

Inspection of 84 caves and rock shelters showed that 36 such locations have potential as human occupation sites; among these, 12 seem particularly attractive as occupation sites. Augering three deposits gave inconclusive results but test-excavations at four others revealed archaeological material at Tunnel Cave and Witchcliffe Rock Shelter (Figure 5.2). Each of these sites was found to contain large quantities of artefacts, burnt bone fragments and other faunal remains, and hearths. The test-excavations showed that where archaeological remains exist, they are found in finely stratified layers that have the potential to give long, detailed records of human occupation.

Chapter 6 Dating episodes of human occupation at Leeuwin-Naturaliste Region sites

The aim of this chapter is to present the distribution of archaeological remains in each of four radiocarbon-dated deposits, to show when human occupation occurred in relation to changes in vegetation. The archaeological deposits were first identified in this study (Chapter 5) or previously (Dortch and Merrilees 1971, Lilley 1993).

Archaeological evidence of human occupation includes, among other things, stone artefacts, hearths (intensely burnt sediments), and some classes of faunal material. Such remains are (1) exotic to the site itself and only people occupying the site would have carried them there, or (2) they are common in and around the site but only people could have modified them. Details of stone artefacts and those faunal remains that derive from human occupation are given in Chapters 7 and 8. Hearths are discussed in this chapter.

Numerous agents move or re-work small archaeological remains within deposits (Hiscock 1985, 1990; Schiffer 1983). These remains can move laterally across surfaces, or vertically through layers (Richardson 1992, Theunissen *et al.* 1998, Villa 1982). In contrast to small artefacts and fragmentary faunal material, hearths reliably indicate human occupation in a layer, since if the component lenses or layers within hearths are mostly intact, they cannot have been re-worked or moved great distances, laterally or vertically.

Hearths are defined as patches or lenses of apparently intensely burnt sediments that are similar to positively identified modern hearths. Both modern and archaeological hearths typically include sharply-defined, circular lenses of ash and charcoal resulting from intense burning of wood and underlying sand deposit. They are identified as artificial because only people make a fire that is both spatially restricted and intense (Bellomo 1993). I give the distribution of hearths at two sites where they are abundant, Tunnel Cave and Witchcliffe Rock Shelter, which suggests the age of at least some of the human occupations of these sites. In addition to hearths, I discuss the following aspects of each site:

1. the stratigraphic associations of small, potentially re-worked archaeological remains such as stone artefacts or exotic faunal remains. I present evidence that these remains were not re-worked from other layers, and radiocarbon dates from those layers give the ages of the archaeological items.
2. the radiocarbon age estimates themselves. Most series of radiocarbon age estimates are in stratigraphic order, and individual estimates are presumably correct, but occasional inverted estimates create some uncertainty. I present statistical and stratigraphic arguments to show which estimates are similar or mutually consistent, and which estimates are probably incorrect.

Devil's Lair

Devil's Lair provides the oldest and most detailed stratified evidence of human occupation in south-western Australia (Turney *et al.* 2001). The site provides an extremely important regional archaeological and palaeontological record, as shown by the number of publications on all aspects of the site (Allbrook 1976; Baird 1993; Balme 1978, 1979, 1980a, 1980b; Balme *et al.* 1978; Baynes *et al.* 1975; Bednarik 1997; David 1993; Davies 1968; C.E. Dortch 1974, 1976a, 1976b, 1979a, 1979b, 1979c, 1980, 1984, 1986a; Dortch and Dortch 1996, 1997; Dortch and Merrilees 1971, 1973; Freedman 1976; Glover 1974, 1979; Lundelius 1960, 1966, 1983, 1989; Merrilees 1975, 1979a, 1979b, 1984; Shackley 1978; Turney *et al.* 2001).

The Western Australian Museum research team, which carried out the bulk of excavations, recovered artefacts, bone fragments, and other remains from some 80 thin, discontinuous sand layers, all excavated separately wherever they could be distinguished. If not distinguishable, layers were excavated as mixtures of two or more layers, *e.g.*, "layers 4 and 5, mixed" (following Balme *et al.* (1978), these mixed units are treated here as layers in their own right). The WAM excavations were in several locations: a "main excavation" (Dortch and Dortch 1997), comprising ten adjoining trenches originally placed on the site of a small pit made by unknown excavators ("Small Excavation" of Dortch and Merrilees, 1971, 1973), trench 1 on the site of Lundelius' 1954 excavation (Lundelius 1960), and trenches 3, 4, and 6 in previously unexcavated areas in or near the cave's present entrance (Figure 6.1). Because trench stratigraphies are not easily reconciled across the floor of the cave, the following analysis draws on the findings in the main excavation, which comprises adjoining trenches 2, 5, 8_2, 8_7, 8-9, 9, and 10. The latter two trenches are further divided into north and south portions (Figure 6.2). Figures 6.3 and 6.4 show section diagrams of the deposit in the main excavation.

Stratigraphic context of evidence for human occupation

C.E. Dortch (1979a) proposes that evidence for human occupation comprises stone artefacts, emu eggshell, marine mollusc shell, hearths (ash and charcoal lenses), and large quantities of burnt bone fragments. Balme *et al.* (1978) argue that the presence of hearths or any of these kinds of material in a layer denotes a human occupation in or near the cave at some time during the formation of that layer.

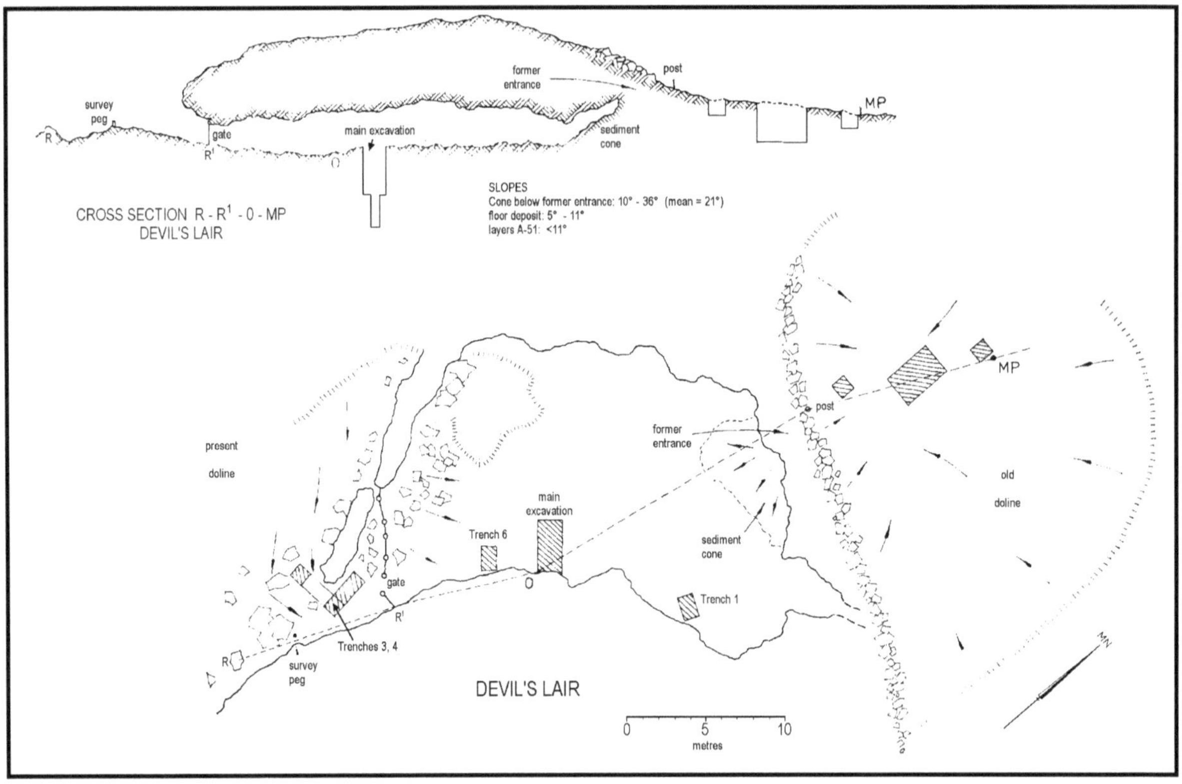

Figure 6.1 Devil's Lair plan and cross-section, showing positions of excavations, after C.E. Dortch (1979a)

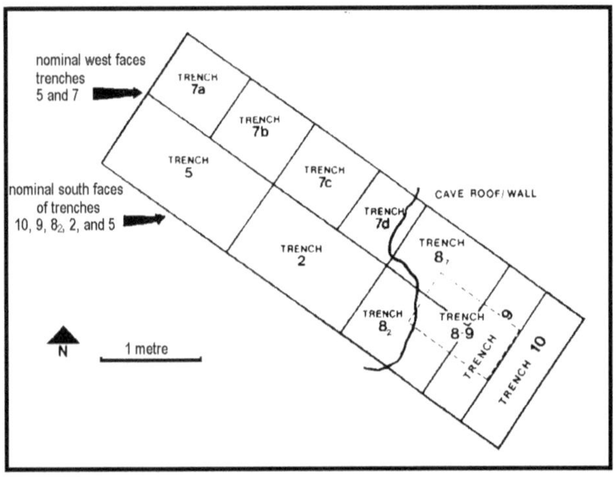

Figure 6.2 Trench lay-out in Devil's Lair main excavation, after C.E. Dortch (1979a)

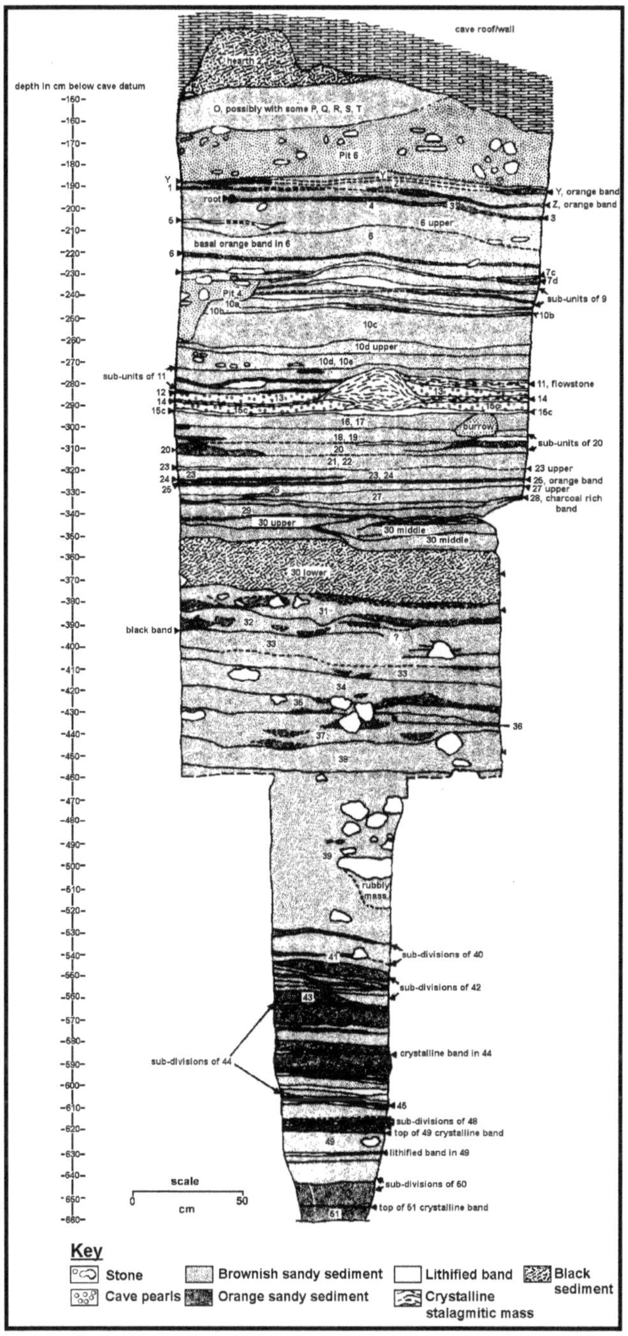

Figure 6.3 **Devil's Lair main excavation, nominal east section of Trench 9, after Balme *et al.* (1978)**

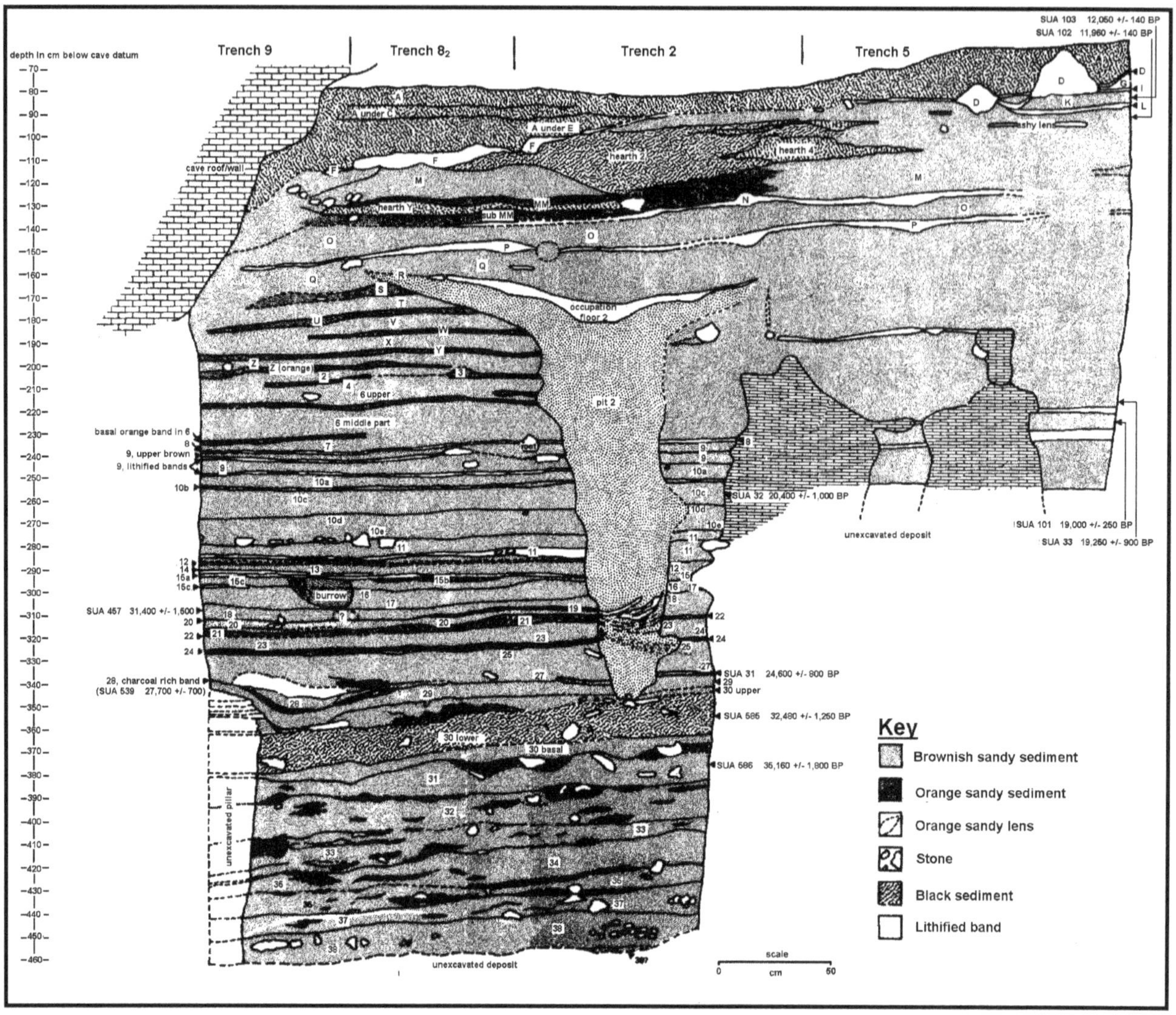

Figure 6.4 Devil's Lair main excavation, nominal south section, after Balme *et al.* (1978)

Interpretations of sedimentation patterns at the site justify the argument (Shackley 1978). Layers were probably deposited rapidly and intermittently rather than slowly and gradually, and artefacts could have been deposited likewise intermittently during the relatively long period of a layer's exposure at the surface, and then quickly buried by deposition of the next layer.

The evidence for rapid, intermittent deposition is as follows: sand grains show little alteration from transport or diagenesis in the cave, suggesting a short period for their movement from outside the cave to their present position (Shackley 1978). The overall rate of sedimentation, as suggested by the depth and age range of the entire deposit, is less than 1 mm per year, which is too slow to be explained by continuous gradual sedimentation. One slender stalagmite was apparently buried in bursts (Dortch and Merrilees 1973). At various intervals along its length, it has lateral extensions of flowstone which would have formed on a ground surface, suggesting that conditions in the cave alternated between long undisturbed periods suited for flowstone formation (flowstone formation is limited in an open cave), and bursts of sand deposition. Flowstone layers throughout the deposit can only have formed slowly, but cannot have formed during the periods of deposition of sand grains and other clastic material. Clastic deposition must have often ceased for long periods.

The uppermost layer A contains much humic material suggesting a rapid influx of sediment washed in from the present forest floor (Balme *et al.* 1978), and therefore suggests a potential means of deposition for some other layers of similar dark brown colour (notably layer 30-lower). Shackley (1978) also notes that the size of the sand grains and the probable shape of the former cave entrance eliminates wind as the transporting medium, hence the sand-grains were probably deposited by water flowing into the cave after heavy rainfalls.

Sand and humic material arrived in intermittent, discrete episodes throughout the period represented by the excavated Devil's Lair deposit. Distinctive orange sand layers suggest single episodes of sand deposition, brown sand layers may represent a summation of several sand and humic material depositions (Figures 6.3, 6.4). As flowstone formed during periods of no sedimentary activity, the cave entrance must have been periodically constricted. Shackley (1978) notes that flowstone formed around objects already in the cave. Artefacts and bones found within flowstone are therefore as old or older than the flowstone.

An important and continuing diagenetic process in the cave, relevant to interpreting the ages of occupational remains, is the gradual lithification of all sand layers by addition of calcitic cement precipitated from solutions entering the deposit. The cause of lithification, whether entry of solutions richer in calcium carbonate, or changes in temperature or water flow, is unclear, but several layers are advanced in this process, to the point that they are termed "lithified bands" (Balme *et al.* 1978).

At Devil's Lair it is unlikely that artefacts or faunal remains would have moved through lithified bands during or after their lithification. The few lithified bands that show signs of having been broken by disturbances show them very clearly (e.g., Pit 2 in trenches 2 and 8_2, burrow in layer 15, trench 9: Figures 6.3, 6.4). I therefore presume that most artefacts and faunal remains date to the stratigraphic interval between the lithified bands or flowstone layers above and below them. There may have been disturbance of layers in between lithified bands, or movement of objects across bands before they became fully lithified, but gradual lithification of the bands, along with less extreme cementation of sand layers, would have limited artefact movement overall.

Because of long exposure of successive surfaces, humans or animals could have excavated or trampled the same surface many times (Balme *et al.* 1978). Some sand layers show signs of mixing, making them unsuitable for this analysis (cf Balme *et al.* 1978). Layer A contains both modern and ancient artefacts and fauna, indicating modern disturbance, and is only included here to indicate its stratigraphic position. Water-channelling is evident in the lowermost layers and Balme *et al.* (1978) assign thick stratigraphic units here, to avoid potential problems from mixing.

In this analysis, I group layers according to their position relative to lithified bands and flowstone layers, which suggest major stratigraphic boundaries. Each group of layers is capped by one of these boundaries and is here termed a Period, indicated by Roman numerals I to XIV. In Table 6.1, I present all layers with the 40 or so mixed and other layers of Balme *et al.* (1978) combined into larger groups in a way consistent with their stratigraphic positions. I established these combinations from field notes, published section diagrams (Balme *et al.* 1978; C.E. Dortch 1979a, 1984; Dortch and Dortch 1996), and use of a Harris matrix (Appendix 3). Some of the excavators' combined units cross the Period boundaries because of trench section cleaning, or because the boundary (lithified band or flowstone unit) was not continuous across the excavated area (the boundary still indicates a stratigraphic division because a lithified band or flowstone unit indicates that sand was not being deposited at that time). Combined units that cross boundaries are excluded, and hence some information about artefacts and excavated volumes is discarded.

In Table 6.1 the flowstone layers and lithified bands that divide the deposit are indicated by medium and light grey shading, respectively. The number of artefacts is divided by the volume excavated as a way of comparing layers of different volumes. This calculation is sometimes referred to as a "density value" (Jerardino 1995).

Dark grey shaded bands in the table indicate two major stratigraphic divisions. These are layers 30-lower and A, thick bands of black or dark brown sediment. Layer A contains humic material, and both are thought by Balme *et al.* (1978) and Shackley (1978) to be the remains of a sudden influx of earth and humic material from the forest floor outside the cave. These influxes are probably relate to the sudden opening or widening of a cave entrance located just below ground level outside the cave. For layer A, the entrance in question is the present entrance located at the south-west end of the chamber. The slope orientation of layer 30 lower and that of inferior and superior layers indicate that deposition then was through an entrance located at the north-east end of the cave, that was closed by rubble fall and stalagmite growth.

Table 6.1 shows that layers below layer 30-lower, that is, half of the excavated deposit, contain very few occupational remains. In contrast, there are few artefact-free layers above layer 30-lower. Volume estimates suggest that the layers above 30-lower contain broadly similar concentrations of artefacts. Arguably, since the time of deposition of layer 30-lower, and before the formation of flowstone D, the only periods that people were absent from Devil's Lair were during the periods of deposition of flowstone layers and lithified bands (but not all of the latter), and sand layers 23, 19, 17-13 (the artefacts in this interval could derive from artefact-rich layer 18), parts of 9, H.P., and H.

While the upper part of the Devil's Lair deposit represents a long record of intermittent human visits to the site, the lower part seems only to indicate long periods of water-borne sand deposition. This contrast suggests no people visited the site locality before the time of layer 37, and no people entered the site until the time of layer 30-lower (Turney *et al.* in press). In the passage quoted below, Balme *et al.* (1978: 58-59) show that the depositional sequence is key to this interpretation. The age estimates derive from radiocarbon dates presented in Balme *et al.* (1978; see also Table 6.2).

Table 6.1 Distribution of Devil's Lair archaeological remains (continued next page)

Material from Trenches 2 and 5 is excluded because the upper parts of these trenches lack detailed stratigraphic information. Volume estimates are from Balme *et al.* (1978). Many layers and layer groupings used here lack published estimates of excavated volume (litres or *l*) and the column entry is given as "na" (not applicable). The number of artefacts is the total number of flaked stone, exotic stone, and bone points. Hearths (identified as definite or probable) and emu eggshell weights are not included in this total as they are not comparable to artefact numbers. The figure at the base of the right hand column is the number of all artefacts divided by the total of known volume estimates.

DEVIL'S LAIR TRENCHES 7a, 7b, 7c, 7d, 8_2, 8_7, 8-9, 9, 10

Period	Layer	Hearth present	Flaked stone	Ochre, granite, other exotic stone	Bone points	Emu eggshell (weight in g)	Volume excav. (*l*)	No. artefacts//
XIV	A and units below A		6	1		0.63	290	0.02
XIII	D						90	0.04
	mixed units below D		3		1		na	
XII	F						50	
	G & mixtures with G		1			1.7	na	
	H						100	
XI	I						na	
	Hearth x	probable	3				na	
	K (including occupation floor)		23	1			60	0.40
X	L and brown-grey layer atop L		3		1		40	0.10
	Mixed Hearth 2 & up		1			0.1	na	
	Hearth 2	definite	16	2	3		60	0.35
	M (inc.orange & brown mottling)		31				450	0.07
	H.P.						10	
	Small lithified mass		2			0.35	na	
	Hearth z	probable	1			12	10	0.10
	Below H.P.						10	
	MM		5				120	0.04
	Hearth y & sub MM	probable	4			0.1	70	0.06
IX	N						50	
	O		33	1			340	0.10
VIII	P		6		1	0.02	40	0.18
	Q		54		3		440	0.13
VII	Occupation Floor 2		4			0.1	na	
	R		2			0.45	na	
	Mixtures of R, S, T, Pit 2		10			0.44	na	
	S mixed		22		1	0.15	na	
	T		11				20	0.55
	Mixtures of U, V, Pit 1		35	1	1	1.36	na	
	Mixtures of Pit 6 & Y to 6		8			0.5	na	
	Pit 6		15				na	
	Mixtures of U, V, W, X		23			0.15	90	0.26
	U		1				10	0.10
	V		24	2			120	0.22
	W		1			0.5	10	0.10
	Mixtures of W, X, Y		14			5.73	30	0.47
	Pit 1 fill		2				na	
	X		47	1			100	0.48
	Mixtures of X with hearth & Pit 2		6			0.8	na	
	Y		11				70	0.16
	Y and Z mixed		3			1.64	na	
	Z		22				200	0.11
	Mixtures of Z, 1, 2		2			0.69	10	0.20
	1		2				40	0.05
	2		12				150	0.08
	2 & 3 mixed		5				30	0.17
	3		8	1			60	0.15
	3 & 4 mixed		17				70	0.24
	4		64	2			na	
VI	5		12				40	0.30
	5 & 6 mixed		1				na	
	6		174	5	3		690	0.26
	7		116	6	2	0.63	280	0.44
V	8		2			0.05	50	0.04
	8 & 9 mixed		1				10	0.10
	9, upper brown part		1				120	0.01
	9, lower		12			0.45	160	0.08
IV	9, lithified bands						na	
	10 & sub-units		27	2		7.91	730	0.04

Table 6.1 **Distribution of Devil's Lair archaeological remains (continued from previous page).**

DEVIL'S LAIR TRENCHES 7a, 7b, 7c, 7d, 82, 87, 8-9, 9, 10								
Period	**Layer**	**Hearth identification**	**Flaked stone**	**Ochre, granite, other exotic stone**	**Bone points**	**Emu eggshell (weight in g)**	**Volume excav. (*l*)**	**No. artefacts/*l***
III	**11**		2				260	0.01
	Mixtures of 11, 12, 13		2			0.25	60	0.03
	13					0.15	90	
	14						50	
II	**15**						100	
	Mixtures of 15, 16, 17, 18, 19		3		1	0.01	na	
	16						70	
	17						40	
	18 (including hearth)	definite	24				110	0.22
	19						220	
	20		12				180	0.07
	21		3				80	0.04
	22		1				40	0.03
	23						80	
	24		7				80	0.09
	25		2		1		100	0.03
	26		13				110	0.12
	27 (including hearth)	probable	8				30	0.27
	Mixtures of 26, 27, 28, 29		5				na	
	28 (including hearth)	definite	33	1			70	0.49
	29		10		1		150	0.07
	Porter's Hearth	probable					100	
	Mixtures of 29,30-upper, 30-lower						150	0.05
	30-upper		8				na	
I	**30-lower**				1		510	0.002
	31						230	
	31 & 32 mixed						10	
	32						530	
	32 & 33 mixed						20	
	33						240	
	33 & 34 mixed			2			50	0.04
	34		2				510	0.004
	35		2				na	
	35 & 36 mixed						30	
	36						na	
	37		1				250	0.004
	38 through to 51						na	
	TOTAL	3 (definite)	1047	28	20	36.86	10,310	**0.11**

"In the light of evidence presented or cited above, we conceive the following sequence of events, beginning 35 000 years ago or more in a cave very much longer and deeper than the present Devil's Lair. This cave had an opening to the surface (which may have been substantially higher than it is now) about 16 m north of the site of our excavations [main excavation].

The entrance was probably shaft-like, not easily negotiable by human beings, but wide enough for owls to fly through. It must have opened from a depression in the surface, and have been capable of receiving an in-wash of sand, perhaps only occasionally during exceptionally heavy rain. At some still-earlier time, it may have acted as a pit trap, sampling the contemporary fauna through individuals falling into it and being unable to get out. Alternatively, it may have functioned as a rubbish disposal chute for bands of human beings camped near it. For these or other reasons, a deposit accumulated near this old entrance which contained remains of Sthenurus brownei, S. occidentalis and other mammal species which were also represented in the Mammoth Cave deposit.

We visualize owls bringing predominantly small mammal prey into this cave, occasionally with juveniles of larger mammal species, and regurgitating pellets containing predominantly unbroken bones of these small or juvenile mammals. From time to time, relatively large quantities of sand were washed into the cave, enveloping and redistributing the owl-pellets, and building up a talus cone extending radially from the entrance, into the deeper recesses of the cave. The successive surfaces of this cone were extensively channelled and filled with sand of a brighter orange colour than the main mass. The water movement responsible for the channel cutting was sufficiently energetic to move material from the old Sthenurus deposit down the talus cone so that such material came to lie side by side with younger bone from the owl pellets. Artifacts derived either from the Sthenurus deposit or from the surface outside were also washed into juxtaposition with this mixture of older and younger bone from time to time. In this way we visualize the accumulation about 33 000 years ago of what we have called layers 39-31.

Layer 30-lower is lithologically very distinctive and presumably marks some unusual event. By analogy with layer A, we take this to be associated with some major change in the nature of the entrance, allowing the sudden ingress of large quantities of unmodified dark humic soil from the surface. Meantime, the owl pellets had continued to accumulate inside the cave, and the Sthenurus deposit to contribute small quantities of the coated bone fragments we have described as "possibly reworked".

This "layer 30-lower event" of about 32 000 years ago, possibly related to the end of channelling, was followed by a resumption of sedimentation producing layer 30-upper, with little contribution from the Sthenurus deposit. However, the Sthenurus deposit was not yet buried and continued to supply very small quantities of old bone to the developing talus cone over the next few thousand years.

A new phase in the history of the cave begins with layer 29. This may relate to the alteration in the character of the entrance denoted by layer 30-lower. It may have become possible or easy for human beings to enter the cave, and they appear to have done so from time to time, beginning some 30 000 years ago and continuing for the next 22 000 years or so. It is possible that small bands or family groups lived in the cave for a few days or a few weeks at a time at more or less regular intervals, or only sporadically, or that larger groups used the cave very sporadically. The bones of their prey, sometimes charred, and sometimes fashioned roughly or more carefully into implements, together with stone implements, fine chips resulting from the fashioning of stone implements, and occasional items of adornment, were left lying on successive surfaces of the talus cone. Deep and shallow pits were dug and fires were lit.

In the intervals between human occupancy of the cave, or possibly more or less continuously, owls and devils and perhaps other small carnivores visited or lived in the cave, contributing broken or unbroken bones of their prey species to the scatter left by the human occupants. It is possible that devils were attracted by only partly eaten carcases left by human visitors, and converted initially largely unbroken bones into small fragments. It is equally possible that the human visitors broke up the bones of their prey species, including devils, as part of the cooking or eating process.

In any case, we visualize sandy surfaces in the cave littered with bone fragments, and from time to time receiving accessions of fresh sand or of carbonate deposited in irregular masses from solutions dripping from the roof and walls, or in more regular sheets from surface films of solution or in the interstices of the sand from percolating solutions. Calcareous deposition may have been continuous, if seasonal, but deposition of sand seems to have been sporadic. Thus the bone litter on a given surface may have received contributions from many episodes of human occupation before becoming buried by sand. Some of our narrower layers may represent single episodes of sand deposition but multiple episodes of bone deposition.

From about layer L up to and including layer D, deposition of carbonate matched or exceeded deposition of sand, and the thick flowstone D, formed later than about 7 000 years ago, probably represents a more or less complete sealing of the entrance to the cave as well as of the deposit in it. The cave may have remained unoccupied, through lack of access, for several thousand years.

Then, only a few hundred years ago, a, large segment of the roof and central west wall of the cave collapsed, forming the existing doline and the present entrances to Devil's Lair and Nannup Cave. What we now call Devil's Lair was left by this collapse as a small remnant of what was once a much more extensive cave. This event is marked in the depositional succession by layer A. This seems to be largely unmodified, black humic soil from the forest floor which suddenly fell into the cave following the opening of the new entrance. Human and animal use of the cave remnant would again have been possible, but seems to have been minimal."

(Balme *et al.* 1978: 58-59)

Chronology of human occupation

The following analysis identifies the radiocarbon estimates that are inconsistent in the chronology of the Devil's Lair deposit and determines the duration of Periods containing evidence for human occupation. Luminescence-based and electron-spin resonance estimates for the age of Devil's Lair sediments and marsupial teeth have been obtained, but although they agree with the radiocarbon estimates (Turney *et al.* 2001), they are unsuitable for the statistical analyses employed here, which require age probability estimates to be normally distributed.

Devil's Lair radiocarbon estimates have been obtained from 1955 to the present day (Table 6.2), resulting in considerable variation in sample types, dating methods, precision, and hence reliability. The age estimates are more or less in correct stratigraphic order, but they do not give a precise duration and age for each Period and several Periods contain inverted series of estimates. Therefore, in each Period, I identify the estimates that are most inconsistent with adjacent estimates.

In Table 6.2, all the Devil's Lair radiocarbon estimates are shown in stratigraphic order. Table 6.2a shows estimates from the main excavation. Tables 6.2b and 6.2c show estimates from trenches 1 and 6, which have not been stratigraphically related to the main excavation. By virtue of their depths below the uppermost flowstone ('D' in main excavation), the estimates broadly agree with those from the main excavation, but they are not considered further here.

Table 6.2 Devil's Lair radiocarbon determinations (continued next page)

a) Main excavation. See next page for explanation of codes and abbreviations.

Period	Sample number	Ref.	Depth (cm) below cave datum	Trench	Layer	Method	Pre-treatment	Sample type	Date	1 σ (±)
XIV	SUA 342	1	70	7b	A	LSC	Alkali sol.	Charcoal	320	85
XII	SUA 364	1	86	7c	G	LSC	Alkali sol.	Charcoal	6,490	145
	OZD 320	9	86	7b	G	AMS	ABA-BC	Charcoal	12,500	100
	AA 19695	8, 9	82	7a	G	AMS	HCl sol.	Eggshell	13,580	110
XI	OZD 321	9	89	7b	I	AMS	ABA-BC	Charcoal	13,050	80
	SUA 103	6	80	5	K (incl. occ. floor)	LSC	None	Charcoal	11,870	150
	SUA 103/2	4	80	5	K (incl. occ. floor)	LSC	None	Charcoal	12,050	140
X	OZD 323	9	96	na	Hearth 2	AMS	ABA-BC	Charcoal	12,950	110
	OZD 322	9	102	na	Hearth 2	AMS	ABA-BC	Charcoal	13,300	120
	SUA 102	6	78-95	5	M	LSC	None	Charcoal	11,960	140
	SUA 102/2	4	78-95	5	M	LSC	None	Charcoal	12,000	180
IX	GX 7249	4	121-137	7d	O	LSC	Alkali sol.	Charcoal	13,975	450
VIII	SUA 975	4	148-151	9	Q	LSC	Alkali sol.	Charcoal	17,560	460
	SUA 1315	4	154	7d	Q2	LSC	None	Charcoal	16,970	620
VII	SUA 976	4	167	8_2	V	LSC	Alkali sol.	Charcoal	19,160	380
	SUA 1248	4	171-172	9	V	LSC	Alkali sol.	Charcoal	17,370	290
	AA 19691	8, 9	184	9	Y	AMS	HCl sol.	Eggshell	19,835	75
	GX 7252	4	189-192	8_2	Z	LSC	Alkali sol.	Charcoal	21,270	620
	SUA 977	4	203-208	8_2	4	LSC	Alkali sol.	Charcoal	21,820	480
VI	GX 7253	4	196-206	7b, 7c	6b	LSC	Alkali sol.	Charcoal	17,100	810
	SUA 101	6	201-212	5	7 lower	LSC	None	Charcoal	19,000	250
V	SUA 33	6	222	5	9	LSC	None	Charcoal	19,250	900
	OZD 324	9	237	na	9	AMS	ABA-BC	Charcoal	23,050	250
	AA 19690	8, 9	240	7d, 8_7	9, upper brown pt	AMS	HCl sol.	Eggshell	24,930	335
	SUA 1316	4	237-238	8_2	9, lower	LSC	None	Charcoal	24,200	1400
IV	AA 19689	8, 9	247	8_2	10a	AMS	HCl sol.	Eggshell	25,500	275
	SUA 32	6	249-267	2	10c and 10d	LSC	None	Charcoal	20,400	1000
	OZD 325	9	266	na	10d	AMS	ABA-BC	Charcoal	21,850	210
III	ANUA 10002	9	293-302	na	11a	AMS	ABOX-SC	Charcoal	26,590	+370 -350
	OZD 326	9	276	na	11b	AMS	ABA-BC	Charcoal	25,900	300
II	SUA 457	1	293-302	8_7, 9	18 (incl. hearth)	LSC	Alkali sol.	Charcoal	31,400	1500
	SUA 31	6	340	2	28	LSC	None	Charcoal	24,600	800
	SUA 539	1	340	8_2, 8_7	28	LSC	None	Charcoal	27,700	700
	OZD 330	9	335	na	28	AMS	ABA-BC	Charcoal	40,500	1750
	OZD 328	9	345	na	28	AMS	ABA-BC	Charcoal	NDFB	
	OZD 329	9	345	na	28	AMS	ABA-BC	Charcoal	41,500	2000
	ANUA 10003	9	345	na	28	AMS	ABOX-SC	Charcoal	41,460	+1400 -1190
	GX 7255	4	353	9	Porter's Hearth	LSC	Alkali sol.	Charcoal	30,590	+2220 -1420
I	SUA 585	1	350	2, 8_2, 9	30-lower	LSC	None	Charcoal	32,480	1250
	ANUA 11502	9	365	8-9	30-lower	AMS	ABOX-SC	Charcoal	45,470	+1420 -1210
	SUA 586	1	375	2, 7d, 8_2, 9	31	LSC	None	Charcoal	35,160	1800
	ANUA 11512	9	397-405	8-9	33	AMS	ABOX-SC	Charcoal	46,730	+2190 -1720
	SUA 546	1	438-460	8_2, 8_7, 9	38	LSC	None	Charcoal	31,800	1400
	GX 7251	4	436-453	2, 7d	38	LSC	Alkali sol.	Charcoal	24,685	1150
	ANUA 11511	9	465	8-9	39	AMS	ABOX-SC	Charcoal	48,130	+2590 -1960
	OZD 331	9	468-470	8-9	39	AMS	ABA-BC	Charcoal	40,400	1900
	OZD 332	9	480-488	8-9	39	AMS	ABA-BC	Charcoal	NDFB	
	SUA 698	3	483-496	8_2, 8_7, 9	39	LSC	None	Charcoal	37,750	2500
	OZD 333	9	495-501	8-9	39	AMS	ABA-BC	Charcoal	NDFB	
	ANUA 11507	9	496	8-9	39	AMS	ABOX-SC	Charcoal	>52,000	
	OZD 334	9	546	8-9	42	AMS	ABA-BC	Charcoal	NDFB	
	ANUA 11510	9	562-505	8-9	44	AMS	ABOX-SC	Charcoal	>55,000	

Table 6.2 Devil's Lair radiocarbon determinations (continued from previous page)

b) Trench 6

Period	Sample number	Ref.	Depth (cm) below cave datum	Trench	Layer	Method	Pre-treatment	Sample type	Date	1 σ (±)
	Wk 5494	2	110-120	6	Hearth 1	LSC	Alkali sol.	Charcoal	12,660	240
	SUA 34	6	245-269	6	brownish earthy layer	LSC	not known	Charcoal	17,400	350

c) Trench 1

Period	Sample number	Ref.	Depth (cm) below cave datum	Trench	Layer	Method	Pre-treatment	Sample type	Date	1 σ (±)
	O 654	5, 7	c. 100	1	below flowstone (D?)	-------not known--------		Charcoal	8,500	160
	O 658	5, 7	224-233	1	rubbly layer	-------not known--------		Charcoal	12,175	275

NOTES:

===== Stratigraphic boundary - "lithified band", flowstone, or thick humic layer

Laboratory codes (sample number prefixes):

AA: National Science Foundation - Arizona AMS Laboratory
ANUA: Australian National University AMS facility
GX: Geochron Laboratory
O: Geochemical Laboratory, Humble Oil & Refining Company
OZD: Australian Nuclear Science and Technology Organisation AMS facility
SUA: Sydney University Radiocarbon Laboratory
Wk: Waikato Radiocarbon Laboratory

References (Ref):

1: Balme *et al.* (1978)
2: S. Burke, Centre for Archaeology, UWA, pers. comm.
3: C.E. Dortch (1979a)
4: C.E. Dortch (1984)
5: Dortch and Merrilees (1971)
6: Dortch and Merrilees (1973)
7: Lundelius (1960)
8: G. Miller, Institute for Arctic and Alpine Research, University of Colorado, pers. comm.
9: Turney *et al.* in press

Trench:

Follows system of Balme *et al.* (1978) except that OZD and ANUA samples were collected from trench walls, and not all their trench provenances were determined.

Method:

LSC: Liquid Scintillation Count
AMS: Accelerator Mass Spectrometry

Pre-treatment:

ABA-BC: acid-base-acid pretreatment with bulk combustion (conventional AMS pre-treatment)
ABOX-SC: acid-base-wet oxidation pretreatment with stepped combustion (Turney *et al.* 2001)
Alkali sol.: washed in alkali solution to remove humic acids
HCl sol.: for emu eggshell only: treated with HCl solution (cf. Miller *et al.* 1997)
None: sample too small or fragile for alkali solution pre-treatment

Sample type:

Charcoal: charcoal collected *in situ* or from sieve residues
Eggshell: Eggshell fragments from Emu (*Dromaius novaehollandiae*)

Date:

NDFB: "Not distinguishable from background" (radiation)

Table 6.3 Assessment of Devil's Lair radiocarbon estimates (see Appendices 4, 5)

Sample numbers of acceptable estimates used in further calculations (Appendix 4) are shown in bold.

						Status of Null Hypothesis - *p* (H0) - "dates are statistically identical"		Compare with other periods - accept or reject?	Age BP at 95% confidence	
Period	Sample number	Depth below datum	Layer	Date	1σ (±)	For largest apparent grouping	For alternative, smaller grouping (aiming for higher *p* [H0])		Min.	Max.
XIV	**SUA 342**	70	A	320	85	NA	none possible	**Accept**	150	490
XII	**SUA 364**	86	G	6,490	145	No chance	none possible	Accept	6,200	6,800
	OZD 320	86	G	12,500	100	Reject - *p* (H0) = 0.4×10^{-13}	none possible	**Accept**	12,300	13,500
	AA 19695	82	G	13,580	110			**Accept**		
XI	**OZD 321**	89	I	13,050	90	Reject - *p* (H0) = 0	none possible	**Accept**	12,900	13,200
	SUA 103	80	K (incl. occ. floor)	11,870	150		Do not reject - *p* (H0) = 0.38	Reject		
	SUA 103/2	80	K (incl. occ. floor)	12,050	140			Reject		
X	SUA 102	78-95	M (incl. mottling)	11,960	140	Do not reject - *p* (H0) = 0.86	none possible	Reject	12,600	13,600
	SUA 102/2	78-95	M (incl. mottling)	12,000	180			Reject		
	OZD 323	96	Hearth 2	12,950	110	Reject - *p* (H0) = 0.031	none possible	**Accept**		
	OZD 322	102	Hearth 2	13,300	120			**Accept**		
IX	**GX 7249**	121-137	O	13,975	450	NA	none possible	**Accept**	13,100	14,900
VIII	**SUA 975**	148-151	Q	17,560	460	Do not reject - *p* (H0) = 0.44	none possible	**Accept**	15,800	18,400
	SUA 1315	154	Q2	16,970	620			**Accept**		
VII	**SUA 976**	167	V	19,160	380	Reject - *p* (H0) = 7.97×10^{-16}	Do not reject for estimates AA 19691 and SUA 976 - *p* (H0) = 0.08	**Accept**	18,400	20,200
	SUA 1248	171-172	V	17,370	290			Reject		
	AA 19691	184	Pit 6	19,835	75			**Accept**		
	GX 7252	189-192	Z	21,270	620	Do not reject - *p* (H0) = 0.48	none possible	**Accept**	20,200	22,800
	SUA 977	203-208	4	21,820	480			**Accept**		
VI	GX 7253	196-206	6b	17,100	810	Reject - *p* (H0) = 0.03	none possible	Reject	dated by Period VII & V	
	SUA 101	201-212	7 lower	19,000	250			Reject		
V	SUA 33	222	9	19,250	900	No chance	none possible	Reject	22,200	26,000
	OZD 324	240	9, upper brown	23,050	250	Reject - *p* (H0) = 3.8×10^{-5}	Do not reject for SUA 1316 and OZD 324 - *p* (H0) = 0.42, nor for AA19690 and SUA 1316 - *p* (H0 = 0.61)	**Accept**		
	AA 19690	237	9	24,930	335			**Accept**		
	SUA 1316	237-238	9, lower	24,200	1400			**Accept**		
IV	**AA 19689**	247	10a	25,500	275	Reject - *p* (H0) = 1.5×10^{-24}	none possible	**Accept**	24,950	26,050
	SUA 32	249-267	10c and 10d	20,400	1000		Do not reject - *p* (H0) = 0.16	Reject		
	OZD 325	266	10d	21,850	210			Reject		
III	**ANUA 10002**	276	11a	26,590	360	Do not reject - *p* (H0) = 0.14	none possible	**Accept**	25,200	27,200
	OZD 326	276	11b	25,900	300		none possible	**Accept**		
II	**SUA 457**	293-302	18 (incl. hearth)	31,400	1500	Reject - *p* (H0) = 3.5×10^{-41}	See GX 7255, SUA 585	**Accept**	28,400	34,400
	SUA 31	340	28	24,600	800		Reject - *p* (H0) = 0.004	Reject	37,400	44,800
	SUA 539	340	28	27,700	700			Reject		
	OZD 330	335	28	40,500	1750		Do not reject - *p* (H0) = 0.89	**Accept**		
	OZD 329	345	28	41,500	2000			**Accept**		
	ANUA 10003	345	28	41,460	1300			**Accept**		
	GX 7255	353	Porter's Hearth	30,590	1810		Do not reject for SUA 457, GX7255,SUA 585- *p*(H0)=0.67	Reject		
	SUA 585	350	30, lower	32,480	1250			Reject		
I	**ANUA 11502**	365	30	45,470	1315	Do not reject for ANUA 11502, ANUA11511, ANUA 11512 - *p* (H0) = 0.58		**Accept**	43,000	51,600
	SUA 586	375	31	35,160	1800	No chance	none possible	Reject		
	ANUA 11512	397-405	33	46,730	1960	See ANUA 11502		**Accept**		
	SUA 546	438-460	38	31,800	1400	No chance	none possible	Reject		
	GX 7251	436-453	38	24,685	1150	No chance	none possible	Reject		
	OZD 331	468-470	39	40,400	1900	Do not reject - *p* (H0) = 0.40	none possible	Reject		
	SUA 698	483-496	39	37,750	2500			Reject		
	ANUA 11511	465	39	48,130	2280	See ANUA 11502		**Accept**		

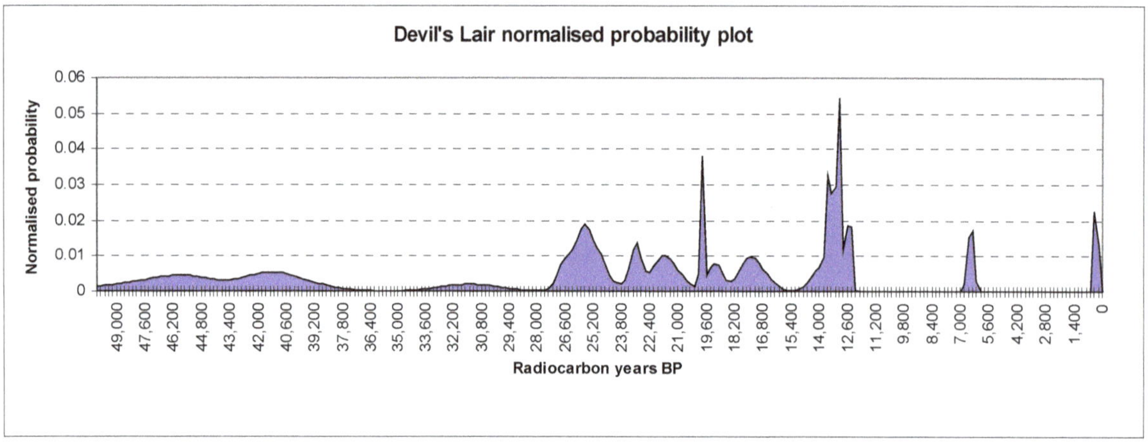

The peaks in this graph indicate the mean age for each set of similar age estimates. The central 95% of each peak's area gives the 95% confidence interval. For accurate results the sets of similar age estimates are graphed separately in Appendix 4 (note that the vertical scale in this figure indicates an probability for all the estimates considered together, not the probability of each group of estimates).

Figure 6.5 Normalised probability curve for acceptable Devil's Lair radiocarbon estimates.

Statistical comparisons of estimates from within Periods show which groups or pairs of estimates are most similar, as an indication of their reliability. Single estimates that are inconsistent with multiple estimates from adjacent Periods are assumed to be incorrect due to laboratory error or contamination, and rejected. In a few cases the only reason to suspect laboratory error is that the estimate was made decades ago, and the original analysis did not eliminate all modern contaminants. Modern contamination of some samples processed in the 1970s is likely because only two of the more recently obtained estimates are anomalous (OZD 325, OZD 331), and the 1970s estimates were all obtained by liquid scintillation counts (LSC), which require relatively large samples, and therefore prohibit the vigorous de-contamination possible with accelerator mass spectrometry (AMS; cf. Turney *et al.* 2001). Modern contamination possibly affected, in stratigraphic order, anomalous estimates SUA 103, SUA 103/2, SUA 102, SUA 102/3, SUA 1248, SUA 33, and SUA 32. All of these estimates are younger than, but broadly agree with, adjacent estimates above them. Modern contamination was probably a serious problem for samples older than *c.*30,000 BP, in which the quantity of radioactive carbon is very small and biased by the slightest modern contamination (as suggested by the result of OZD 331). Seriously affected estimates probably also include SUA 31, SUA 539, GX 7255, SUA 585, SUA 586, SUA 546, GX 7251, and SUA 698. No explanation can be offered for the very young age estimate for OZD 325; it is younger than 30,000 BP and there is no reason to suspect mixing of the deposit around the sample.

In the other cases of discrepant estimates, one can suspect mixing of the deposit. C.E. Dortch (1984) shows that the prehistoric excavation of Pit 2 from layer Q would have introduced material up to 10,000 years younger into layer 28 (Figure 6.4), and possibly affected samples SUA 31 and SUA 539. Dortch and Dortch (1996) affirm this rejection, noting that these dates are "anomalously different from one another and from the statistically identical dates, SUA 457 and GX 7255, that bracket layer 28". Recently obtained samples from layer 28 were collected from or near a hearth in layer 28, which is not penetrated by Pit 2. Another source of mixing is that limited amounts of material below layer 30-lower may have rolled down a talus slope from the ancient "*Sthenurus* deposit" posited by Balme *et al.* (1978). Mixing is also suspected because of evidence for water channelling and down-cutting in layers 31-38, all within Period I (Balme *et al.* 1978, C.E. Dortch 1979a). The possibility of contamination by older material applies to estimates from layers 31 to 39. Samples from above layer 30-lower are probably unaffected by such mixing, especially where obtained from hearths (SUA 457 from hearth in layer 18; OZD 327, 329, 330, 331, and ANUA 10003 from layer 28 and the hearth in 28).

Other sources of contamination are probably minor. Some small, fragile samples were not pre-treated by washing in an alkali solution to remove humic acids (C.E. Dortch 1984: Table 2). However, none of these samples are greatly discrepant from others, except where stratigraphic mixing is already suspected. Another error derives from a Sydney University laboratory miscalculation in processing SUA samples with numbers less than 975, but the error is no more than a few centuries (Temple and Barbetti 1981), and insignificant, given the age of the deposit.

In Table 6.3, I compare the estimates to determine groups of statistically similar estimates, using the method of Ward and Wilson (1978; see Appendix 4). Having identified the groups of statistically similar estimates, I reject single, discrepant estimates from further analysis. Single estimates that are still feasible are retained, and in two cases suggest long periods of deposition within each Period. SUA 364, 6,000 years younger than two other estimates in layer G, is either contaminated or indicates long exposure of layer G, as perhaps confirmed by O 654, which also suggests the overlying flowstone D formed after 8,500 BP and that very little sediment accumulated in this early Holocene time. SUA 457 from layer 18 suggests c.10,000 years elapsed during Period II. Since no other age estimates bracket either of these two estimates, one cannot show that the estimates are discrepant, and possibly, their respective Periods are very long.

The age ranges of grouped dates and the few isolated feasible dates indicate the age of the Periods. For grouped dates, the probability of the Null Hypothesis (H0) that dates in the group are identical, is generally less than 0.95 and more than 0.05. In these cases, H0 can neither be accepted nor rejected and it would be incorrect to follow Ward and Wilson (1978) further and derive mean pooled ages from these groups of similar dates (as there is a greater than 5% chance that dates have an age outside the pooled age range). Instead, I infer minimum and maximum ages for every Period from the central 95% of area under a normalised probability curve representing all the age estimates in the group (Appendix 4), using the method of Holdaway and Porch (1996). The results for each group of estimates are graphed in Appendix 4. The resulting estimate of minimum and maximum ages for each Period is given in Table 6.3 (right-hand column). Figure 6.5 gives the normalised probability curve for all the acceptable estimates from the main excavation.

Summary

At Devil's Lair, almost every layer above 30-lower contains archaeological remains, so human visits to the cave occurred at some time in every radiocarbon-dated Period in the long depositional sequence between layer 30-lower and flowstone D. These Periods represent the time elapsed between the sudden widening of the cave's former entrance *c.*45,000 BP, and the closure of the same entrance by *c.*6,500 BP. Human occupation began in the cave 41,000 ± 3,700 BP; it probably occurred near the cave even earlier, as indicated by in-washed artefacts in layers capped by the thick humic layer 30-lower, which are estimated at 47,000 ± 4,300 BP (Turney *et al.* in press). Human occupation may have ceased as early as 12,000 BP, however, as few artefacts are found in the thin layers above layer G. Because of this lack of detailed evidence the final date of human occupation at Devil's Lair cannot be resolved.

For the bulk of the archaeological deposit, Table 6.4 and Figure 6.5 show that the Devil's Lair radiocarbon dates as interpreted here provide a continuous sequence of overlapping or adjacent error margins. Most parts of the Devil's Lair deposit, and all human occupations, fall within broad age estimates that have a 95% chance of being correct. Of the layers that accumulated after first human occupation and before closure of the cave entrance, the only ones undated are layer P (top of Period VIII), layers 12-17 (Period III and top of Period II), and layers 19-27 (middle of Period II). These undated layers, which all contain artefacts, represent gaps of, respectively, 900 years, 1,200 years, and 3,000 years. These gaps are of relatively short duration given the age of the deposits.

The analysis of radiocarbon age estimates does not permit any identification of discrete episodes of human occupation or absence at Devil's Lair in the time between 47,300 ± 4,300 BP and 6,490 ± 145 BP. Nevertheless, human visits must have been highly intermittent, if only because the cave would have been uncomfortably wet during the several periods of flowstone formation (Balme *et al.* 1978: 58, C.E. Dortch 1979a: 275). These periods were probably significant, almost certainly exceeding hunter-gatherers' normal absences of months or years, but radiocarbon dating is insufficiently precise to determine their duration. Because sand layers containing artefacts probably also formed over long periods, this site can provide little information as to whether layers containing very few artefacts indicate a phase of occupation or of abandonment. The sequence of rapidly accumulated hearths at Tunnel Cave gives more scope to answer this question.

Tunnel Cave

Tunnel Cave has a large, dry entrance area containing a 3.5 m deep deposit with evidence for intensive prehistoric hearth-building (Figures 6.6-6.9). Fifteen hearths, identified in a 6 m^3 test-excavation, form the basic record of human occupations at this site (J. Dortch 1994, 1996). The test-excavation is on a gently sloping part of the sandy deposit in the entrance area (Figure 6.6.).

Stratigraphic context of hearths and artefacts

The Tunnel Cave deposit, which is divided into hearths, hearth layers, and non-hearth layers, comprises quartz sand, fine carbonate sediment, fragments of bone, charcoal, emu eggshell, and mollusc shell, and artefacts. Field observations of water flow and surface sediments suggest that thin sheets of sand are periodically washed into the site during winter. This effect was obvious during the 1993 field season as tourist traffic along a footpath had de-vegetated parts of the forest floor above Tunnel Cave, increasing the volume of sediment deposited by surface run-off (Appendix 5, Photographs 8, 9).

Other observations support deposition by surface run-off, and also wind. Throughout much of the Tunnel Cave deposit, sand grains are sub-angular to rounded (Appendix 5: Photographs 10, 11). A small proportion of the sediment appears to comprise fine, angular grains, probably including carbonaceous material, which may

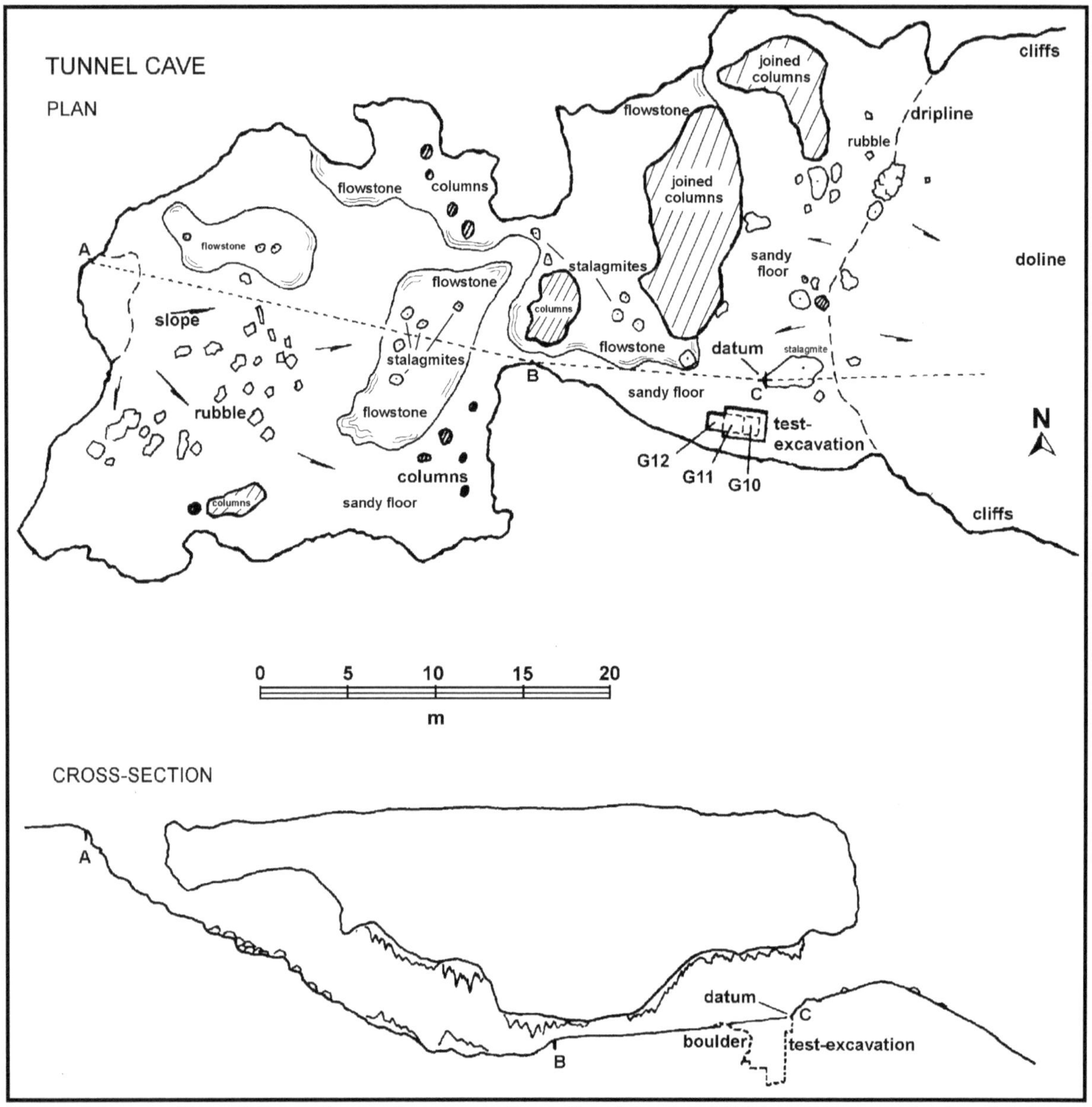

Figure 6.6 **Tunnel Cave plan and cross-section, after J. Dortch (1996) and B. Loveday (no date)**

have been deposited by wind or from the cave roof. In the trench sections, the layers are undulating but still well-defined, as if moved a short distance by water action (they are still largely intact and not incised, suggesting that the water flow is not turbulent or rapid).

Considering the large entrance and wide overhang at Tunnel Cave, sand may have been deposited more continuously and less sporadically than at Devil's Lair. Human occupation within the excavated area may have periodically declined or ceased, but natural sedimentation probably continued at all times.

Identification of hearths as evidence for occupation.

Hearths are composed largely of the sediments they are built on, so a hearth lens cannot have been re-worked or moved very far, and it is contemporaneous with the layer it is found in. Hearths suggest episodes of human occupation (from building a campfire to a prolonged residence in the cave of days or weeks) that are too short-lived to be dated by radiocarbon age estimates, whose error margins amount to centuries.

The age estimates bracketing entire layers that contain hearths also tend to fall within radiocarbon error margins. These "hearth layers" are richest in archaeological remains. Between these layers are large sections of deposit with few artefacts, which accumulated over millennia. Thus the Tunnel Cave hearths are the principle indicators of occupations, and estimating the age of every hearth is a means of inferring the minimum and maximum ages of occupations. Comparison of episodes of human occupation with environmental records is then relatively straightforward.

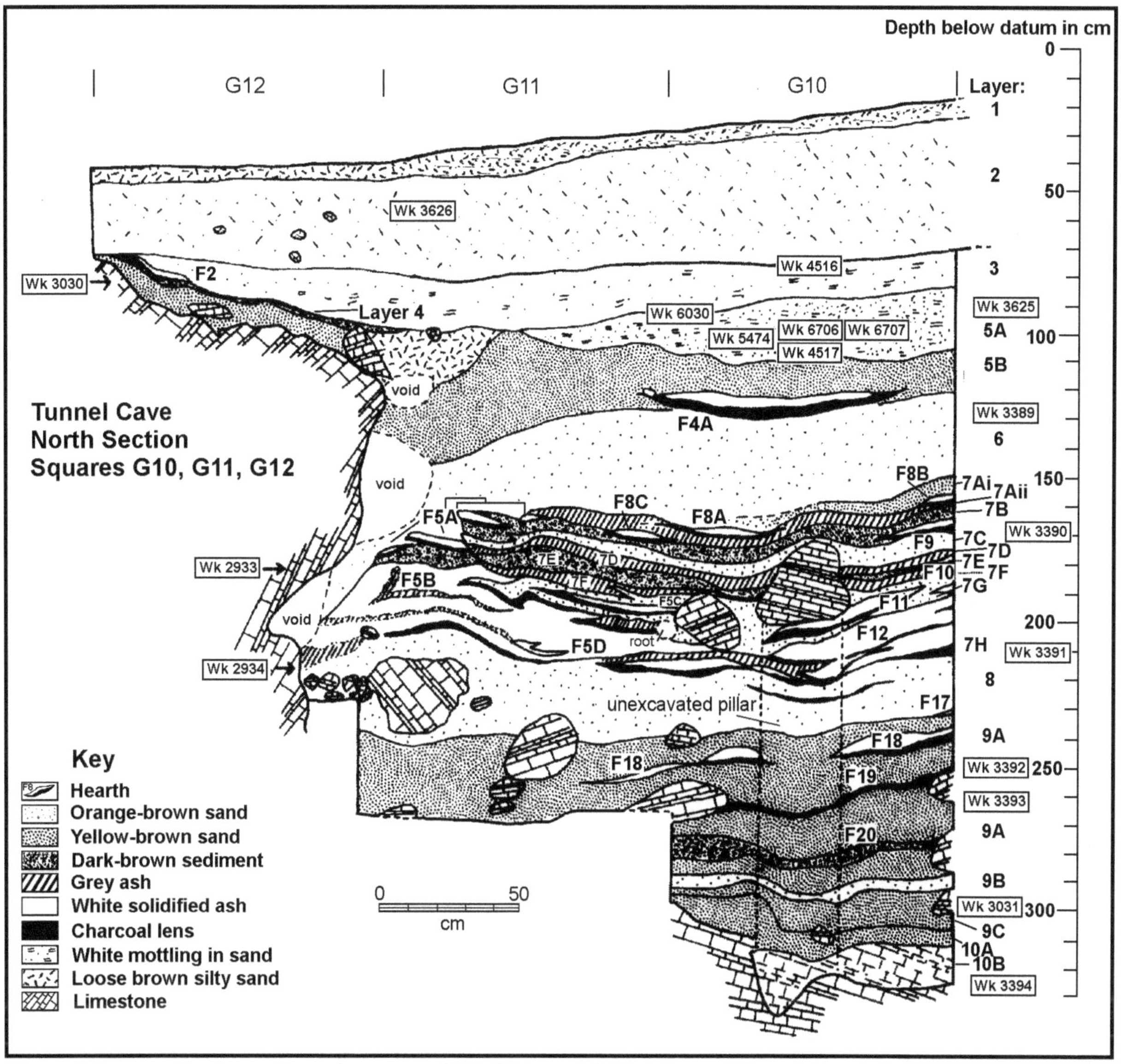

Figure 6.7 **North section of squares G10, G11, and G12 at Tunnel Cave (after J. Dortch 1996)**

Below, I identify Tunnel Cave hearths and assess the propositions that *(i)* hearth-containing layers contain the bulk of archaeological remains and *(ii)* they aggraded rapidly. Non-hearth layers aggraded slowly, or at least do not permit identification of episodes of rapid deposition. They contain little evidence of human occupation, and although they cover hearth layers, they can be distinguished from them, showing no sign of having been mixed into them. These propositions imply that all archaeological and non-archaeological remains are in reliable stratigraphic context, even though they may have been perturbed within the layer they were found in.

Many possible functions of hearths follow from human occupations. The association of archaeological material and hearths with remains of certain animal species suggests that these remains derive from people's meals (see Chapter 8), possibly cooked in the hearths. Other functions for hearths are plausible, *e.g.*, providing warmth, for several people, as suggested by large hearths (e.g. Feature 12 at Tunnel Cave), or a single, sleeping person (e.g., small hearths such as Features 8a-c; cf. Binford 1983, Gould 1977a, Nicholson and Cane 1991). In ethnographically recorded campsites, people built hearths for warmth, cooking, or tool manufacture, and in many social and utilitarian contexts: cooking and eating food; in small "dinner-time" camps or in long term base-camps (cf Binford 1980, Meehan 1988); for warmth at night (Binford 1983, Gould 1980); for making or repairing tools (Bindon 1986); and as a centre-point in ritual sites and campsites (Hammond 1933), the word "hearth" in some hunter-gatherer languages meaning also the family or the home (Hallam 1975, Walters 1988).

In the excavations, hearth lenses, along with various other sediment lenses and pits, were designated "Features" and numbered with the prefix "F". Table 6.5 shows where the Tunnel Cave features are located and which are hearths (shaded lines). Features and layers appear in stratigraphic diagrams (Figures 6.7-6.9).

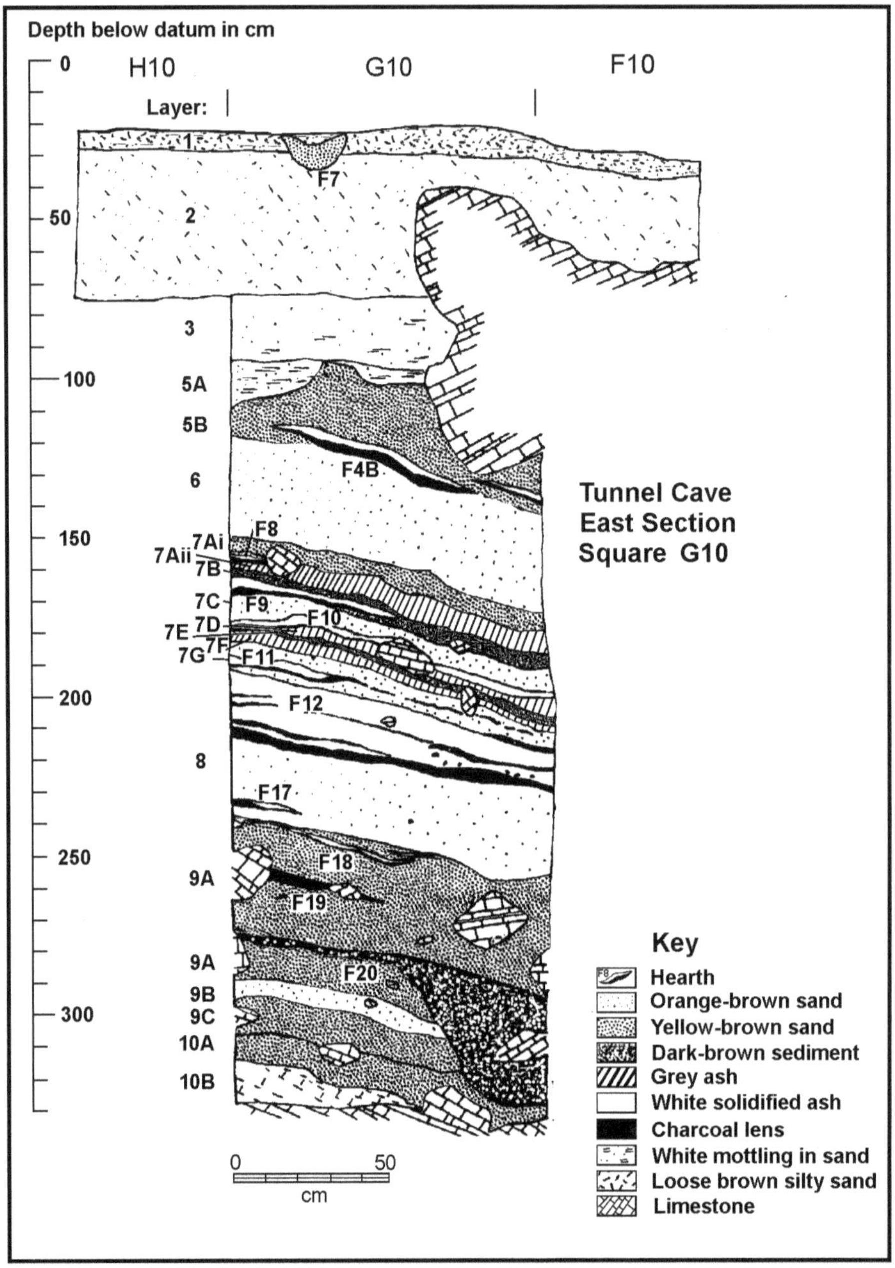

Figure 6.8 **East section of square G10 at Tunnel Cave, after J. Dortch (1994)**

Fifteen Tunnel Cave features are identified as hearths on the basis of the following observations:

1) They comprise white 'ash' lenses overlying black 'charcoal' lenses, in turn overlying scorched sand, exactly the same in appearance and texture to modern campfires in limestone rock shelters and in open-air campsites (Appendix 5: Photographs 6, 7). At one recent camp-fire that had burnt only six months before I sectioned it, the ash and charcoal lens showed a degree of compaction and organisation of ash and charcoal layers comparable to the Tunnel Cave hearths.

2) Ash and charcoal lenses as seen in the Tunnel Cave trench section are identified in photographs of trench sections of limestone cave floor deposits in Europe, the Middle East, and North America (Bar-Yosef *et al.* 1992: Figure 8, Bellomo 1993: Figure 10, Courty *et al.* 1989: Figure 10.3).

3) Other features of the Tunnel Cave hearths match descriptions of other hearths (Courty *et al.* 1989, Bellomo 1993, Nicholson 1993), as follows:

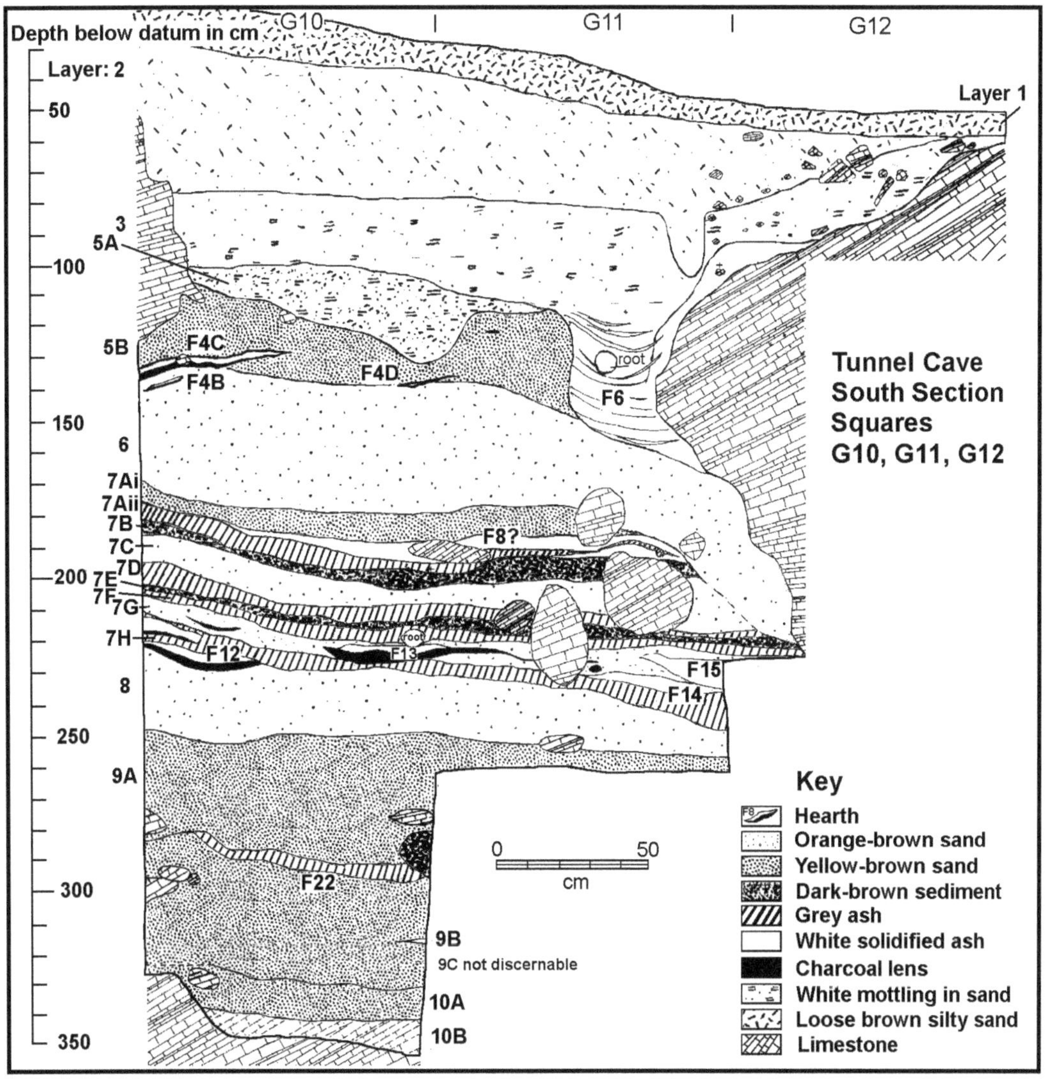

Figure 6.9 **South section of squares G10, G11, and G12 at Tunnel Cave (previously unpublished)**

a) Hearths, representing high temperature, localised fires, tend to show organised layers of ash, charcoal, and scorched earth. Convection draws up sediment from below the base of the fire and mixes it with ash (produced at high temperatures) above the base, making a dense ash layer (Bellomo 1993). The base of the fire is cooler so here wood burns to charcoal rather than ash. The charcoal layer has much more charcoal than surrounding sediments. The sediment below the charcoal layer may be scorched to a depth of two centimetres, and have a dark reddish brown colour indicating oxidisation. All three hearth sub-units - ash lens, charcoal lens, and scorched earth - are visible in most Tunnel Cave hearths.

b) In alkaline sediments (as at Tunnel Cave: pH in all layers ~ 9, 9.5), ash layers tend to become cemented (Courty *et al.* 1989: 110-111). Calcium carbonate in the ash layer is dissolved and then re-precipitated in the ash layer itself. Re-firing of hearths also leads to cementation of pre-existing ash layers, as re-firing melts calcite (calcium carbonate) crystals in the underlying ash. Fine particles, probably of calcium carbonate, are more common in hearth layers than in the predominantly sand non-occupation layers (Appendix 5, photographs 8, 9).

b) Bone fragments in the ash lens tend to be smaller, coated with white ash, or turned blue-white as a result of high-temperature burning (cf Nicholson 1993, Walters 1989). In accordance with the observations on temperature differences, bone fragments from the cooler-burning charcoal lens are blackened or coloured dark brown or dark grey, as if only scorched.

Other evidence of occupation

Hearth-containing layers contain large numbers of artefacts and exotic faunal remains. Only one hearth layer, Layer 1, has few archaeological remains. The sparsity of all classes of archaeological material in other layers suggests infrequent human use of the site at those times. Table 6.5 lists the total number of various archaeological items standardised against the weight of excavated deposit (hence a "density value"); Figure 6.10 graphs this data.

Table 6.4 Archaeological features (hearths, shaded lines) and natural features at Tunnel Cave

Feature numbers, assigned in the course of progressive excavation of squares, are not in stratigraphic order. Asterisks indicate features not preserved in section and therefore absent from section diagrams.

No.	Layer(s)	Identification
F1	**1**	**hearth**
F2	**4**	**hearth**
F3*	3	pits formed by soil slumping
F4	**5B**	**hearth, possibly dissected by water action, hence 4 sections (F4A, B, C, D)**
F5	**7B-H**	**large, deep hearth, disturbed by soil slumping, 4 sections (F5A, B, C, D)**
F6	3	pit formed by soil slumping
F7	1	small pit probably formed by stalactite drips
F8	**7A-B**	**series of small hearths or one dissected hearth (three small hearths or one broken-up hearth: sections F8A, 8B, 8C)**
F9	**7B**	**hearth**
F10	**7E**	**hearth, sheet of ash, no charcoal lens**
F11	**7G**	**hearth**
F12	**7H**	**large, deep, highly compacted hearth**
F13	**7G**	**hearth**
F14	**7H**	**hearth**
F15	**7G**	**hearth, very loose ashy sediment, possibly disturbed by soil slumping**
F16*	8	concentration of tiny bones in small pit
F17	**9A**	**hearth, very thin, dispersed**
F18	**9A**	**hearth**
F19	**9A**	**hearth, charcoal lens only**
F20	9A	lens of dark brown sand with charcoal fragments, possibly an eroded hearth
F21*	9A	poorly defined shallow pit filled with grey-brown sediment, possibly a hearth
F22	9A	very thin, inclined lens of grey sediment with charcoal fragments, possibly an eroded or slumped hearth

Table 6.5 Stratigraphic distribution of archaeological material in square G10, Tunnel Cave.

Layers with hearths are shaded. The right-hand column indicates which layers have as many archaeological items per kilogram of deposit as layer 9A, which after layer 1 is the hearth layer poorest in archaeological remains. Layer 4 and hearth F2 do not appear in square G10. Hearth F12 is listed separately from the two layer 7 units around it, as it is too large to assign to either one. F21 is not definitely a hearth.

Layer	No. of hearths	No. of flaked stone artefacts	No. of exotic stone pieces	No. of ochre pieces	No. of bone points	No. of emu eggshell fragments	No. of aquatic mollusc shell fragments	Kilograms sediment excavated	Total items divided by kg sediment	Higher or lower than value obtained for hearth layer 9A?
1	1	3					1	159.0	0.03	Lower
2		1		1				273.2	0.01	Lower
3		2						230.5	0.01	Lower
5A		15				2	5	201.0	0.11	Lower
5B (incl. F4)	1	236		2		86	7	368.0	0.90	**Higher**
6		43	1			25	13	327.0	0.25	Lower
7 indeterminate		43		17		16	6	73.0	1.12	Higher
7A1		22	1	9		10		28.0	1.50	**Higher**
7A2	1	72		38		23	3	48.5	2.80	**Higher**
7B	1	130	1	23		59	2	92.5	2.32	**Higher**
7C		130	1	8		22	1	78.5	2.06	**Higher**
7D	1	167		9		17	2	40.5	4.81	**Higher**
7E	1	125		3		36	33	30.5	6.46	**Higher**
7F	1	22				9		7.0	4.43	**Higher**
7G	1	278		22	3	81	10	126.0	3.13	**Higher**
F12	1	21		2		7	29	97.5	0.61	**Higher**
7H	2	15	3	20		3	9	28.0	1.79	**Higher**
8		19		1		9	12	235.5	0.17	Lower
9A (incl. F18, F19)	3	80	1	15	2	60	22	608.5	0.30	Same
9B (incl. F21)		4				3	1	206.0	0.04	Lower
9C		3				1	2	182.5	0.03	Lower
10A		6				1	1	162.5	0.05	Lower
10B						1	2	91.0	0.03	Lower
Total	14	1437	8	170	5	471	161	3694.7	1.22	

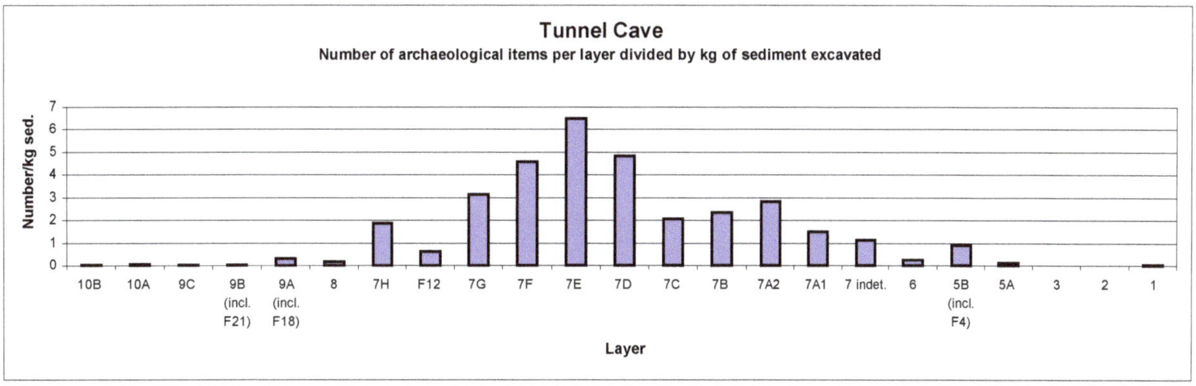

Figure 6.10 Number of archaeological items per kilogram of sediment excavated from each layer in square G10, Tunnel Cave

Materials classed as archaeological, listed across the top of Table 6.5, are either humanly modified, or could only have entered the cave deposit from human transportation. With the exception of layer 1, all layers containing hearths have more artefacts per kilogram of excavated deposit than all layers without hearths. After layer 1, the hearth layer poorest in archaeological remains is layer 9A (0.3 items/kg). This threshold divides the Tunnel Cave layers into roughly equal sized groups of layers (the median is a similar value, 0.61 items/kg). In Table 6.5, the layers with a higher density value than layer 9A are indicated by the entry "higher" in the right hand column, and the hearth layers by shading. The two indications of human occupation, hearths and artefacts, conform well, the exception being layer 1. Almost all the archaeological material lies in hearth-containing layers, while faunal remains from various potential sources are distributed throughout all layers. This distribution could hardly have resulted from random perturbance of layers from water-flow or other agents. Continuous, slow natural deposition is a backdrop to the isolated instances of localised re-working represented by hearths, which during combustion probably drew up sediment from underlying layers. The following section identifies the age and duration of occupation and non-occupation episodes.

Chronology of human occupation

For square G10, I have modified previous stratigraphic observations and age estimates (J. Dortch 1996), taking into account more recent analyses and the view that hearth layers formed within short periods while deposition of non-hearth layers continued over long periods. The following discussion shows that every hearth layer (and hence most archaeological items) falls within an age estimate based on statistically indistinguishable radiocarbon determinations from that layer.

Table 6.6, below, shows the complete list of Tunnel Cave radiocarbon estimates, including previously unpublished AMS estimates obtained on emu eggshell fragments in an ongoing analysis by G. Miller, University of Colorado. No estimates are converted to calendar years, because no calibration curve exists for estimates greater than 18,500 radiocarbon years, such as several of those presented here (Stuiver and Reimer 1993).

At Tunnel Cave, all but two of 22 radiocarbon estimates are in stratigraphic order, suggesting the deposit is little disturbed. The exceptional estimates are Wk 3031 and the determination from a trial at the UWA Geography Department dating laboratory, which I reject from further analysis. Wk 3031 derives from F22, a sloping and possibly eroded feature that appears to intrude into the lower part of layer 9 from its upper part. If F22 is eroded then it may be contaminated with younger material and so register as erroneously young. The UWA trial estimate is in broad agreement with stratigraphically superior dates from layer 9A, but it is statistically different from other dates at a similar depth (Wk 3393, AAL 8205). It may have an apparently young age because of failure to eliminate a slight modern contamination of the vials used to contain the sample (W. Wilson, Geography Department, UWA, pers. comm.).

As at Devil's Lair, the method of Ward and Wilson (1978) allows one to determine which estimates are statistically identical. Almost all hearth layers are dated to short periods by virtue of the fact that they contain, or are surrounded by, similar radiocarbon estimates. There are four groups of similar estimates, each associated with an episode of occupation (Table 6.7).

In these groups of estimates, the probability of the Null Hypothesis (H0), that estimates in the group are identical, ranges from 0.70 to 0.88. As at Devil's Lair, H0 can neither be accepted nor rejected in all four cases of similar dates. Therefore, I infer an age range for every layer (Table 6.7, second column from right) from the ages falling within the central 95% of area under a normalised probability curve representing all the dates in the pool (Appendix 4; cf. Holdaway and Porch 1996). The most probable ages are the peaks on the curve (Figure 6.11). The groups of similar estimate in Table 6.7 suggest a re-grouping of spits into new layers, especially in layers 5A,

5B, sub-divisions of layer 7, and 9A. Radiocarbon dates Wk 6706 and Wk 6707 show that deposition of the upper part of layer 5A dates from 10,000 years BP, while dates Wk 4517 and Wk 5474 show that the lower part (which contains a few archaeological items, including a human deciduous tooth) is contemporaneous with all of layer 5B.

Layer 5A does not extend to square G12, so it has no stratigraphic relation to layer 4 and hearth F2 in that square. However, a radiocarbon date from F2 (Wk 3030), which is a hearth on the top of layer 4, indicates that layer 4 predates layer 5A, and is contemporaneous with layer 5B. The radiocarbon determinations that bracket layers 4 and 5B (Wk 3030, Wk 3389, Wk 4517, Wk 5474) provide no statistical evidence that they are from samples of different age (p [H0: dates are identical] = 0.20). The deposit between spits 18 and 33 could have aggraded during one occupational episode, or several. The estimate given in Table 6.7 is the maximum period of deposition for the lower part of layer 5A, and layers 4 and 5B, including the hearth F4, that is, some time between 13,300 and 12,000 years BP. Apparently more precise, alternative estimates based on other combinations of these dates are not stratigraphically consistent. Hearth F4 and its sub-units therefore represent a major occupation within this 1,300 year period; there may also have been later occupations within layer 5B and this period.

Layer 7, comprising 14 hearths and sub-units of hearths (F5A to F15) and sub-layers 7A to 7H, contains more than half of Tunnel Cave's excavated archaeological material. There is no reason to suspect the layer 7 sub-layers are mixed because they are clearly defined by their different colours, degrees of compaction, and composition (i.e., predominantly sand, ash, or charcoal). The ash and charcoal layers composing hearths are highly compacted, and often quite thick, so there is little chance that material could have moved through these layers. Therefore, the radiocarbon determinations from the layer 7 hearth complex probably derive from material in primary stratigraphic position. Small inversions within the layer 7 radiocarbon determinations probably derive from slight modern contamination during excavation or analysis of samples, or from the use of different dating laboratories, methods, and sample materials.

Table 6.6 Radiocarbon age estimates from Tunnel Cave.

Sample number	Depth below datum (cm)	Square	Spit	Layer	Sample type	Feature	Date, years BP	±
Wk 3626	55	G12	3	1 (base)	CL	F1	1,370	40
Wk 4516	74-80	G10	11	3 (top)	C, 5 mm	none	4,280	60
Wk 6030*	90-93	G10	15	5A (top)	C, 5 mm	none	6,900	60
Wk 3625	93-98	G10	16	5A (top)	C, 5 mm	none	8,270	80
Wk 6706*	98-101	G10	17	5A (mid)	C (K), 5 mm	none	9,780	80
Wk 6707*	98-101	G10	17	5A (mid)	C (B), 5 mm	none	9,940	110
Wk 5474*	101-105	G10	18	5A (mid)	C, 5 mm	none	12,890	250
Wk 4517	106-110	G10	20	5A (base)	C, 5 mm	none	12,840	90
Wk 3030	80	G12	13	4	CL	F2	12,590	180
Wk 3389	128	G10	33	5B (base)	CL	F4A	12,400	240
Wk 3390	173	G10	64	7B	CL	F8	16,080	90
AA 22900*	179-182	G10	68	7B	AMS-EE	F9	17,380	105
Wk 2933	177	G11	32	7 (mid)	C, # 1	none	17,110	250
Wk 3391	210	G10	86	7G	CL	F11	16,850	110
Wk 2934	213	G11	49	7 (near base)	C, # 1	F5D	17,010	260
Wk 3392	251	G10	112	9A (top)	C, # 1	F18	19,300	650
AA 22901*	257-258	G10	117	9A (top)	AMS-EE	none	19,735	130
Wk 3393	263	G10	121	9A (mid)	C, # 1	F19	21,110	220
UWA trial*		G10	130-131	9A (base)	C, 5 mm	none	20,100	360
AA 22902*	276-278	G10	132	9A (base)	AMS-EE	F20	21,215	165
Wk 3031	300	G10	146	9B	C, # 1	F22	19,110	460
Wk 3394	325	G10	164	10A	C, 5 mm	none	22,410	850

Asterisks indicate previously unpublished dates; other dates are given in J. Dortch (1994, 1996).

Laboratory codes:
- Wk Waikato Radiocarbon Laboratory
- AA University of Arizona AMS
- UWA UWA Department of Geography

Sample types:
- CL charcoal-rich sediment from charcoal lens of hearth
- C, 5 mm charcoal fragments from 5 mm mesh sieve
- C (K), 5 mm charcoal identified as from Karri (*Eucalyptus diversicolor*), from 5 mm sieve
- C (B), 5 mm charcoal identified as from a banksia (*Banksia* sp.), from 5 mm sieve
- C, # 1 charcoal fragments collected *in situ* (the number is the charcoal collection for that spit).
- AMS-EE AMS date on emu eggshell fragments collected from 5 mm sieve,l courtesy of G. Miller, Institute for Arctic and Alpine Research, University of Colorado, with support from the US National Science Foundation, Climate Dynamics Program, Division of Atmospheric Sciences.

Table 6.7 Assessment of Tunnel Cave radiocarbon dates

Samples are listed in stratigraphic order. Layers 2, 6, and 8 have no radiocarbon dates and are not shown. Column headed "New layer" indicates layer divisions used in subsequent chapters, although new layer 5-upper is potentially divisible by its radiocarbon determinations into component spits 15, 16, and 17. For new layers 5-lower and 7-lower, age estimates indicate the age range for the largest possible grouping. Shaded lines indicate hearth-containing layers.

					Status of Null Hypothesis (H0): "dates are statistically identical"		Age BP at 95% confidence		
Sample	Depth in cm below datum	Layer and position within it	Date	1σ (±)	For largest apparent grouping	For alternative, smaller grouping (aiming for higher *p* [H0])	Min.	Max.	New layer
Wk 3626	**55**	**1 (F1)**	**1,370**	**40**	**No chance**	**none possible**	**1,290**	**1,450**	**1**
Wk 4516	74-80	3 (top)	4,280	60	No chance	none possible	4,160	4,400	3
Wk 6030	90-93	5A (top)	6,900	60	No chance	none possible	6,780	7,020	5-upper
Wk 3625	93-98	5A (top)	8,270	80	No chance	none possible	8,110	8,430	
Wk 6706	98-101	5A (mid)	9,780	80	Do not reject - *p* (H0) = 0.24	none possible	9,600	10,200	
Wk 6707	98-101	5A (mid)	9,940	110		none possible			
Wk 5474	**101-105**	**5A (mid)**	**12,890**	**250**	**Do not reject - *p* (H0) = 0.20**	**Do not reject - *p* (H0) = 0.88**	**12,000**	**13,300**	**5-lower**
Wk 4517	**106-110**	**5A (base)**	**12,840**	**90**					
Wk 3030	**80**	**4 (in G11)**	**12,590**	**180**		**Do not reject - *p* (H0) = 0.53**			
Wk 3389	**128**	**5B (base)**	**12,400**	**240**					
Wk 3390	**173**	**7B (top)**	**16,080**	**90**	**No chance**	**none possible**	**15,900**	**16,260**	**7-upper**
AA 22900	**179-182**	**7B (base)**	**17,380**	**105**	**Do not reject - *p* (H0) = 0.32**	**Reject - *p* (H0) = 0.01**	**16,500**	**17,600**	**7-lower**
Wk 2933	**177**	**7 (top)**	**17,110**	**250**		**Do not reject - *p* (H0) = 0.78**			
Wk 3391	**210**	**7G (mid)**	**16,850**	**110**	**Do not reject - *p* (H0) = 0.57**				
Wk 2934	**213**	**7 (base)**	**17,010**	**260**					
Wk 3392	**251**	**9A (top)**	**19,300**	**650**	**Do not reject - *p* (H0) = 0.51**	**none possible**	**18,200**	**20,450**	**9-upper**
AA 22901	**257-258**	**9A (top)**	**19,735**	**130**		**none possible**			
Wk 3393	**263**	**9A (mid)**	**21,110**	**220**	Do not reject - *p* (H0) = 0.33	**Do not reject - *p* (H0) = 0.70**	**20,700**	**21,600**	**9-middle**
AA 22902	**300**	**9A (mid)**	**21,215**	**165**					
Wk 3394	325	10A (top)	22,410	850		none possible	20,710	24,110	9-lower, 10

The peaks in this graph indicate the mean age for each set of similar age estimates. The central 95% of each peak's

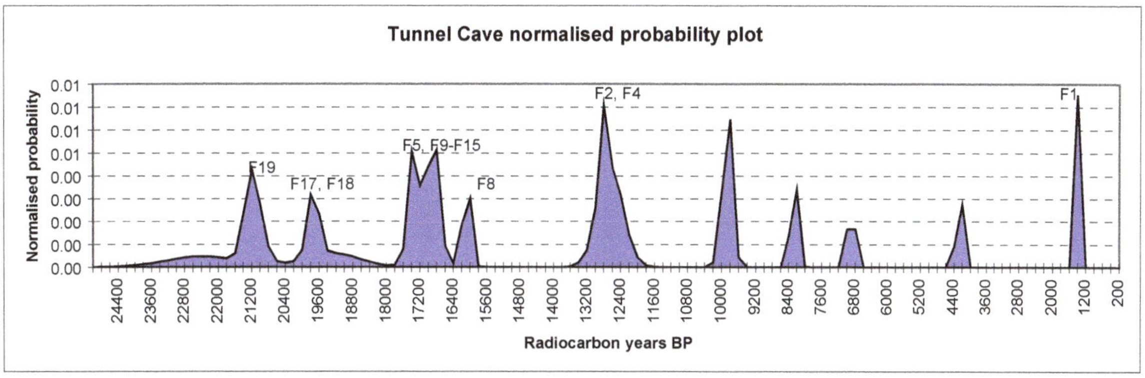

area gives the 95% confidence interval. The vertical scale in this figure indicates a probability for all the estimates considered together, not the probability of each group of estimates. For accuracy the sets of similar age estimates are graphed separately in Appendix 4. "F" numbers indicates age estimates for hearths; other age estimates are from non-hearth charcoal.

Figure 6.11 Normalised probability curve for Tunnel Cave radiocarbon dates.

Table 6.8 Sediment accumulation rates in hearth layers versus non-hearth layers (layers grouped according to radiocarbon dates)

Hearth layers	Maximum period of deposition (years)	Kilograms sediment excavated	Kilograms sediment/ millennium	Non-hearth layers	Maximum period of deposition (years)	Kilograms sediment excavated	Kilograms sediment/ millennium
1	1,450	198.5	137	2	3,110	301.2	97
5-lower	1,300	450	346	3	2,860	163	57
7-upper	360	242	672	5-upper	3,420	119	35
7-lower	1,100	408	371	6	4,340	327	75
9-upper	2,250	243.5	108	8	3,950	235.5	60
9-middle	900	251.5	279	9-lower, 10	3,400 or more	755.5	222?
9-middle	900	251.5	279	9-lower, 10	3,400 or more	755.5	222?

The layer 7 radiocarbon determinations suggest that the hearth complex accumulated between 17,600 and 16,000 years BP. The stratigraphically uppermost and lowermost dates (Wk 3390, Wk 3391) from layer 7 suggest that the period of deposition for layer 7 was at least 300 years, which is the difference between these dates at two standard errors, or a confidence level of 95%. Determination AA 22900 is older than all the other layer 7 determinations, perhaps because of differences in the sample, the dating method, or the laboratory. It suggests that layer 7B is at least 17,100 radiocarbon years old. The lowermost determination from layer 7 is Wk 2934, which suggests the basal part of the complex is no older than 17,530 radiocarbon years. Thus, layer 7 represents a complex of hearths that was first built between 17,500 and 16,700 years BP. Hearth building in layer 7 continued for some time, at least until 16,260 years BP, but no later than 15,900 BP. In short, the period for deposition of 60 cm of deposit was no more than 1,600 years, and more likely, less than 1,000 years.

In layer 9A there are artefacts and hearths (F17, F18, F19) and three hearth-like features (F20, F21, F22) that are possibly eroded or disturbed hearths. Feature 18 near the top of Layer 9A is dated 19,300 years BP. Radiocarbon estimates Wk 3393 and Wk 3031, which date features F19 and F22 respectively, are inverted, but Wk 3031 is discounted, due to possible contamination (see above). Wk 3393 is statistically similar to AA 22902 (p [H0] =0.70) and to Wk 3394 (p [H0] =0.33). As the similarity with Wk 3394 is probably due to the large error margin on that date, I prefer the alternative grouping of dates that separates the middle and basal parts of layer 9A from lower parts of the deposit. This grouping gives the layers "9-middle" and "9-lower and 10".

Table 6.8, based on the age ranges inferred above, suggests that sediment accumulated in hearth-containing layers more rapidly (in terms of kilograms of sediment per millennium elapsed), than sediment in non-hearth layers. All hearth-containing layers appear to have accumulated in such a short time that the entire layer falls within the error margins of a single radiocarbon date or pool of dates. Some hearth layers and hearths could have aggraded during a single human occupation.

These figures cannot be precise, since layers 2, 6, and 8 are dated only by estimates from adjacent layers, layer 10 has no basal age estimate, and age estimates for layers 9-upper and 9-lower have large error margins. Moreover, one cannot determine whether a layer aggraded slowly over the entire period shown above, or was deposited quickly in one or more episodes in that period. For these reasons, archaeological items are standardised against weight of deposit, not period of deposition.

Despite the imprecision, there is a significant difference in the deposition rates of hearth and non-hearth layers. According to a Fisher test, p (H0: "the rates of deposition of all layers are indistinguishable") is less than 0.039. The hearth layers presumably aggraded by non-natural means as well as natural ones. For example, prehistoric human traffic would have pushed sand from higher parts of the sloping floor deposit to the location of the future test-excavation, and people would have carried in fire-wood to be converted to large quantities of ash and incorporated with the sand beneath hearths.

Summary

Sediment was deposited in several ways at Tunnel Cave: water-flow moving sand; deposition of fine carbonaceous sediment from the cave roof; deposition of bone and coprolites by animals; deposition of bone, charcoal, and organic material, and scuffage of sand, during periods of human occupation; and creation of charcoal and ash from burning wood in hearths.

The first two, natural agents are probably relatively slow and perhaps continued during all periods. Deposits by animals are probably also small in terms of total volume of deposit, although highly significant within the class of faunal remains. The anthropogenic agents are rapid and relatively short-lived. The radiocarbon dating of occupational episodes at Tunnel Cave suggests that hearths and occupation layers represent punctuation points in a continuum of relatively slow sand and carbonate deposition.

Six main occupations at Tunnel Cave each fall within the error margins of an individual radiocarbon date, or within a group of statistically similar radiocarbon dates (Table 6.7). With future analysis of other parts of the Tunnel Cave deposit, or by means of more accurate dating, it may be possible to identify other occupations outside of the restricted area of the test-excavation, or within the large error margins on the older dates. The inferred occupations are as follows:

1. Hearth F1 built between 1290 and 1450 BP (leaving minimal occupational remains).

2. Hearths F2 and F4 built between 12,400 and 13,000 BP. After the building of these hearths there may have been other occupations within layer 5B and the lower part of layer 5A, still within this period. There are occupational remains, not associated with a hearth, from stratigraphically higher positions dated to the same age.

3. Hearth F8A built between 15,900 and 16,100 BP; hearths F8B and F8C appear to date to either the same period or a slightly earlier one.

4. Hearths F5D, F9, F10, F11, F12, F13, F14, and F15 built between 16,800 and 17,500 BP. Within this period, there may be two shorter occupations. Hearths F9, F10, F11, and the hearth complex comprising F5D, F12, F13, F14, and F15, represent several hearth-constructions in

this period. The smaller hearths F9, F10, or F11 may be part of the same occupation during which people built new hearths on old ones but layer 7G (orange sand) separates them from the hearth complex, suggesting a short period of natural sand deposition between the hearth complex and the smaller hearths.

5. Hearth F18 built at sometime between 18,000 and 20,000 BP.

6. Hearth F19 and possible hearth F20 built between 20,700 and 21,545 BP, possibly in separate episodes.

Some sparse occupational material may extend back in age to 22,000 BP or more, but it is not associated with any feature readily identifiable as a hearth.

Witchcliffe Rock Shelter

Witchcliffe Rock Shelter, adjacent to Witchcliffe Cave, overlooks a small valley cut by Boodjidup Brook (Figures 5.2, 6.12). Archaeological evidence suggests that people avoided camping in the cave and occupied the shelter mainly in the last millennium (J. Dortch 1996).

Like Rainbow Cave, discussed below, Witchcliffe Rock Shelter differs from Devil's Lair and Tunnel Cave because it contains a recent archaeological record that is relatively detailed. In this discussion, I assess the proposition (J. Dortch 1996, Lilley 1993) that these deposits record several occupations within a short period, all slightly more recent than the last recorded occupation at Tunnel Cave, as represented by hearth F1 there.

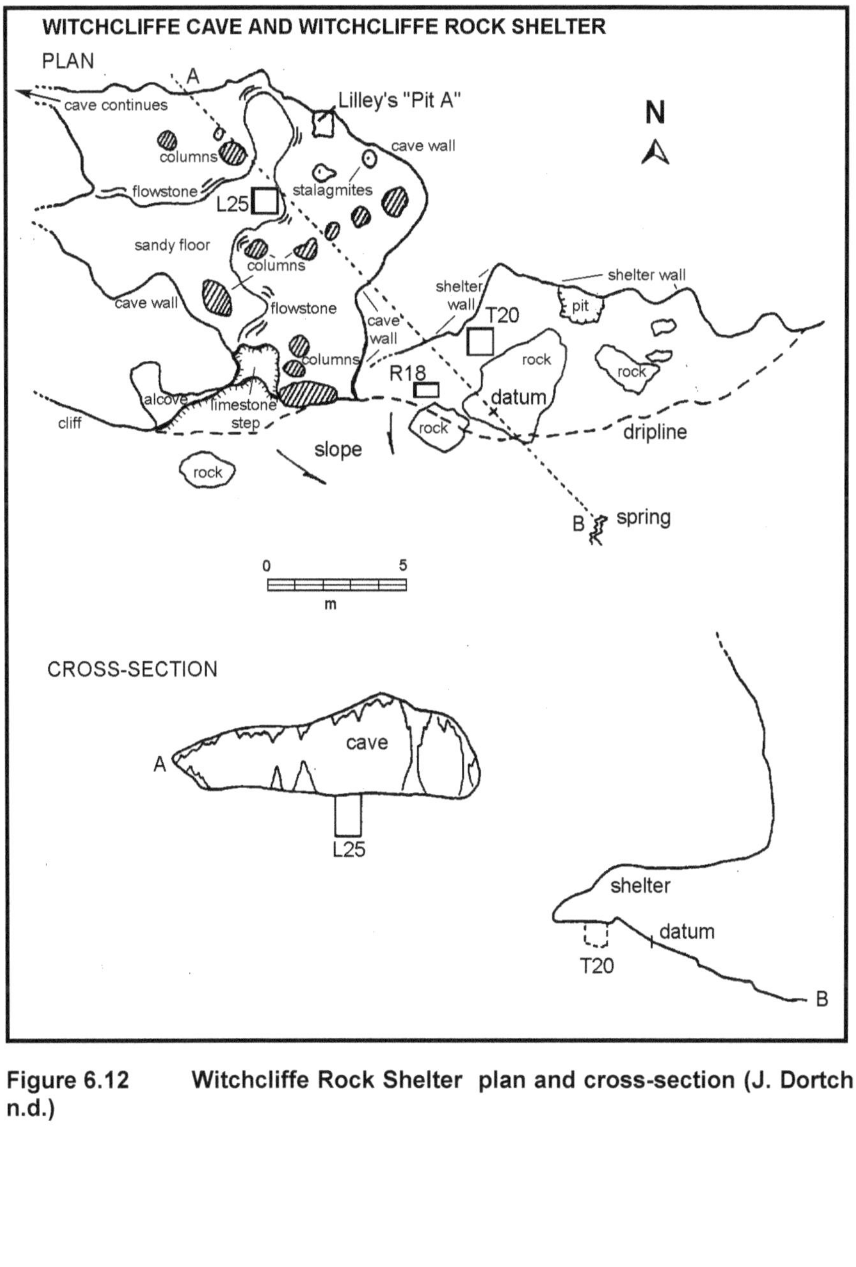

Figure 6.12 **Witchcliffe Rock Shelter plan and cross-section (J. Dortch 1996, Lilley, pers. comm., Loveday, n.d.)**

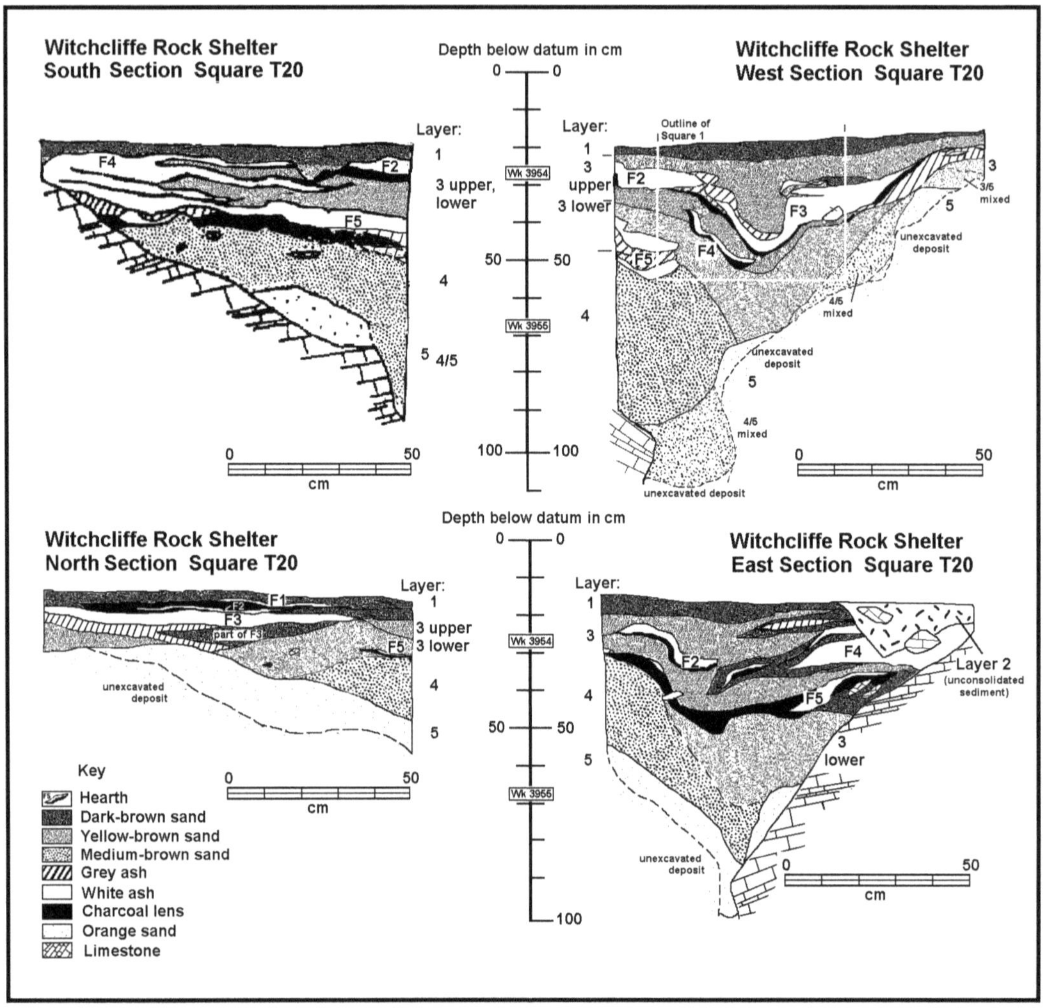

Figure 6.13 Section diagrams of Witchcliffe Rock Shelter test-excavation, square T20 (south section after J. Dortch 1996, others previously unpublished)

Table 6.9 Archaeological remains from Witchcliffe Rock Shelter

Layers with hearths are shaded grey. Modern artefacts are not included in the standardised total of archaeological items.

Layer	Hearths	Modern artefacts	Flaked stone	Feldspar, granite-gneiss	Ochre pieces	Other exotic stone	Fish bone	Fish scales	Crust-acean shell	Fish oto-liths	Emu egg-shell	Marine mollusc	Fresh-water mollusc	Kg deposit excav.	Arch'l items per kg deposit	More items/ kg than F4?
1	1	53	307	2	1	2	6	10	1	3	45	12	2	35.0	11.2	Yes
2			9								2	1		1.5	8.0	Yes
3-upper	1	8	743	3	9	7	17	25			112	27		65.5	14.7	Yes
F3	1		60		1						12		3	11.5	6.7	Yes
F4	1	1	142		1						6			29.0	5.1	Yes
3-lower			307	2	2		2	1	3		37	3	1	16.0	22.4	Yes
F5	1	2	301	1	2		3	3			26	3		33.5	10.2	Yes
4-upper		1	415		9		5	2			36	2	1	62.0	7.6	Yes
F6	1		71		3		1				7			5.5	14.9	Yes
F7			11								3			2.0	7.0	Yes
4-middle			311	6	8		6	5			58	8	6	93.5	4.4	No
F8, mixed	1		117	2	3	3	5		1		5	2	1	18.5	7.6	Yes
4-lower			124		4	1	4	1			6	1	1	72.0	2.0	No
4-lower/5			104	3	1		1	1			40	1	1	154	1.0	No
5			34			1					3			49.5	0.8	No
TOTAL	7	65	3056	17	44	16	50	48	5	3	398	60	16	649	5.8	

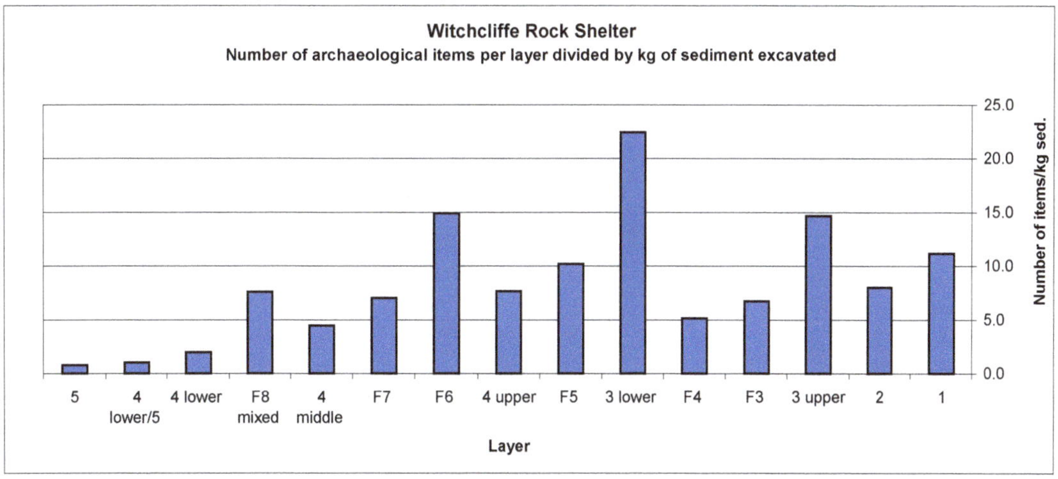

Figure 6.14 Distribution of archaeological remains from Witchcliffe Rock Shelter.

Evidence for human occupation

The occupation deposit in Witchcliffe Rock Shelter consists of dark organically stained quartz sand, rich in charcoal, and containing hearths (Figure 6.13), similar in form to those at Tunnel Cave, as well as numerous artefacts (Table 6.9). As at Tunnel Cave, artefacts could have moved between layers, and only hearth-containing layers are unequivocal occupation layers.

Archaeological remains include flakes and fragments of quartz, pieces of ochre and granite-gneiss, bird bone, emu eggshell fragments, fragments of freshwater and marine bivalve shells, and fish bones and scales (almost certainly marine species since they are larger than those of the very small south-western Australian freshwater fish). Unlike Tunnel Cave, many non-hearth layers have a high density of archaeological items (Figure 6.14). All layers above layer 4-middle have a higher density of archaeological items than the hearth poorest in these remains, F4 (see right-hand column, Table 6.9).

F1 is a thin, flat hearth lying within 5 cm of the modern-day surface. It lies beneath layer 1, which contains modern artefacts and may have been disturbed by 20th C occupiers of the shelter. Layer 1 and F1 also contain many artefacts of apparent prehistoric origin (Table 6.9). Since F1 is compacted and well-defined, showing no signs of disturbance, I assume that layers beneath it are undisturbed. Table 6.10 shows all the archaeological and other features, of which only Feature 7 (F7) is not identified as a hearth.

Witchcliffe Rock Shelter hearths F3 and F5 may be contemporaneous and part of the same hearth-building event. F3 and F5 also show faint brown or grey lenses suggesting partial combustion of charcoal lenses within the hearth, which may indicate refiring or re-use of the same hearth. The west section of square T20 (Figure 6.13) suggests that F3 was created by an excavation into the layers below and filled with ashy sediment, which was in turn excavated again.

Below the lowest hearth, Feature 8, there are c.250 quartz artefacts in loose brown sand called layer 4-lower. Layer 4 lower is charcoal-poor, in contrast to the compacted dark brown, charcoal-rich sand of superior layers, and not being compacted, the artefacts in this layer could have been more easily moved into it from superior layers.

At this depth in the excavation, the brown sand deposit sits between a limestone boulder on one side and a bank of partially cemented yellow sand on the other (Figure 6.13). This yellow sand, called layer 5, lacks artefacts and bone fragments, and is similar in colour, texture, and degree of cementation to sand exposed in a pit dug by persons unknown in another part of the shelter (Figure 6.12). This sand appears to be intercalated with the calcarenite. If layer 5 is part of this apparently ancient sand, it pre-dates human occupation. Given that the lateral extent of layer 4-lower is very limited, and that the layer contains no occupation features and is loose enough for artefacts to move through, it is probably not an occupation layer either.

Table 6.10 Archaeological and natural features at Witchcliffe Rock Shelter

Feature numbers are in stratigraphic order. Asterisks indicate features not preserved in section.

No.	Layer(s)	Identification
F1	1	small hearth, thin layers of ash and charcoal
F2	3 upper	small hearth, probably part of F3
F3	3 upper	thick, large hearth, intruding into lower layers, possibly the ash fill of a fire pit
F4	3 lower	small hearth
F5	4 upper	thick, large hearth
F6*	4 upper	hearth, very small, perhaps coeval with F5
F7*	4 upper	unidentified white mottling
F8*	4 middle	hearth, perhaps slumped part of F5

F5 seems to represent the lowermost undoubted occupation feature. F8 is lower but it may not be significantly older, as it is located immediately below the lowest part of the large, dipping lens of F5. It is very small for a hearth and may be a piece of indurated sediment separated from the bottom edge of F5. F6 is adjacent to F5, rather than below it.
As the upper side of F5 extends upwards to 10 cm below the surface, immediately below hearths F1, F2, F3, and F4, occupation at Witchcliffe Rock Shelter perhaps took place either during a period of little sedimentation, or a very short period. Before the known period of hearth-building, and to a limited extent during it, natural sedimentation was probably by means of water-flow washing in sand from the hillside above the rock shelter.

The Witchcliffe Rock Shelter archaeological remains suggest pre-European hunter-gatherer occupation. But also found in the uppermost layer (layer 1) and on the surface are 20th C artefacts: cigarette butts, pieces of plastic, and copper rimfire 0.22" calibre rifle cartridge cases (including some with the stamp "ICI"), suggesting manufacture since World War I (Kass, n.d.; G. Vines, Melbourne's Living Museum, pers. comm.). The presence of some of these remains in lower layers also suggests some mixing or disturbance. This problem is discussed in the next section.

Chronology of human occupation

The two radiocarbon estimates from Witchcliffe Rock Shelter indicate the young age of the entire archaeological deposit (Table 6.11).

Table 6.11 Witchcliffe Rock Shelter radiocarbon estimates
Sample and type codes explained in Table 6.7.

Sample number	Depth below datum	Square	Spit	Layer	Sample type	Feature	Date	1 σ (±)
Wk 3954	26-27 cm	T20	7	3	CL	F1	400	50
Wk 3955	66 cm	T20	37	4A	C, #1	F8	680	90

As indicated by hearths, there were at least seven occupations throughout the period represented by the two age estimates, which at two standard deviations are adjacent in time: 860-500 years BP and 500-300 years BP, and do not overlap significantly (p [H0] = 0.006). By virtue of their stratigraphic position, the age estimates for hearths F1 and F8 provide age estimates for the large hearths F3 and F5 and the small hearths F2, F4, and F6. Wk 3954 shows that F1, even considering the large number of modern artefacts in the overlying layer 1, is a prehistoric hearth. The modern artefacts have perhaps infiltrated layer 1 and, to a limited extent, lower layers because of trampling or shallow digging of the deposit by modern occupiers of the shelter. However, every hearth lens appears to be intact, suggesting that this kind of perturbation has not disturbed layers or hearths from F1 downwards. Perturbation is a likely cause of modern artefacts being mixed with prehistoric ones for parts of layers 1 and 3-upper located above F1 and F2, and for the upper part of F4, intersected by a recent post-hole excavation from the surface (indicated by "Layer 2", interpreted as loose fill, in Figure 6.13, T20 east section).
A few tiny fragments of copper cartridge case from layer 4, below F4, may have been inadvertently introduced into the excavation because I excavated square T20 in two stages, the first stage being a 35 cm deep, 50 by 50 cm trial excavation ("Square 1") within the 1 by 1 m square T20 (see Figure 6.13, T20 west section). Although I attempted to prevent the mixing of archaeological material, by covering the base of Square 1 during excavation of the remainder of T20, and not including spilled sediment with analysed material, small quantities of surface sediments were possibly accidentally knocked into Square 1, to be later excavated from its base, below which there are no modern artefacts.

The quantity of modern material in the archaeological sample is small. Fifty-three modern artefacts were found within the uppermost layer 1 (5 cm below surface), another eight were recovered from layer 3-upper (30-50 cm below surface), and only four others were found in lower layers. The chances of modern artefacts being introduced into the archaeological sample because of slight mixing during excavation are probably very small below layer 3-upper.

Despite these problems, the Witchcliffe Rock Shelter radiocarbon estimates are probably reliable. Charcoal sampled from F8 was probably burnt in that feature. Moreover, there are no modern artefacts in the layers immediately above or below it. The radiocarbon sample from F1 is even more reliable, as it derives from an entire charcoal lens formed beneath a hearth, rather than loose charcoal fragments.

Summary

People occupied Witchcliffe Rock Shelter in prehistoric times from little earlier than 860 years BP to about 300 years BP. More recent Aboriginal occupation is not identifiable in the excavation. Occasional occupation by people continues in the present day, resulting in disturbance of the layers above the highest hearths, F1 and F2, and the archaeological sample is slightly contaminated with introduced modern artefacts. However, the dates for hearths F1 and F8 are reliable since these hearths, like the others, show no signs of disturbance, and the samples dated were taken from charcoal lenses of the hearths. These hearths are on either side of the large hearths, F3 and F5. All derive from occupations within a period no longer than six centuries, and possibly much shorter, so that periods between occupations cannot be identified. As at Tunnel Cave and Devil's Lair, most of the occupation deposit belongs to a radiocarbon-dated period.

Rainbow Cave

Lilley (1993) excavated seven test-pits at Rainbow Cave in 1990 and 1991 (Figure 6.15). Stone artefacts and faunal remains are analysed in Honours dissertations (Cocks 1993, Jackson 1992) and an unpublished student report (Adie *et al.* 1990). Lilley aimed to identify Aboriginal occupation sites in or near the Margaret River valley. Of the three deposits that he excavated, only Rainbow Cave contained archaeological remains. Most of these derive from excavation square D22.

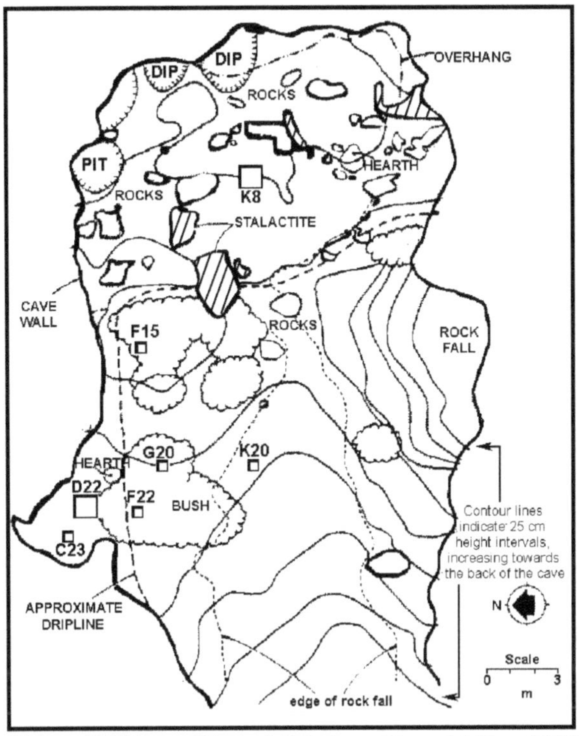

Figure 6.15 Plan of Rainbow Cave (Lilley 1993)

The deposit in square D22, where the original analysis concentrated, is divided into an upper section 0-40 cm below surface (Figure 6.16), containing most of the artefacts, articles inferred to have been carried in by people, and charcoal fragments (Lilley 1993: Figure 4), and a lower section poor in artefacts, argued to represent a period of natural sand deposition with no human occupation (Jackson 1992, Lilley 1993).

Evidence for human occupation

The upper section includes layers that appear to be hearths, identified by Lilley (1993: 37) as "ash and charcoal-rich bands of sand". Given the similar appearance between these bands in photographs of Rainbow Cave excavations (Cocks 1993: Figures 4.2, 4.3, 4.5, 4.6, 4.8) and the photographs of hearths at Tunnel Cave (Appendix 5) and at Devil's Lair (C.E. Dortch 1984: photograph on p. 43), I consider that these bands of ash and charcoal are hearths. Figure 6.16 shows his section drawings from the excavation square D22, which I have shaded to indicate brown sand (grey shading), ash layers (white), and charcoal layers (black). It can be seen from the section drawing for the south face of this square, and the photographs of this section, that ash layers overlie charcoal layers, just as they do at Devil's Lair, Tunnel Cave, and Witchcliffe Rock Shelter.

Archaeological remains at Rainbow Cave are similar to those at Witchcliffe Rock Shelter: quantities of burnt and unburnt bone, including fish bone, marine mollusc shell, charcoal, flakes and fragments of quartz, and layers of ash and charcoal (Cocks 1993, Lilley 1993, Jackson 1992). Some of the bone fragments from terrestrial animals and all the fish bones are attributed to human occupation. None of the analysts of the site reports finding emu eggshell.

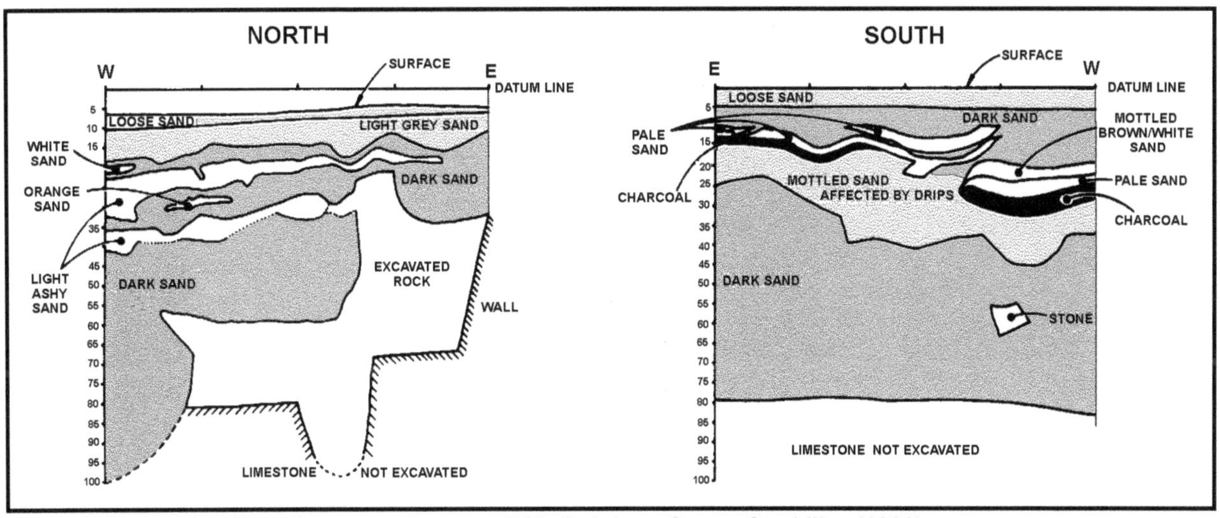

Figure 6.16 Section diagrams of square D22, Rainbow Cave, after Lilley (1993)

In the grey-brown sand below 40 cm depth, the presence of small quantities of artefacts, burnt bone, and charcoal probably derives from the downward migration of these items through trampling and mild disturbance at the surface. This scenario is similar to that inferred for the Witchcliffe Rock Shelter floor deposit. I inspected the floor deposit at Rainbow Cave, as revealed there by shallow pits of unknown origin (Figure 6.15), and concur with Cocks (1993), that the deposit comprises unconsolidated sand, similar to the sand-box deposits in which Cocks' trampling experiments revealed substantial movement of a small proportion of all the artefacts initially placed in one layer. Undulations in the ash and charcoal lenses shown in Figure 6.16 suggest that there may have been some later re-working or contemporary disturbance within the occupation layer indicated by the labels "white sand", "pale sand", "orange sand", "light ashy sand", and "charcoal [-rich layer]". This re-working might have caused the small degree of mixing inferred between occupation and non-occupation layers.

Chronology of human occupation

Lilley (1993: Table 1) obtained the following radiocarbon age estimates from Rainbow Cave (Table 6.12):

Table 6.12 Radiocarbon age estimates from Rainbow Cave

Collection methods are not given in Lilley (1993), but are potentially from the sieve or from *in situ*. All samples are reported as charcoal fragments dates using LSC.

Sample number	Depth below surface (cm)	Square	Date	1 σ (±)
Wk 1875	5-10	D22	340	45
Wk 1876	25-30	D22	790	50
Wk 1877	35-40	D22	830	45
Wk 1878	70-75	D22	4150	70
Wk 1879	25-30	C23	1030	50
Wk 1880	95-100	K8	2600	90

The age estimates of interest here are those from excavation square D22, which contained most of the occupational remains that were located and analysed. Of these, only Wk 1878 dates a non-occupation layer, for reasons given above. Wk 1879 and Wk 1880 date very sparse archaeological remains in other excavation squares, and are considered no further here. The remaining estimates, Wk 1875, Wk 1876, and Wk 1877 suggest that the major human occupation at Rainbow Cave took place on at least two occasions sometime between 920 and 250 BP, which are the outermost error margins of the series at a 95% confidence level. The youngest and oldest estimates of this series are statistically different (*p* [H0: they are the same age] = 1.4×10^{-14}). However, dates Wk 1876 and Wk 1877 are very similar (*p* [H0] = 0.55), and hence provide mutual support for an age estimate of no earlier than 900 BP for first recorded occupation of the site. Neither of these age estimates is distinguishable from the age estimate for first recorded occupation of Witchcliffe Rock Shelter, as indicated by Wk 3955 (680 ± 90 years BP; for Wk 3955 and Wk 1876, *p* [H0] = 0.28; for Wk 3955 and Wk 1877, *p* [H0] = 0.13).

Summary

The Rainbow Cave deposit and its archaeological contents (hearths; burnt and unburnt faunal material, from both marine and terrestrial species; abundant flaked quartz artefacts) are similar to those from Witchcliffe Rock Shelter, located only two kilometres to the south-east. The radiocarbon ages of several human occupations are virtually identical, at 800-400 BP. The major differences between the deposits, to be discussed in Chapter 8, are to do with the types of exotic faunal remains in them, and the location of the Witchcliffe Rock Shelter deposit slightly further away from the coast.

Conclusion

The four sites discussed here provide detailed evidence of well-defined periods of human occupation in the Leeuwin-Naturaliste Region from 47,000 BP to 300 BP. Given research indicating that there were major climatic changes in the world at c.22-18,000 BP, and that in south-western Australia, vegetation boundaries shifted at this time and afterwards, these sites have the potential to show how human occupation has altered in reaction to changes in vegetation type.

Devil's Lair was occupied many times, albeit intermittently, throughout the period from 45,000 BP to 12,000 BP, and perhaps as late as 6,500 BP. Uncertainties about the final date of human occupation mean that the site can provide little detailed information about human reactions to vegetation change at what is probably a key time, the Pleistocene-Holocene transition at *c.*10,000 BP (Balme *et al.* 1978, Burke 1997).

Tunnel Cave was occupied by people building hearths and leaving varied archaeological evidence at c. 20-18,000 BP, 17-16,000 BP, 16,000 BP, 13-12,000 BP, and perhaps briefly at 1,400 BP. This record suggests some prolonged absence of people from the site throughout most of the Holocene. The question of whether these possible absences are related to vegetation changes is dealt with in subsequent Chapters.

Witchcliffe Rock Shelter and Rainbow Cave were occupied by people who also left hearths and abundant archaeological evidence, but in a relatively short period, the last 800 years. These sites suggest that people made full use of these sites at a time when vegetation was perhaps little different from today, although it may have been slightly more open (McNicol 1999).

Chapter 7 Stone artefacts and occupation intensity

In this chapter I aim to identify how intensively people occupied sites at different times. Intensive occupation refers to occupation episodes that are particularly long or frequent or involve many people. I infer changes in occupation intensity by analysing long-term trends in stone artefact characteristics. People tend to produce and maintain stone tools more carefully during intensive occupation of sites when raw material for making tools is more quickly exhausted or cannot be replenished by visits to stone quarries that are distant from the sites (Bamforth 1986; Callow 1986; McNiven 1994; O'Connell 1977; Rolland and Dibble 1990; Roth and Dibble 1998; Torrence 1992; Veth 1993; Walsh 1998). In this chapter I identify changes in raw material choice and artefact production, use, maintenance, and discard as indicators of raw material conservation. I compare this evidence with changes in raw material access probably caused by sea-level fluctuations, and determine whether changes in access or occupation intensity provide the best explanation for changes in raw material conservation. In Chapter 10, I compare the inferred trends in occupation intensities to changes in vegetation.

Evidence for raw material conservation is potential evidence for occupation intensity, but many other variables affect the appearance of stone artefacts also. One can envisage additional variables that must be controlled for, at every stage in the life of a stone artefact: quarrying, manufacture, use, and discard. These stages, which I assess at each of the sites identified above, are those of the *chaîne opératoire*, or "operational sequence" (Bar-Yosef *et al.* 1992, Grace 1997).

Basic to the concept of the *chaîne opératoire* is the assumption that tools are objects used or intended for a task. The *chaîne opératoire* concept is often used in inferring stone tool production, but it also applies to stone tool makers' and users' techniques from "the acquisition of raw material... to the final abandonment of the desired and/or used objects" (Bar-Yosef *et al.* 1992: 511). It is a theoretical history of a stone tool, with stages of quarrying, manufacture, use and maintenance, and discard. It is also a framework for studying by-products of the quarrying, manufacture, and use stages. The by-products, which are much more abundant in archaeological sites than tools, are *debitage*, meaning waste, although debitage is informative to archaeologists. Moreover, debitage, discarded but still potentially useful, can re-enter the *chaîne opératoire* at any stage as recycled raw material.

Regional variations in lithological and geological parameters and cultural practices sometimes lead to distinct regional sets of tool types and manufacturing techniques (Bar-Yosef *et al.* 1992, Inizan *et al.* 1992, Mellars 1971, Walsh 1998). In south-western Australia, as in other regions, people made some distinctive tools, but stone flaking techniques are not known to have differed from elsewhere in Australia (C.E. Dortch 1977, cf. Flenniken and White 1985). Holdaway (1995) notes that much of the apparent uniformity in Australian stone technology and typology probably stems from the continued application of the same theories and analytical methods, but at the same time, the number of ways a person can flake and use stone are circumscribed by lithology (Grace 1997). One can therefore make similar arguments about raw material conservation here as elsewhere.

In the following discussion I identify links between stone artefact production and occupation intensity. The links derive from ethnographic observations of stone tool users and the assumption that these observations would hold true for prehistoric stone tool users.

Flaked stone artefacts, the main category of stone artefacts analysed here, are the result of a series of choices made by people about what to quarry and how to quarry it, what to make with the raw material and how to make it, what task to carry out and how to make and maintain the tool for that task, and finally, when and where to discard the tool (Grace 1996, Jones and White 1988). Every choice informs the others. A person handling a quantity of stone is aware of the other choices that they will make, or have made, about that stone (Kuhn 1992). The impact of this knowledge is that stone tools tend to be made more efficiently, made to be more robust, used more intensively, or maintained longer in direct proportion to the scarcity or inaccessibility of good-quality raw material (Grace 1997; Hiscock 1993, 1994, 1996a; O'Connell 1977).

Changes in raw material supply

Stone raw material supply becomes more limited when group territories become smaller, when groups visit quarries less often, or when access is physically limited (Bamforth 1986; Callow 1986; McNiven 1994; O'Connell 1977; Rolland and Dibble 1990; Roth and Dibble 1998; Torrence 1992; Veth 1993; Walsh 1998). Physical access to stone and occupation intensities are of greatest interest here.

The probable quarries for the raw materials analysed here are either located within the Leeuwin-Naturaliste Region, or very close to it, probably in locations now submerged by glacio-eustatic sea-level rise (Glover 1984; see below). These locations were not so far from the sites as to be outside of fairly small group territories, and Glover (1984) indicates that there was little geographical movement of raw material, so territorial restriction (if any) was probably not a factor encouraging raw material conservation. However, the quarries were probably located far enough from the sites analysed here to discourage rapid consumption of stone. These sites are all located on the Leeuwin Ridge, the 10 km wide dune complex mentioned in Chapter 2. Except in incised river valleys (not located near the sites analysed here) there are no exposures of rock suitable for flaking on the Leeuwin Ridge. It is assumed here that return journeys from cave and rock shelter sites to stone sources, perhaps requiring one or two days travel, were long enough to encourage some degree of raw material conservation – except when occupations were very short, and imminent arrival at new sources of stone could be anticipated. In short, people would not have visited quarries as often if they were intensively occupying sites on the Leeuwin Ridge.

Physical access to rock outcrop is also an important issue in the Leeuwin-Naturaliste Region. As in other parts of coastal south-western Australia, a major stone type for artefacts in older sites is fossiliferous chert[1], a rock whose outcrops were probably inundated by mid-Holocene sea-level rise (Glover 1984).

Only in glacial periods of low sea-level could people obtain fossiliferous chert by travelling directly to outcrops. Contact with the glacial-period shoreline is confirmed by fragments of marine shell at both Devil's Lair and Tunnel Cave, dated to times when the coastline was 30-40 km west of the present one, 18,000-15,000 BP (Balme *et al.* 1978, J. Dortch 1996). Like the visitors to another occupation site on a alternately receding and transgressing marine shore, La Cotte de St Brelade on the Isle of Jersey (Callow 1986), the supply of chert (flint at that site) largely depended on peoples' ability to reach the previously submerged sources or outcrops. The rate of sea-level rise, and hence submergence of chert outcrops, after the LGM may have been fastest after *c.*12,000 BP (Chappell and Shackleton 1986). Human use of chert outcrops, at least on the Indian Ocean coast of south-western Australia, ended with the mid-Holocene inundation of much of the continental shelf (Ferguson 1980a, Pearce 1978).

Fossiliferous chert nodules may have been larger, providing more material for knapping and more potential for making a range of tool sizes, than quartz cobbles, which were probably widely available from pebble-beds or veins in granite-gneiss at all times (Glover 1984). Quartz artefacts in south-western Australia often appear to be smaller than chert ones (Dortch and Kelly 1997, Pearce 1978, Schwede 1990), perhaps because large chunks of polycrystalline vein quartz are often incohesive (Glover 1984). At Witchcliffe Rock Shelter and Rainbow Cave, the quartz artefacts are small and suggest flaking by bipolar technique (Coventry 1998), as often used on small-sized raw material (Roth and Dibble 1998). The prevalence of this technique also depends on raw material availability and group mobility (Hiscock 1994, 1996a), factors related to occupation intensity.

Chert artefacts seemed to have travelled only small distances to their find spots. On the Indian Ocean coast of south-western Australia, they are uncommon in sites more than 10 km east of the present coastline (Glover 1984, cf. Pearce 1978), and they probably travelled equally short distances up and down the coastline. Glover and Lee (1984) show that the proportions of trace elements in chert artefacts are correlated with the latitude of their discovery. The proportions form a north-south gradient probably because of variations in oceanic conditions at the time of chert formation, in the Eocene. As people could hardly replicate this gradient with hundreds of scattered artefacts on a north-south axis if there were only a few chert quarries, the gradient suggests that chert artefacts derive from many quarries scattered from north to south on the submerged part of the coastal plain. These quarries were perhaps also scattered east-west across the coastal plain. Thus a decrease over time in the proportion of chert artefacts from some time before 7000 BP at Dunsborough (Ferguson 1980a) is perhaps attributable to glacio-eustatic sea-level rise gradually submerging quarries from west to east across the continental shelf.

At the Dunsborough site, Ferguson (1980a) proposes a "behavioural response" to the gradual restrictions in chert supply, which would have begun, he reasons, *c.*12,000 BP. After 6000 BP, chert outcrops would have been fully submerged, and fresh sources inaccessible, but Ferguson suggests that people continued to use ancient chert artefacts until they had exhausted this source also, until the final use of chert at 4,500 BP, well after sea-level stabilisation at 6,000 BP. In support of his hypothesis, Ferguson notes evidence for chert artefact re-use at Quininup Brook, a nearby site occupied from 18,000 BP to at least 10,000 BP (Ferguson 1980b). The re-used artefacts are too few to form a sequence of increasing re-use, however.

At Walyunga (near Perth), the only site with fossiliferous chert that was contemporaneous to Dunsborough, there is no evidence of declining chert supply (Pearce 1978, 1981). Like Dunsborough, chert use at Walyunga terminates at 4,500 BP, but throughout older layers, the proportion of chert artefacts changes little. Pearce (1981) argues there is no need to propose a behavioural explanation. Chert outcrops were perhaps not slowly submerged by 6000 BP, but were suddenly covered by a later geomorphological change, such as large dunes forming on top of coastal chert outcrops. But at Tunnel Cave and Devil's Lair, there is evidence for a gradually declining proportion of chert artefacts after the LGM (see below). The Leeuwin-Naturaliste Region sites old enough

[1] At Tunnel Cave, about 90% of the stone artefacts are of fossiliferous chert (Dortch 1996); at Devil's Lair, there are equal numbers of fossiliferous chert and quartz artefacts (Dortch 1984: Table 3). Almost all of the Witchcliffe Rock Shelter and Rainbow Cave artefacts are made of quartz.

to contain chert artefacts provide little support for Pearce's proposed sudden cut-off of chert supply.

In this region, responses to raw material scarcity might stem from raw material inaccessibility as much as intensive occupation. Ferguson (1980a) has already argued that responses to gradual restrictions in access would have included a take-up of quartz as an alternative raw material, and re-use of ancient artefacts. Below, I discuss other possible responses to raw material scarcity, as a means of distinguishing restrictions in access from intensive occupation.

One can distinguish the two potential causes of "scarcity responses" if some or all responses occur when there are no restrictions in physical access, that is during periods of low sea-level. If, at these times of good chert access, there are several contemporaneous trends towards conserving chert, occupation in the site was probably intense. I aim to identify multiple responses because of the manifest uncertainties in this type of analysis.

One possible counter-argument can be dealt with in advance of the analysis. Possibly, fluctuations in sea-level (which control access to chert) also influenced occupation patterns in Leeuwin-Naturaliste Region sites, since lower sea-levels would have allowed people to travel west of the Leeuwin Ridge. However, opportunities for greater population dispersal would not be expected to lead to more intensive occupations in the sites on the Leeuwin Ridge. Conversely, in periods of moderately high sea-level, access to chert is restricted, and populations are more constricted, if anything. If chert is used freely at these times, occupation was probably not intense. The two potential causes of raw material scarcity considered here would not cause the same response at the same time.

Characteristics of stone artefacts that can suggest behavioural responses to raw material scarcity are as follows:

1. The practice of using debitage pieces as adventitious tools (because they have a suitable edge angle or profile: Flenniken and White 1985, Hayden 1977, Kamminga 1982, White *et al.* 1977) intensifies as raw material becomes scarcer (Grace 1996). Such adventitious tools can be detected from use-wear or residues on their edges (Hayden and Kamminga 1979). The rate of production of retouched tools such as scrapers (found at Devil's Lair: C.E. Dortch 1984: 47, 55) and the frequency that the same tool is used, likewise increases with scarcer material (Dibble 1987; Hiscock 1996a, 1996b; Rolland and Dibble 1990; Walker 1978).

2. Hiscock (1993) and Burton (1980) suggest that additional means of conserving raw material are to produce artefacts with greater precision (e.g. if long cutting edges are desired, to attempt to make flakes more elongate) and with greater efficiency (more flakes or potential tools per core). Hiscock (1993) notes additional means of improving knapping efficiency: preparing platforms to provide better flaking angles (hence more flaked and facetted platforms), reducing overhang on platform edges (leading to more overhang-removal flakes), and heating the raw material, which makes it brittle and easier to fracture (Cotterell and Kamminga 1990).

3. If tools and cores are knapped for longer, because supplies of the stone are limited, one can expect to find more abundant and finer debitage (Byrne 1980, Gould 1977b, Wright 1977). Tunnel Cave artefacts are particularly small - 67% are less than one centimetre long and 90% are less than two centimetres long. Only nine of 1,646 recorded artefacts are identifiable as tools (Appendices 8, 9). The large proportion of small debitage at Tunnel Cave is possibly a small sample bias (the excavation square G10)[2], but could also result from the effects of various intra-site behaviours, such as removing large stone debitage from hearth areas where people sit or sleep, or recycling large artefacts for other purposes (cf. Newcomer and Karlin 1987, Theunissen *et al.* 1998). Whatever the case, these small artefacts still offer profitable avenues of study (Burton 1980, Newcomer and Karlin 1987), because all other factors being equal, increases in the proportions of small-sized artefacts imply that fresh raw material is scarce.

4. Studies of artefact edges can further identify intensity of use of raw material. For example, microscope examinations of the edges of several Devil's Lair and Tunnel Cave flakes indicate the presence of fine, incomplete fractures. These finely fractured edges would not have sustained much use or even been damaged after deposition, because otherwise the fractures would have been completed, removing a flake from the edge (J. Kamminga, pers. comm.). The more numerous the unused flakes with these "fresh" edges, the less conservation of raw material is implied.

5. Similar principles apply to use-wear and residues acquired during an artefact's use as a tool (Grace 1996, Hardy and Garufi 1998). Use-wear and residues are observable on edges of an artefact used as a tool, or on

[2] In the analysis of any archaeologically excavated sample its representativeness is in question. Possible reasons for bias in the artefact samples analysed here include: the location of the excavated material on a prehistoric surface, which may have been divided during occupation into activity areas according to the site's physical shape and other factors; the small size of the sample in relation to the unknown total population of artefacts; differential preservation of artefacts and use-wear on the artefacts according to spatially variable factors, such as soil chemistry or re-working of sediments within the cave. These factors cannot be easily controlled in a small test-excavation. Without extending the test-excavations I am forced to assume that each test-excavation is a random and roughly representative sample of the site.

debitage before it was detached from a tool. The proportion of used artefacts (or parts of used artefacts) should increase with increasing scarcity of raw material.

6. In south-western Australia, re-use of ancient artefacts is proposed at Quininup Brook (Ferguson 1980b, 1981) and at Lake Jasper (Brock 1998), near Point D'Entrecasteaux (Figure 2.1). The presence of different use-fractures on a tool or use-flakelet, or on different edges of the same artefact, indicate the number of uses. The number of tool rejuvenation flake scars on an artefact indicates the degree of tool edge rejuvenation. These measures both increase with increasing raw material scarcity.

7. The same principle applies to cores as turning the core and striking off the old platform rejuvenates a core. An old platform detached in this manner is evinced by the truncated flake scars around its perimeter, most of which initiate from its dorsal side (the former platform surface). A platform rejuvenation flake may also have a tabular appearance. Rejuvenated tools and cores, and rejuvenation flakes from tools and cores, should theoretically become more common during periods when fossiliferous chert is scarce (Rolland and Dibble 1990).

The last events that happen to artefacts tend to obscure all of the above evidence, and must be assessed carefully before moving on to cultural determinants of artefact form. After artefacts are discarded or lost, they can potentially become incorporated into hearths, burnt, and made brittle and prone to breakage; damaged by trampling of the deposit's surface by people; or weathered *in situ* (Hiscock 1985, 1990; McBrearty *et al.* 1998, Shea and Klenck 1993).

Weathering is probably unimportant at these sites, as it is not observed at Devil's Lair and Tunnel Cave (fine, unweathered chert flakes are found in the oldest occupation deposits), nor at Witchcliffe Rock Shelter and Rainbow Cave, where the archaeological deposits are young, and the quartz artefacts are harder and resist chemical decomposition (Cotterell and Kamminga 1990). However, post-depositional breakage and burning could have affected chert artefacts, at least. Post-depositional burning, which can occur when artefacts in a deposit are incorporated into hearths, or when they are discarded directly into hearths, is chemically and petrologically indistinguishable from heat-treatment (Griffiths *et al.* 1987). I attempt to distinguish between such "pre-depositional" and post-depositional burning by identifying chert flakes whose bulbar or ventral faces are coloured white or grey (the discolouration of chert indicating heating). Flakes with burnt bulbar faces were probably heated after they were produced (J. Kamminga, pers. comm.).

Artefacts whose edges are chipped by trampling and re-working of sediments can look like well-used artefacts (McBrearty *et al.* 1998, Shea and Klenck 1992). This damage appears as snaps of very fine edges, and conchoidal fractures along thicker ones (Cotterell and Kamminga 1990, Lawrence 1979). Conchoidal fractures may reach lengths up to 3 mm (Shea and Klenck 1992). In extreme cases, conchoidal flake scars from trampling damage reach lengths of 5 to 10 mm, and are a form of pseudo-retouch flake scars (McBrearty *et al.* 1998: Figures 4-10). However, if artefact edges with such large flake scars also lack small edge-fractures, then the scars probably derive from fine retouch or use, since most damaged artefacts also acquire many smaller scars, less than 3 mm long. Moreover, post-depositional damage does not cause bending fractures on moderate or large edge angles, or on thick edges. Bending fractures are initiated only by pressure from a soft indenter, e.g, in pressure retouch with a soft flaking tool, or use of a tool on soft material.

To summarise all of the above, raw material scarcity (caused by loss of access to stone sources or intense occupation of cave and rock shelter sites) would have several responses. These would be manifested in collections of stone artefacts as follows:

1. a high frequency of platform preparation, core rotation, heat treatment, and bipolar technique,

2. large proportions of formal tools, used artefacts, and flakes relative to other debitage,

3. a high frequency of retouch, use, or rejuvenation of tools, as may be evident on tool edges or debitage deriving from tool edges. Hence one would expect to find more use-wear, retouch, or tool rejuvenation flakes, as tools are used more intensively, retouched more often, or rejuvenated by re-flaking,

4. abundant evidence for continued knapping or use of poor quality or exhausted tools or cores, as provided by

a) the prevalence of step and hinge fractures;

b) changing size of debitage, as shown by measurement of pieces' dimensions and weights;

c) changing production indices of debitage, as shown by ratio of flake area to platform area, flake length to width, and flake length to thickness.

In the following sections I discuss my methods and present results in terms of the *chaîne opératoire*: raw material preferences, tool manufacture and debitage production, and tool use, re-use, and rejuvenation.

Methods

Materials and equipment

I examined every stone artefact from Tunnel Cave and Devil's Lair under a binocular microscope at magnifications ranging from x 6 to x 40. I measured each one to the nearest 0.01 mm using Mitsumi scientific callipers and was able to weigh artefacts to the nearest 0.01 g on Sartorius Laboratory electronic scales. I estimated the weight of artefacts weighing less than 0.01 g as 0.005 g. I noted attributes on recording sheets and a desktop computer, recording the data on a spreadsheet program (Microsoft Excel, version 5.0).

Types of evidence

Edge-fractures and other characteristics of Tunnel Cave and Devil's Lair artefacts are identifiable at ×6 to ×40 magnification. There is little evidence for polish or residues, and the low-magnification is sufficient to identify initiations and terminations on the smallest of artefacts and their edge-fractures.

The evidence that I examined does not fall naturally into exclusive categories. Kamminga (1982) notes that observations of use-fractures form a continuum between extremes. One use-wear flake produced during use of a tool will not show all of the evidence of use that might be seen on an identical flake produced from the same tool used in the same way. Conversely, the same types of evidence may be seen on more than one type of artefact and even in more than one type of use-wear. Thus only bending initiated fractures can be considered as undeniably use-fractures, since conchoidal fractures can be initiated by artefact damage, by retouch, or by use.

The types of evidence expected for various artefact types and knapping techniques are set out in Appendix 6. Use-fracture interpretations derive from observations of use-fractures on chert and quartz artefacts used experimentally on hardwoods, softwoods, bone, and skin (Ho Ho Committee 1979, Kamminga 1982, Lawrence 1979). I consider that these observations apply to fossiliferous chert and quartz, artefacts, that is 99% of the analysed artefacts from Tunnel Cave, Witchcliffe Rock Shelter, and Rainbow Cave; and 95% of the analysed artefacts from Devil's Lair. The terminology for use-fractures follows that of the Ho Ho Committee (1979: Figure 1). Other terms derive from general references (Holdaway 1995, Inizan *et al.* 1992, McCarthy 1967, Prentiss 1998).

Stone artefact recording procedure

For every artefact, I recorded the following observations:

- Provenance (excavation square, spit, layer, and if applicable, feature and "small find" number: see Chapter 5).
- Raw material.
- If a tool (retouched piece): what type of retouch - simple, notched, or denticulate (Appendix 6).
- If debitage: whether a single platform core; multi-platform core; bipolar core; flake; bipolar flake; broken flake (proximal end); broken flake (distal end); fragment; bipolar flake; bipolar fragment; pebble, granule, or other unmodified manuport (Appendix 6).
- Dimensions: oriented length, width, thickness; platform width and thickness, to the nearest tenth of a millimetre.
- Weight: to the nearest hundredth of a gram.
- Initiation: conchoidal; bending; bipolar (or wedging type).
- Platform type: flat (in the computer record, the "default" entry), gull-wing, facetted, focal.
- Evidence of core rotation: number of times rotated.
- Termination: feather; step; hinge; retroflex hinge.
- Location of use-fractures and residues: platform; lateral margins; distal end.
- Scars and other features on dorsal face.
- Scars and other features on ventral face.
- If burnt or heat-crazed, and whether on bulbar (ventral) or dorsal surfaces.
- If potlidded, and whether on bulbar (ventral) or dorsal surfaces.
- Percentage of artefact with cortex remaining, if any.
- Any sediment or other material adhering to artefact.

Devil's Lair

Most artefacts from the main excavation are assignable to one of the fourteen radiocarbon-dated Periods (Chapter 6). Artefacts from mixed units that cross periods are excluded from the analysis. Also excluded is Period XIV, representing layer A. Layer A may include material from earlier periods, since pieces of what is believed to be flowstone D were found in layer A (Baynes *et al.* 1975). No artefacts were found in undisturbed parts of layer A (C.E. Dortch 1979a).

Post-depositional alteration

This analysis shows whether events such as trampling or hearth-building on the floor of the site could have broken artefacts or burnt artefacts already deposited. Breakage and burning would increase the apparent numbers of artefacts, especially broken flakes and fragments (Hiscock 1985). Either process would affect the analysis of responses to raw material scarcity. Table 7.1 lists all artefact characteristics that might derive from post-depositional events.

All the burnt artefacts are chert, presumably because quartz only reacts to heat at higher temperatures (*c.*800° C) than campfires produce (*c.*3-400° C; Courty *et al.* 1989). Most burnt artefacts were burnt on their bulbar surfaces, suggesting that they were burnt after manufacture, and therefore, probably after deposition. Since burning already-deposited chert artefacts in campfires can produce fragments that are indistinguishable from knapping fragments (Hiscock 1985), and burning is common in Devil's Lair artefacts, it is important to find out whether there are changes in the proportion of artefacts getting burnt, and if so, whether the changes matter.

Table 7.1 indicates large fluctuations in total sample size from Period to Period, but the Kolmogorov-Smirnov (K-S) tests used below are sufficiently robust to allow for such variations in the counts of each artefact type (Appendix 7; cf. Shennan 1988: Table 5.2). K-S tests on the cumulative proportions drawn from the columns in Table 7.1 show that artefacts with burnt bulbar surfaces (burnt after deposition) comprise a disproportionately small proportion of the artefacts in Period VI and earlier Periods (Table 7.2).

Table 7.1 Post-depositional alteration of Devil's Lair artefacts

Double horizontal line indicates presence of flowstone, lithified band, or other major stratigraphic boundary.

Estimated ^{14}C age, years BP Min.	Max.	Period	Burnt on bulbar surfaces	Burnt on other surfaces	Broken flakes, proximal	Broken flakes, distal	Snapped artefacts	Water-rolled artefacts	Edge-damaged artefacts	Post-excavation damage	Total altered	Total artefacts
150	490	XIV	4	1							5	7
?6,200	13,500	XIII										9
?6,200	13,500	XII		3							3	
12,900	13,200	XI	12			1					13	27
12,600	13,600	X	16	3		4					23	68
13,100	14,900	IX	12	1	1	1	1	2	1	1	20	45
15,800	18,400	VIII	18	9	1	5	1		1	1	36	56
18,400	22,800	VII	140	16	7	34	9	6	7	7	226	434
18,400	26,000	VI	49		7	23			2	2	83	316
22,000	26,000	V	3			2					5	11
24,950	26,050	IV	7			5	1	1	1	1	16	47
25,200	27,200	III	1	7							8	4
37,400	44,800	II	5		1	18	1	2	1	1	29	145
43,000	51,600	I										13
		Total	267	40	17	93	13	11	13	13	467	1182

Table 7.2 K-S test results on comparisons of cumulative proportions of artefact characteristics indicating post-depositional alteration at Devil's Lair

The Null Hypothesis (that proportions of artefacts with different characteristics do not vary) is rejected if P (H0) < 0.05.

Where the Null Hypothesis should be rejected, the Period containing the greatest difference between proportions is given.

Reject H0?	Burnt on bulbar surfaces	Burnt on other surfaces	Broken flakes, proximal	Broken flakes, distal	Snapped artefacts	Water-rolled artefacts	Edge-damaged artefacts	Post-excavation damage
versus all other artefacts	Yes - Period VI	No	No	No	No	No	No	No
versus all other flakes	not tested	not tested	No	No	not tested	not tested	not tested	not tested
chert alone - versus all other chert artefacts	No	Yes - Period VI	No	No	No	No	No	No

In other words, burning of artefacts already in the deposit was apparently less intense up to and including Period VI. There may have been more frequent hearth-building in Period VII onwards, although few hearths are recorded anywhere in the excavation (the hearths that burnt the artefacts may be present in adjacent sections of unexcavated deposit). If other analyses show a large proportion of fragments from Period VII onwards, it might not be important for inferring a response to raw material availability.

Other causes of artefact breakage include people walking on artefacts, water-flow that re-works deposits and artefacts in them, and breakage of thin edges or thin artefacts during knapping. K-S tests suggest that the respective proportions of broken, snapped, edge-damaged, and water-rolled pieces in the entire assemblage do not vary throughout the deposit (*p* [H0, that there is no variation] > 0.05). Various causes may have broken artefacts after their deposition, and burning was apparently more intense after Period VI, but the overall proportion of broken artefacts does not vary. With the possible exception of fragments, changes in artefact numbers probably do not relate to changes in post-depositional processes.

Variations in post-excavation damage (identified from fresh or recent-looking edge-fracture, or crumbling of burnt and friable pieces) are not significant (*p* [H0] > 0.05).

Raw material

At Devil's Lair, the changes in the relative numbers of artefacts made of chert, quartz, and other rock can be shown in more detail than previously (C.E. Dortch 1984: Table 3) because in Chapter 6 I have made finer divisions of the deposit than previously attempted. Table 7.3 shows these revised groupings.

Thirty small granules of quartz, and all the "other rock" listed in Table 7.3 (fragments of ochre, granite-gneiss, and feldspar), are not flaked, although they were almost certainly introduced into the deposit by people. These items are not analysed in the following arguments for identifying occupation intensity from flaked stone.

K-S tests of the respective proportions of chert, quartz, and calcrete artefacts indicate that the period where each raw material is most disproportionate to its overall representation is Period VI (Table 7.4). In this Period fewer artefacts are made of chert and calcrete than would be expected from the overall representation of each raw material.

Table 7.3 Materials used for stone artefacts from main excavation at Devil's Lair
Double horizontal line indicates presence of flowstone, lithified band, or other major stratigraphic boundary.

Layers	Estimated ^{14}C age, years BP Minimum	Maximum	Period	Number of stone artefacts: Foss. chert	Quartz	Calcrete	Other rock	Total
A and sub-units of A	150	490	XIV	6	1			7
D and units below D	?6,200	13,500	XIII	2	7			9
F and units below F	?6,200	13,500	XII					
I to K (incl. Hearth x)	12,900	13,200	XI	18	7	2		27
L to sub-MM	12,600	13,600	X	37	26	4	1	68
N and O	13,100	14,900	IX	26	18	1		45
P and Q	15,800	18,400	VIII	38	16	1	1	56
Occupation Floor 2 to 4	18,400	22,800	VII	291	122	18	3	434
5 to 7	18,400	26,000	VI	66	238	7	5	316
8 to 9,-lower	22,000	26,000	V	6	4	1		11
9 lithified bands and 10	24,950	26,050	IV	18	19	8	2	47
11 to 15	25,200	27,200	III		4			4
16 to 30,-lower	37,400	44,800	II	2	141	1	1	145
31 to 38	43,000	51,600	I		9	4		13
			Total	510	612	47	13	1182

Table 7.4 K-S test results on comparisons of cumulative proportions of Devil's Lair artefact raw materials
The Null Hypothesis (that proportions of artefacts with different characteristics do not vary) is rejected if *p* (H0) < 0.05. Where the Null Hypothesis should be rejected, the Period containing the greatest difference between proportions is given.

Reject H0?	Foss. Chert	Quartz	Calcrete
Foss. chert	---------------------------	Reject - Period VI	Reject - Period VI
Quartz		---------------------------	Reject - Period VI
Calcrete			---------------------------

The low representation of calcrete in Period VI is perhaps due partly to the minor importance of this raw material, and partly to the difficulty of identifying conchoidal fracture on its friable and often weathered surfaces (C.E. Dortch 1979a). Calcrete is a secondary deposit commonly found on calcarenite surfaces in the Leeuwin-Naturaliste Region (Lowry 1967). It would be easily quarried in or near any limestone cave site, but microscope examination suggests that calcrete artefacts would make poor tools since they have coarse and friable edges, unlikely to last long. People perhaps used calcrete mainly because of the ease of obtaining it. The distribution of calcrete artefacts throughout the Devil's Lair deposit indicates that small quantities of calcrete were flaked at all times. Possibly, the large quantity of quartz brought into the site during Period VI made calcrete even less attractive as a raw material at that time.

The amount of chert is disproportionately low from Period I to Period VI inclusive, and high afterwards. This finding suggests that chert availability was improved at times of low sea-level and maximum exposure of chert outcrops, as posited by Glover (1984). Sea-level fall associated with the last glacial maximum began by 25,000 BP, and terminated by 18,000 BP (Chappell and Shackleton 1986). Large quantities of chert appeared in the Devil's Lair record after 22,800 BP (the maximum age inferred for Period VII). Interpretations of occupation intensity that follow here will allow for the possibility of changing availability of chert raw material. After *c.*23,000 BP, people at Devil's Lair may have used chert more freely.

Technology

Analysing tool and debitage categories within raw material classes suggests that changes in the overall proportions of tools and other categories relate to changes in raw material (Tables 7.5-7.7). When chert is abundant, so are tools. This is potentially a misleading situation. Many quartz flakes were produced at times of chert scarcity, so quartz tools must have been manufactured. However, I failed to identify any, perhaps because the material is difficult to analyse. The interlocking crystal faces in polycrystalline (vein) quartz, comprising most quartz artefacts, make it hard to identify tool characteristics such as retouch.

For example, C.E. Dortch (1984: Table 3) indicates that the Devil's Lair excavations yielded 22 quartz tools. I examined nine of these tools (all those from the main excavation), and identified six as debitage. One notched flake (WAM registration number B3768) may have been flaked accidentally in the deposit; artefact B1544 has a rough surface that makes it difficult to identify putative retouch scars; two quartz pieces (B5086, B3668) are probably a bipolar core and a bipolar flake; and two artefacts (B3668, and an unregistered artefact from trench 7c, layer 26) are unmodified flakes[3]. Quartz seems to be either an unsatisfactory material for making tools, or more likely, producing recognisable tools. However, most of C.E. Dortch's (1984) quartz tools, and most quartz artefacts, pre-date Period VI, when access to chert was probably poor.

There appears to have been a transition from Period VII (18,400-22,800 BP), from manufacturing many quartz flakes (probably for use as tools) to manufacturing many chert flakes. This transition was possibly due to a realisation, after *c.*23,000 BP, that chert was superior to quartz for tool manufacture and use. It was by then available in larger quantities, encouraging less conservative use of it. Table 7.8 shows that in Period VIII (and in no other period) the proportion of chert tools to chert debitage is lower than would be expected from their overall representations. Period VIII is dated 18,400-15,800 BP, that is, the period of lowest sea-level. Possibly, the improved supply of fossiliferous chert required less careful use of it for tools.

In no other raw material are variations in artefact types significant, nor are variations in other chert artefact types significant (p [H0] > 0.05). This finding suggests that apart from the number of chert tools in Period VIII, people produced more or less the same numbers of artefact types within each raw material class throughout time.

Chert artefacts also indicate changes in knapping technique, again suggesting changes in raw material treatment (Tables 7.9, 7.10).

K-S tests indicate that there are significantly fewer chert bipolar pieces and step or hinge terminated flakes in Periods VII and VIII (Table 7.10). These changes at a time of low sea-level (*c.*20,000-15,000 BP) indicate less sustained reduction, and hence less intensive knapping, of cores, at the probable time of greatest chert availability (cf. Hiscock 1993, 1996a). Among quartz artefacts, minor changes in these knapping techniques (Table 7.11) are not significant (p [H0] > 0.05).

[3] The remaining three artefacts (B1853, B3796, and an unregistered piece from trench 7d, layer 10d) and two others not discussed by Dortch (1984) were probably used on the core, so are not technically tools. They are discussed in the section on use and discard.

Table 7.5 Types of chert artefacts at Devil's Lair

Notched flakes include a notched denticulate flake in layer 21 (Period II). Double horizontal line indicates presence of flowstone, lithified band, or other major stratigraphic boundary.

Period	Scrapers	Notched flakes	Cores	Bipolar cores	Flakes	Broken flakes, longit.	Broken flakes, prox.	Broken flakes, distal	Fragments	Bipolar flakes & fragments	Potlids	Total chert
XIII	1	1										2
XII												
XI	1			1	7	1		1	5	1	1	18
X	3			3	22	1		2	4	2		37
IX	2	1			15	3	1		4			26
VIII					24	3		4	7			38
VII	3	1	2	1	192	6	5	25	46	9	1	291
VI	1	2		1	31	5	2	6	17		1	66
V					4			2				6
IV					7			2	8	1		18
III												
II		1						1				2
I												
Total	11	6	2	6	302	19	8	43	91	13	3	504

Table 7.6 Types of quartz artefacts at Devil's Lair

Double horizontal line indicates presence of flowstone, lithified band, or other major stratigraphic boundary. K-S tests show that no quartz artefact types are found in any Period out of proportion to their overall representation in the entire deposit (p [H0] > 0.05).

Period	Scrapers	Notched flakes	Cores	Bipolar cores	Flakes	Broken flakes, longit.	Broken flakes, prox.	Broken flakes, distal	Fragments	Bipolar flakes & fragments	Potlids	Total quartz
XIII			1	1	3				1	1		7
XII												
XI				1	2				1	3		7
X				2	7	1		2	6	6		24
IX				1	5	1		1	9	1		18
VIII					8		1	1	3	2		15
VII			1	4	38	2	2	9	48	7		111
VI			1	6	96	1	5	17	90	15		231
V					2				2			4
IV					6			2	11			19
III					2							2
II			1	7	47	3	1	17	44	17		137
I									2			2
Total			4	22	216	8	9	49	217	52		577

Table 7.7 Types of calcrete artefacts at Devil's Lair

Double horizontal line indicates presence of flowstone, lithified band, or other major stratigraphic boundary. K-S tests show that no calcrete artefact types are found in any Period out of proportion to their overall representation in the entire deposit (p [H0] > 0.05).

Phase	Scrapers	Notched flakes	Cores	Bipolar cores	Flakes	Broken flakes, longit.	Broken flakes, prox.	Broken flakes, distal	Fragments	Bipolar flakes & fragments	Potlids	Total
XIII												
XII												
XI					2							2
X				1	1				2			4
IX										1		1
VIII									1			1
VII			2		7	1			1			11
VI		1	2		8				2	1		14
V					1							1
IV			1		5			1		1		8
III												
II					1							1
I					3				1			4
Total		1	5	1	28	1		1	7	3		47

Table 7.8 K-S test results on comparisons of cumulative proportions of chert artefact types at Devil's Lair

The Null Hypothesis (that proportions of artefacts of different types do not vary) is rejected if *p* (H0) < 0.05. Where the Null Hypothesis should be rejected, the Periods containing the significantly disproportionate frequencies (and the direction of the difference) are given. Not applicable (na) - quantities are too small for comparison.

Reject Ho?	Tools (scrapers & notched flakes)	Cores	Bipolar cores	Flakes (inc. broken flakes)	Fragments	Bipolar flakes & fragments	Heat fracture (potlids)
versus other chert artefacts	**Yes - Period VIII (low)**	na	na	No	No	No	na

Table 7.9 Instances of knapping techniques in chert artefacts at Devil's Lair

Period	Overhang removal	Facetted platform	Flaked platform	Focal platform	Precision blow	Core rotation	Bipolar technique	Retouch	Step or hinge termination	Total chert
XIII								3		2
XII										
XI				2	5	1	4	2	3	18
X	8			3	2		2	1	18	37
IX	3			3	1	4	3	4	14	26
VIII	10				2	3			16	38
VII	36	2	2	23	9	9	5	6	87	291
VI	3			6		4		5	9	66
V		1							1	6
IV	2				3				4	18
III										
II	1							2	1	2
I										
Total	63	3	2	37	22	21	14	23	153	504

Table 7.10 K-S test results on comparisons of cumulative proportions of chert artefact types

The Null Hypothesis (that proportions of artefacts of different types do not vary) is rejected if *p* (H0) < 0.05. Where the Null Hypothesis should be rejected, the Period(s) containing the significantly disproportionate frequencies (and the direction of the difference) are given. Not applicable (na) - quantities too small for comparison.

Reject H0?	Overhang removal	Facetted platform	Flaked platform	Focal platform	Precision blow	Core rotation	Bipolar technique	Retouch	Step or hinge termination
versus other chert artefacts	No	na	na	No	No	No	**Yes - Periods VII & VIII (low)**	No	Yes - Period VII (low)

Table 7.11 Instances of knapping techniques in quartz artefacts at Devil's Lair

Precision blow is either behind bulb or crest, or between crests. Step or hinge termination indicates possible core exhaustion.

K-S tests show that no knapping techniques in quartz artefacts are represented in any Period out of proportion to their overall representation in the entire deposit (*p* [H0] > 0.05).

Period	Overhang removal	Facetted platform	Flaked platform	Focal platform	Elongate flake	Precision blow	Core rotation	Bipolar technique	Retouch	Step or hinge termination	Total quartz
XIII	1			1	2		1	2			7
XII											
XI								3			7
X				1	2	1		3			24
IX	1							1			18
VIII	1			1	1	1				1	15
VII	5			4	2	2		7		4	111
VI	9			12	11	4	2	12		7	231
V					1						4
IV					1						19
III	1				1						2
II	6		1	2	8	3		5		6	137
I											2
Total	24		1	21	29	11	3	33	0	18	577

Table 7.12 Results of ANOVA and t-tests of Devil's Lair chert artefact dimensions

ANOVA and t-tests performed on logged values to obtain approximately normal frequency distributions. Periods XIII and II are excluded from calculations because such small samples give misleading results. "Not applicable" (na) indicates that there are no artefacts in the category.

CHERT	*n*		Maximum length			Weight			Flake length		
Period	All	Flakes	Total score	Sum of squares	Mean	Total score	Sum of squares	Mean	Total score	Sum of squares	Mean
XIII	2	0	7.48	30.10	3.74	25.88	471.01	12.94	na	na	na
XII	0	0	na	Na	na	na	na	na	na	na	na
XI	18	7	34.25	85.22	1.90	32.07	330.94	1.78	13.97	34.93	2.00
X	37	22	75.70	186.56	2.05	88.58	603.07	2.39	42.76	97.51	1.86
IX	26	15	57.87	164.81	2.23	93.94	1227.02	3.61	32.77	90.73	2.18
VIII	38	24	50.65	83.77	1.33	25.29	62.73	0.67	30.26	47.49	1.26
VII	291	192	345.63	550.95	1.19	201.84	1341.47	0.69	183.47	240.76	0.96
VI	66	33	71.94	98.33	1.09	21.86	29.87	0.33	34.05	48.29	1.03
V	6	4	8.64	17.82	1.44	4.88	10.35	0.81	6.39	15.27	1.60
IV	18	7	19.00	23.13	1.06	2.62	0.68	0.15	5.82	5.17	0.83
III	0	0	na	na	na	na	na	na	na	na	na
II	2	0	5.33	21.85	2.67	11.65	134.33	5.83	na	na	na
I	0	0	na	na	na	na	na	na	na	na	na
Total	504	302	676	1263	19	509	4211	29	350	581	12
Variance ratio				9.54			7.28			6.20	
P (H0) - F -test				1.86×10^{-13}			1.49×10^{-11}			3.55×10^{-9}	
Reject H0?				Yes			Yes			Yes	
For Periods (t-tests)				Period XI versus Periods VIII, VII, VI, and IV			Period XI versus Periods VIII, VII, VI, and IV			Period XI versus Periods VII, VI, and IV	
				Period X versus Periods VIII, VII, VI, and IV			Period X versus Periods VIII, VII, VI, V, and IV			Period X versus Periods VIII, VII, VI, and IV	
				Period IX versus Periods VIII, VII, VI, V, and IV			Period IX versus Periods VIII, VII, VI, and IV			Period IX versus Periods VIII, VII, VI, and IV	
										Period VIII vs Period VII	

CHERT	*n*		Flake width			Flake thickness			Platform width		
Period	All	Flakes	Total score	Sum of squares	Mean	Total score	Sum of squares	Mean	Total score	Sum of squares	Mean
XIII	2	0	na	na	na	na	na	na	na	na	na
XII	0	0	na	na	na	na	na	na	na	na	na
XI	18	7	14.44	41.57	2.06	14.44	41.57	2.06	1.46	0.51	0.21
X	37	22	41.93	97.71	1.82	41.93	97.71	1.82	22.75	33.51	0.99
IX	26	15	30.14	80.88	2.01	30.14	80.88	2.01	14.12	23.87	0.94
VIII	38	24	32.58	54.60	1.36	32.58	54.60	1.36	19.43	20.35	0.81
VII	291	192	195.43	272.09	1.02	55.35	25.57	0.29	114.02	92.90	0.60
VI	66	33	29.90	33.92	0.91	10.25	4.40	0.31	17.17	13.36	0.52
V	6	4	6.09	11.17	1.52	6.09	11.17	1.52	3.62	4.71	0.91
IV	18	7	5.79	5.38	0.83	5.79	5.38	0.83	3.43	1.90	0.49
III	0	0	na	na	na	na	na	na	na	na	na
II	2	0	na	na	na	na	na	na	na	na	na
I	0	0	na	na	na	na	na	na	na	na	na
Total	504	302	357	598	12	357	598	12	71	60	2
Variance ratio				5.06			2.85			2.24	
P (H0) - F -test				3.02×10^{-7}			1.47×10^{-3}			0.01	
Reject H0?				Yes			Yes			Yes	
For Periods (t-tests)				Period XI versus Periods VII, VI, and IV			Period X versus Periods VII and VI			Period XI vs Period VI	
				Period X versus Periods VII, VI, and IV			Period IX versus Periods VIII, VII, and VI			Period X versus Periods VII and VI	
				Period IX versus Periods VII, VI, and IV						Period IX vs Period VI	
				Period VIII versus Periods VII and VI						Period VIII versus Periods VII and VI	

Table 7.13 Results of ANOVA and t-tests of Devil's Lair quartz artefact dimensions

ANOVA and t-tests performed on logged values to obtain approximately normal frequency distributions. Periods XIII, V, III, and I are excluded from calculations because such small samples give misleading results. "Not applicable" (na) indicates that there are no artefacts in the category.

QUARTZ	*n*		Maximum length			Weight			Flake length		
Period	All	Flakes	Total score	Sum of squares	Mean	Total score	Sum of squares	Mean	Total score	Sum of squares	Mean
XIII	7	3	16.75	51.27	2.39	79.40	5465.90	11.34	4.09	6.09	1.36
XII	0	0	na	na	na	na	na	na	na	na	na
XI	7	2	13.25	26.47	1.89	13.38	34.14	1.91	2.87	4.34	1.44
X	24	7	39.22	79.92	1.57	34.70	142.64	1.39	8.60	15.37	1.23
IX	18	5	18.92	27.16	1.05	8.88	18.08	0.49	4.68	6.09	0.94
VIII	15	8	13.04	16.40	0.87	3.97	7.36	0.26	7.78	11.74	0.97
VII	111	38	112.24	219.34	1.00	268.69	31069.0	2.40	24.96	19.81	0.66
VI	231	95	207.20	264.10	0.90	76.26	216.20	0.33	71.06	70.80	0.75
V	4	2	2.48	1.61	0.62	0.12	0.00	0.03	1.32	0.92	0.66
IV	19	6	13.63	10.94	0.72	2.18	1.28	0.12	3.67	3.34	0.73
III	2	2	1.95	2.01	0.98	0.43	0.15	na	1.53	1.17	0.77
II	137	47	165.34	317.34	1.21	253.60	6620.11	1.89	41.55	64.48	0.94
I	2	0	0.87	0.38	0.44	0.06	0.002	0.03	na	na	na
Total	577	216	651	3226	14	742	43575	21	174	205	11
Variance ratio			6.43			5.26			1.77		
P (H0) - F -test			5.16×10^{-8}			6.43×10^{-8}			0.06		
Reject H0?			Yes			Yes			No		
For Periods (t-tests)			Period XI versus Periods IX, VIII, VII, VI, V, IV, II Period X versus Periods IX, VIII, VII, VI, V, IV, II Period II versus Periods VII, VI, and IV			Period XI versus Periods IX, VIII, VII, VI, V, IV, II Period X versus Periods IX, VIII, VII, VI, V, IV, II Period II versus Periods VIII, VII, VI, V, and IV			not listed due to non-significant result		

QUARTZ	*n*		Flake width			Flake thickness			Platform width		
Period	All	Flakes	Total score	Sum of squares	Mean	Total score	Sum of squares	Mean	Total score	Sum of squares	Mean
XIII	7	3	3.65	4.49	1.22	4.09	6.09	1.36	0.72	0.22	0.24
XII	0	0	na	na	na	na	na	na	na	na	na
XI	7	2	2.24	2.51	1.12	2.87	4.34	1.44	0.37	0.09	0.19
X	24	7	6.58	6.44	0.94	8.60	15.37	1.23	1.24	0.40	0.18
IX	18	5	4.47	4.24	0.89	4.68	6.09	0.94	1.15	0.86	0.23
VIII	15	8	7.66	7.65	0.96	7.78	11.74	0.97	0.54	0.06	0.07
VII	111	38	24.09	19.75	0.63	7.76	2.16	0.20	13.82	6.75	0.36
VI	231	95	65.32	59.11	0.69	23.57	9.79	0.25	39.85	23.28	0.42
V	4	2	0.95	0.52	0.48	1.32	0.92	0.66	0.06	0.00	0.03
IV	19	6	4.19	3.63	0.84	3.67	3.34	0.73	0.73	0.34	0.15
III	2	2	1.61	1.45	0.81	1.53	1.17	0.77	0.67	0.42	0.33
II	137	47	35.42	31.32	0.81	41.55	64.48	0.94	6.97	3.55	0.16
I	2	0	na	na	na	na	na	na	na	na	na
Total	577	216	173	155	10	174	205	11	25	11	2
Variance ratio			1.38			1.42			0.73		
P (H0) - F -test			0.19			0.17			0.71		
Reject H0?			No			No			No		
For Periods (t-tests)			not listed due to non-significant result			not listed due to non-significant result			not listed due to non-significant result		

Analysis of variance (ANOVA) supports the proposition that the major changes in Devil's Lair artefacts occurred as a result of restricted raw material supply, not because of variations in occupation intensity (Tables 7.12, 7.13, see Appendix 7 for method, note that measurements included in ANOVA are normally distributed for each Period). Chert debitage size was proposed to increase at the LGM time of low sea-level and greater abundance of chert. Table 7.12 shows that the variance in chert artefacts' maximum length and weight between periods is several times greater than the variance in chert artefact size within them. This *variance ratio* (Lumsden 1971) is high also for chert flake length, width, thickness, and platform width. To summarise all the results in Table 7.12, chert flakes and other artefacts appear to remain at roughly the same small average size until the end of Period VI. From Period VIII onwards they attain a larger size. In Period VII they have a size intermediate between earlier and later Periods.

Period VIII is approximately the time of lowest sea-level, and during most of the subsequent periods, sea-level remained low. The size differences seem trivial from the averages given in Table 7.12 but they are significant given the small size of artefacts. Also, growth in chert artefact size begins immediately after Period VII, when chert seems to have become most abundant generally.

One response to raw material scarcity would be to make thinner or longer flakes, but there are no apparent variations in chert flake shape. ANOVA shows that ratios of flake length to width (elongation index), flake width to platform width (parallel index), and the product of platform width and platform thickness (platform area) do not change significantly (p [H0] > 0.05; these indices are not tabulated here). There was apparently no change in the shape of flakes produced by chert knappers at Devil's Lair, and no apparent attempt to make thinner flakes in order to conserve raw material.

Table 7.13, based on ANOVA of quartz artefacts, shows a different trend. Variance in quartz flake length, width, thickness, and platform width between periods is not as important as variance within periods. Platform and elongation indices, and platform area, do not change either. Unlike the chert sources, quartz sources (pebbles in river beds and veins on local granite-gneiss outcrops) were probably readily accessible throughout the period of human occupation at Devil's Lair. There was perhaps no need to conserve quartz and the lack of variation in quartz flakes seems to reflect this.

As mentioned above, quartz artefacts, including possible quartz tools, are only numerous before Period VII. Quartz artefacts are on average larger in Periods XI, X, and II, on either side of the LGM (Table 7.13). Possibly, quartz was an alternative raw material carried in to the site in larger quantities and as larger pieces at times when chert supply was restricted.

Given that chert artefact characteristics vary little between Period VII and later Periods, it is conceivable that the large numbers of chert artefacts in Period VII indicate frequent or prolonged occupations in the site at this time (22,800-18,400 BP). Very long occupations would have involved extended reduction of the same cores and tools, and chert artefacts are not greatly reduced (flaked intensively) at this time. Possibly, human visits to Devil's Lair at this time were more frequent.

Use and discard

Tables 7.14 and 7.16 show the incidence of use-wear among chert and quartz artefacts. Most instances of use-wear were use-fractures. Thus calcrete artefacts are not analysed, because the coarseness of the material prohibits identifying use-fractures. Other types of use-wear (edge-rounding, blunting, residue) are represented by only one artefact each and are not shown (but see Appendix 9). The large number of clearly used chert artefacts, relative to quartz ones, is probably due partly to the difficulty of identifying fractures on the rough surfaces of polycrystalline quartz, and perhaps, partly to a greater attraction of chert artefacts for use as tools.

The rate of use of chert artefacts may relate to availability of chert. Chert artefacts seem to have been less often used in Period VIII, the time of lowest sea-level. Table 7.15 shows that in this Period used margins and step or hinge-terminated use-fractures are disproportionately less common in chert artefacts. Use-wear does not vary in quartz artefacts (p [H0] > 0.05). Chert and quartz artefacts seem to have been used for similar tasks, including scraping and cutting of soft and hard materials. In sum, as sea level fell, the total use of quartz declined (quartz artefacts became less common after Period VI and there are no significant changes in the proportion of used quartz artefacts), and by the time sea-level reached its lowest point (Period VIII), chert was used least intensively

As mentioned above, during Period VIII and afterwards, chert was abundant, and fewer chert tools were produced relative to the total amount of chert. The use-wear trends are consistent with the hypothesis that chert was favoured as raw material and it was used less efficiently when it was most abundant.

As for discard behaviour, there is no evidence that people recycled or prolonged their use of chert or quartz artefacts as a result of any restriction in raw material supply. No chert or quartz artefacts show signs of re-use or re-flaking.

Table 7.14 Instances of use-wear in Devil's Lair chert artefacts

Except for one conchoidally initiated flakelet in Period VII, flakes or flakelets produced from use of core have bending initiations

Period	artefact from tool working edge	use-wear on dorsal or bulbar face	use-wear on dorsal edge of platform	use-wear on other margins	whole piece produced from use	use-fracture: conchoidal initiation	use-fracture: bending initiation	use-fracture: feather termination	use-fracture: hinge or step termination	Fresh (un-used) edges	Total chert
XIII											2
XII											
XI				2		1	1	1			18
X	2	3		6		2	6	3	5		37
IX		4		7		3	6	6	5	2	26
VIII				1		2	2		1		38
VII	10	12		15	8	15	15	16	10	4	291
VI	3	1	2	2	1	3	3	4	1		66
V				1			1	1			6
IV										1	18
III											
II				1			1	1			2
I											
Total	15	20	2	35	9	26	35	32	22	7	504

Table 7.15 K-S test results on comparisons of cumulative proportions of instances of use-wear in Devil's Lair chert artefacts

The Null Hypothesis (that proportions of instances of use-wear do not vary) is rejected if *p* (H0) < 0.05. Where the Null Hypothesis should be rejected, the Period(s) containing the significantly disproportionate frequencies (and the direction of the difference) are given.

Reject H0?	artefact from tool working edge	use-wear on dorsal or bulbar face	use-wear on dorsal edge of platform	use-wear on other margins	Whole piece produced from use	use-fracture: conchoidal initiation	use-fracture: bending initiation	use-fracture: feather termination	use-fracture: hinge or step termination	Fresh edges
versus other chert artefacts	No	No	No	Yes- VIII (low)	No	No	No	No	Yes - VIII (low)	No

Table 7.16 Instances of use-wear in Devil's Lair quartz artefacts

Except for one conchoidally initiated flakelet in Period VI, flakes or flakelets produced from use of core have bending initiations.
K-S tests show that no types of use-wear in quartz artefacts are found in any Period out of proportion to their overall representation in the deposit (*p* [H0] > 0.05).

Period	artefact from tool working edge	use-wear on dorsal or bulbar face	use-wear on dorsal edge of platform	use-wear on other margins	whole piece produced from use	Fracture: conchoidal initiation	Fracture: bending initiation	Fracture: feather termination	Fracture: hinge or step termination	Total quartz
XIII										7
XII										0
XI										7
X										24
IX				1			1	1	1	18
VIII										15
VII		2		2		1	1	1	2	111
VI	2	3	2	1	3	4	3	4	3	231
V	1		1				1	1		4
IV	1			1		1		1	1	19
III										2
II	1	1	1		1	1	1	1		137
I										2
Total	5	6	4	4	4	7	6	8	6	577

Summary

Earlier than *c.*23,000 BP, chert was little used at Devil's Lair, but from then on, lowering sea-level probably exposed more outcrops, and the proportion of chert artefacts increased. There is no evidence for chert rationing before or after this date. From 23,000 BP, however, chert seems to have been used more freely: for a given quantity of chert, fewer tools were produced. At *c.*12,000 BP, rapidly rising sea-level probably began to submerge chert outcrops, but human occupation seems to have ceased at this time and so provides no evidence for chert rationing. Tunnel Cave, where human occupation extended a little beyond this time, gives a fuller picture of the importance of the change. Nevertheless, at Devil's Lair, there is no change in chert or quartz rationing that is not attributable to changes in physical access to chert and its availability. Apart from large numbers of chert artefacts in Period VII, which suggest frequent visits, Devil's Lair stone artefacts provide limited evidence for changes in occupation intensity.

Tunnel Cave

Post-depositional alteration

As at Devil's Lair, post-depositional breakage may increase artefact numbers. Many Tunnel Cave artefacts are burnt or produced from heating (Table 7.17). Some of these were burnt and potlidded on their bulbar surfaces, suggesting that they were burnt after deposition. Burnt artefacts and potlids or heat-fractured pieces are more common in hearth-containing layers 5-lower, 7-upper, and 7-lower. However, K-S tests show that the proportion of burnt artefacts to other artefacts does not change significantly (p [H0] > 0.05). This finding and the stratigraphic association between hearths and artefacts shown in Chapter 6 confirm that artefacts are mainly found in hearth layers: they are burnt (in hearths) to much the same extent in each hearth layer.

Like burnt pieces, the proportions of broken, snapped, or edge-damaged pieces remain the same throughout all layers (p [H0] > 0.05). Artefact breakage may have increased the apparent numbers of knapped artefacts but, as at Devil's Lair, it did so more or less uniformly. Also like Devil's Lair, the chert assemblage shows the same trends, probably because chert artefacts comprise most of the artefacts shown in Table 7.17, and the few altered quartz artefacts do not give significant results. Post-excavation damage seems to be minor, perhaps because artefacts were individually wrapped in the field.

Just as at Devil's Lair, post-depositional effects on Tunnel Cave artefacts have remained at a constant level throughout the period of deposition. Artefact movement through the deposit is considered to be negligible, since most artefacts are associated with well-defined occupation layers as indicated by hearths and a wide variety of archaeological remains (Chapter 6). Post-depositional effects have not substantially altered the numbers or appearance of artefacts at any one time.

Table 7.17 Post-depositional alteration of Tunnel Cave artefacts

K-S tests support H0, i.e., no instances of post-depositional alteration are found in any layer out of proportion to their overall representation in the entire deposit (p [H0] < 0.05).

Layer	Burnt or potlidded on bulbar surface	Burnt or potlidded on other surfaces	Potlids & heat fracture	Broken flakes, proximal	Broken flakes, distal	Snapped artefacts	Edge-damaged artefacts	Post-excavation damage	Total altered	Total artefacts
1									1	2
2										1
3										1
5-upper										3
5-lower	20	26	47	13	3		1	1	119	249
6		2	10	1		1			16	43
7-upper	14	25	33	20	9	4	2		122	266
7-lower	50	120	168	50	34	16	3	5	473	869
8			2						4	21
9-upper	1	2	1	2	2				12	37
9-middle			2	3	1				7	35
9-lower, 10			1	1					2	21
Total	85	175	264	90	49	21	6	6	756	1548

Raw materials

Fossiliferous chert artefacts make up most of the assemblage, but Table 7.18 shows a significantly high proportion of chert artefacts in the middle of the deposit (layers 9-middle, 9-upper, 8, 7-lower, 7-upper, and 6). The earliest dated large-scale use of fossiliferous chert is during formation of layer 9-middle (21,700-21,600 BP). This period falls within Period VII at Devil's Lair (22,800-18,400 BP), when the proportion of chert artefacts increased significantly at that site.

Unlike Devil's Lair, the part of the Tunnel Cave deposit dated to *c.*13,000 BP contains a large number of stone artefacts, imparting more confidence to interpretation of statistical results (Table 7.19). Chert ceases to be represented in disproportionately large quantities in layer 5-lower. The trend in the relative proportions of quartz and fossiliferous chert probably has the same cause as shown at Devil's Lair: fossiliferous chert was used more during times of lowest sea-level, from *c.*20,000 to 13,000 BP.

In Table 7.18, "other rock" includes one silcrete flake and one yellow chert flake that has a superficial resemblance to Plantagenet chert, found near Albany (Glover 1984). These artefacts may have been carried from regions well beyond the Leeuwin-Naturaliste Region, but the quantity is too small to be statistically meaningful. Unflaked rock (pieces of granite-gneiss, basalt, and feldspar) is not analysed here either.

Table 7.18 Materials used for stone artefacts from square G10, Tunnel Cave
Other rock includes flaked stone (not fossiliferous chert or quartz) and manuports.

Estimated ^{14}C age, years BP Minimum	Maximum	New layer	Number of stone artefacts Foss. chert	Quartz	Calcrete	Other rock	Total
1,290	1,450	**1**		2			2
1,290	4,400	**2**		1			1
4,160	4,400	**3**			1		1
6,780	10,200	**5-upper**	1	1	1		3
12,000	13,300	**5-lower**	156	92		1	249
12,000	16,260	**6**	33	9		1	43
15,900	16,260	**7-upper**	249	16		1	266
16,500	17,600	**7-lower**	850	9	1	9	869
16,500	20,450	**8**	18	3			21
18,200	20,450	**9-upper**	24	11	1	1	37
21,700	21,600	**9-middle**	27	7		1	35
21,710	24,110	**9-lower, 10**	8	12		1	21
		Total	1366	163	4	15	1548

Table 7.19 K-S test results on comparisons of cumulative proportions of Tunnel Cave artefact raw materials
The Null Hypothesis (that proportions of artefacts with different characteristics do not vary) is rejected if p (H0) < 0.05. Where the Null Hypothesis should be rejected, the layer containing the greatest difference between proportions is given.

Reject H0?	Foss. chert	Quartz	Calcrete
Foss. chert	---------------------------	**Reject - Layer 7-upper**	No
Quartz		---------------------------	No
Calcrete			-------------------------

Technology

Tunnel Cave artefact types are shown in Tables 7.20-7.21. Tools are made exclusively of fossiliferous chert. They are all utilised pieces, showing bending-initiated and conchoidally-initiated use fractures. Only one of them has retouch (Appendix 9). The high ratio of chert tools to debitage, and of chert flakes to other debitage, in layer 7-lower (Table 7.22) confirms the observation in Chapter 6 that numerous hearths in that layer indicate intensive occupation at that time (*c*.17-16,000 BP). The faint internal lenses suggestive of hearth re-firing in layer 7-lower may also indicate several occupations of the site, at the time of greatest chert abundance. Contemporaneous layers at Devil's Lair lack hearths, and show a low ratio of chert tools to debitage. Tunnel Cave is probably not located any further from chert sources than Devil's Lair, so there would not be special reasons to conserve chert at Tunnel Cave. The trend of more efficient tool and flake production at the time of chert abundance is consistent with intense occupation of the site at around the LGM.

Among quartz artefacts, the significantly low quantities of bipolar pieces in layers 7-upper and 6 (Table 7.23, above) reflect the large take-up of quartz in layer 5-lower. Layer 5-lower contains most of the quartz bipolar pieces, so it determines the overall proportion of these artefacts. This trend is consistent with increased use of quartz after *c*.13,000 BP, and a corresponding increase in the bipolar flaking of this intractable material.

Within chert and quartz artefacts, there are no significant changes in knapping techniques (Tables 7.24 and 7.25). The incidence of these techniques reveals no changes in occupation intensity.

ANOVA and t-tests, summarised in Tables 7.26 and 7.27, show that the size of chert debitage at Tunnel Cave does reflect changes in raw material availability. According to ANOVA, chert artefacts decrease significantly in size in layer 5-lower, a time of rising sea-level (Table 7.26). This pattern complements the trend identified at Devil's Lair, where chert artefacts are largest at the time of lowest sea-level. As Tunnel Cave artefacts are even smaller than Devil's Lair ones, the tiny changes in average size are relatively important. There are no significant changes in the size of quartz artefacts (Figure 7.4, Table 7.27), and quartz flakes provide too small a sample to analyse.

ANOVA on chert flakes shows that there are no significant changes in flake shape (elongation and parallel indices and platform area; p [H0] > 0.05). Like Devil's Lair, there seem to be no changes in chert flake shape, and as mentioned above, in knapping technique, revealing no response to raw material scarcity in this aspect.

To summarise, changes in quartz and chert debitage mainly occurred in layer 5-lower time (*c*.13,300-12,000 BP), when sea-level was rising and perhaps gradually submerging fossiliferous chert outcrops. In this period, chert artefacts became shorter, and people flaked more quartz, using the bipolar technique perhaps more than before. These are similar trends to those that took place at Devil's Lair, but a major difference at Tunnel Cave is that the proportions of tools and flakes suggest that the most intense occupation was probably during the hearth-building episodes represented by layers 7-lower and 7-upper, dated 16-17,000 BP. This possibility is confirmed by analysis of use-wear.

Table 7.20 Types of chert artefacts at Tunnel Cave

Layer	Scrapers, retouched flakes	Utilised flakes	Bipolar cores	Flakes	Broken flakes, longit.	Broken flakes, prox.	Broken flakes, distal	Fragments	Bipolar flakes & fragments	Potlids or heat fracture	Total chert
1											
2											
3											
5-upper				1							1
5-lower				23	7	8	1	69	1	47	156
6				15	1			7		10	33
7-upper			3	93	14	19	9	77	1	33	249
7-lower	1	6	1	262	28	50	34	292	8	168	850
8				4	2			10		2	18
9-upper		1		8	4	1		9		1	24
9-middle				8	1	1	1	14		2	27
9-lower, 10			1	5				2			8
Total	1	7	5	419	57	79	45	480	10	263	1366

Table 7.21 Types of quartz artefacts at Tunnel Cave

Layer	Scrapers, retouched flakes	Utilised flakes	Bipolar cores	Flakes	Broken flakes, longit.	Broken flakes, prox.	Broken flakes, distal	Fragments	Bipolar flakes & fragments	Potlids or heat fracture	Total quartz
1				1	1					1	3
2								1			1
3											
5-upper				1							1
5-lower			3	13	1	5	2	46	22		92
6				2	1	1		4	1		9
7-upper				3	1	1		10	1		16
7-lower				2				7			9
8								3			3
9-upper				1		1	2	6	1		11
9-middle						2		5			7
9-lower, 10				1		1		9			11
Total			3	24	4	11	4	91	25	1	163

Table 7.22 K-S test results on comparisons of cumulative proportions of chert artefact types at Tunnel Cave

The Null Hypothesis (that proportions of artefacts of different types do not vary) is rejected if p (H0) < 0.05. Where the Null Hypothesis should be rejected, the layers containing the significantly disproportionate frequencies (and the direction of the difference) are given.

Reject Ho?	Tools (scrapers & utilised flakes)	Bipolar cores	Flakes (inc. broken flakes)	Fragments	Bipolar flakes & fragments	Heat fracture (potlids)
versus other chert artefacts	Layer 7-lower (high)	No	Layers 6, 7-upper (high)	No	No	No

Table 7.23 K-S test results on comparisons of cumulative proportions of quartz artefact types at Tunnel Cave

The Null Hypothesis (that proportions of artefacts of different types do not vary) is rejected if *p* (H0) < 0.05. Where the Null Hypothesis should be rejected, the layers containing the significantly disproportionate frequencies (and the direction of the difference) are given. Not applicable (na) - quantities too small for comparison.

Reject Ho?	Tools (scrapers & utilised flakes)	Bipolar cores	Flakes (inc. broken flakes)	Fragments	Bipolar flakes & fragments	Heat fracture (potlids)
versus other quartz artefacts	na	na	No	No	Yes - layers 7-upper and 6 (low)	na

Table 7.24 Instances of knapping techniques in chert artefacts at Tunnel Cave

Step or hinge terminations indicate possible core exhaustion. Retouch includes retouch flakelets.

K-S tests show that no knapping techniques in chert artefacts are represented in any layer out of proportion to their overall representation in the entire deposit (*p* [H0] > 0.05).

Layer	Overhang removal	Facetted platform	Flaked platform	Focal platform	Precision blow	Core rotation	Bipolar technique	Retouch	Step or hinge termination	Total chert
1										
2										
3										
5-upper										1
5-lower				3	3	1	2	8	5	156
6				1		1		1	1	33
7-upper	8		2	9	7	5	5	14	22	249
7-lower	31	1	3	42	41	23	20	60	82	850
8										18
9-upper	1			1	1		2		4	24
9-middle				1		2		1	3	27
9-lower, 10				1						8
Total	40	1	5	58	52	32	29	84	117	1366

Table 7.25 Instances of knapping techniques in quartz artefacts at Tunnel Cave

Step or hinge terminations indicate possible core exhaustion. Retouch includes retouch flakelets.

K-S tests show that no knapping techniques in quartz artefacts are represented in any layer out of proportion to their overall representation in the entire deposit (*p* [H0] > 0.05).

Layer	Overhang removal	Facetted platform	Flaked platform	Focal platform	Precision blow	Core rotation	Bipolar technique	Retouch	Step or hinge termination	Total quartz
1										2
2										1
3										
5-upper										1
5-lower	1						14	2	3	92
6							1			9
7-upper				1				1		16
7-lower	1								1	9
8						1				3
9-upper	1						1		3	11
9-middle				1						7
9-lower, 10										12
Total	3			2		1	16	3	7	163

Table 7.26 Results of ANOVA and t-tests of Tunnel Cave chert artefact dimensions

ANOVA and t-tests performed on logged values to obtain approximately normal frequency distributions. Layer 5-upper is excluded from calculations because such small samples give misleading results. "Not applicable" (na) indicates that there are no artefacts in the category.

CHERT	*n*		Maximum length			Weight			Flake length		
Layer	All	Flakes	Total score	Sum of squares	Mean	Total score	Sum of squares	Mean	Total score	Sum of squares	Mean
1	0	0	na	na	na	na	na	na	na	na	na
2	0	0	na	na	na	na	na	na	na	na	na
3	0	0	na	na	na	na	na	na	na	na	na
5-upper	1	1	0.67	0.45	0.67	0.05	0.00	0.05	0.59	0.35	0.59
5-lower	156	22	133.83	147.65	0.86	28.09	36.70	0.18	15.22	15.49	0.69
6	33	15	33.95	43.62	1.03	9.70	18.03	0.30	15.83	23.17	1.06
7-upper	249	93	237.89	287.76	0.96	70.18	181.45	0.28	69.34	66.06	0.75
7-lower	850	262	850.27	1131.01	1.00	278.82	1189.43	0.33	241.12	303.32	0.92
8	18	4	16.83	20.05	0.94	3.41	4.55	0.19	3.29	3.18	0.82
9-upper	24	8	27.16	46.98	1.13	14.73	54.74	0.64	6.19	6.36	0.77
9-middle	27	8	30.89	45.45	1.14	7.66	7.79	0.29	10.40	16.47	1.30
9-lower, 10	8	5	10.08	18.28	1.26	7.23	23.71	0.90	5.13	7.54	1.03
Total	1366	418	1342	1741	8.98	420	1516	3.17	367	442	7.92
Variance ratio				2.33			1.11			2.30	
P (H0) - F -test				0.02			0.35			0.02	
Reject H0?				Yes			No			Yes	
For layers (t-tests)			Layer 5-lower versus layers 6, 7-upper, 7-lower, 9-upper, and 9-middle			not listed due to non-significant result			Layer 5-lower versus layers 6,, 7-lower, and 9-middle Layer 7-upper versus layers 7-lower, 9-middle		

CHERT	*n*		Flake width			Flake thickness			Platform width		
Layer	All	Flakes	Total score	Sum of squares	Mean	Total score	Sum of squares	Mean	Total score	Sum of squares	Mean
1	0	0	na	na	na	na	na	na	na	na	na
2	0	0	na	na	na	na	na	na	na	na	na
3	0	0	na	na	na	na	na	na	na	na	na
5-upper	1	1	0.67	0.45	0.67	0.42	0.18	0.42	0.09	0.01	0.09
5-lower	156	22	15.83	14.28	0.72	10.11	5.74	0.46	1.70	0.31	0.08
6	33	15	14.49	17.01	0.97	7.89	5.91	0.53	2.22	1.05	0.15
7-upper	249	93	76.37	80.37	0.82	46.85	34.38	0.51	11.37	8.41	0.12
7-lower	850	262	232.49	274.55	0.88	130.87	101.07	0.50	39.89	52.48	0.15
8	18	4	4.05	5.81	1.01	1.98	1.08	0.50	0.55	0.19	0.14
9-upper	24	8	6.29	7.07	0.79	3.43	2.06	0.43	0.58	0.08	0.07
9-middle	27	8	12.71	24.86	1.59	5.30	4.58	0.66	1.16	0.30	0.17
9-lower, 10	8	5	5.69	10.44	1.14	2.96	2.38	0.59	0.81	0.27	0.16
Total	1366	418	369	435	8.59	210	157	4.59	58	63	1.13
Variance ratio				2.13			1.10			060	
P (H0) - F -test				0.03			0.37			0.78	
Reject H0?				Yes			No			No	
For layers (t-tests)			Layer 9-middle versus layers 5-lower, 6, 7-upper, 7-lower, 9-upper			not listed due to non-significant result			not listed due to non-significant result		

Table 7.27 Results of ANOVA and t-tests of Tunnel Cave quartz artefact dimensions

ANOVA and t-tests performed on logged values to obtain approximately normal frequency distributions. All flakes and layers 1, 2, and 5-upper are excluded from calculations because such small samples give misleading results. "Not applicable" (na) indicates that there are no artefacts in the category.

QUARTZ	*n*		Maximum length			Weight		
Layer	All	Flakes	Total score	Sum of squares	Mean	Total score	Sum of squares	Mean
1	2	1	1.78	1.59	na	0.14	0.01	na
2	1	0	na	na	na	na	na	na
3	0	0	na	na	na	na	na	na
5-upper	1	1	1.02	1.04	1.02	0.08	0.01	0.08
5-lower	92	13	82.32	109.19	0.90	86.95	4778.07	0.96
6	9	2	7.33	7.13	0.81	0.89	0.16	0.10
7-upper	16	3	17.53	35.28	1.10	2.51	1.31	0.16
7-lower	9	2	10.08	15.30	1.01	7.88	44.24	0.79
8	3	0	3.79	3.05	0.76	0.86	0.29	0.17
9-upper	11	1	10.60	14.67	0.96	4.99	10.03	0.45
9-middle	7	0	5.70	6.19	0.81	1.03	0.76	0.15
9-lower, 10	12	1	7.71	4.39	0.55	0.53	0.03	0.04
Total	163	24	148	198	7.93	106	4835	2.89
Variance ratio				1.11			1.05	
P (H0) - F -test				0.36			0.40	
Reject H0?				No			No	
For layers (t-tests)			not listed due to non-significant result			not listed due to non-significant result		

Table 7.28 Instances of use-wear in Tunnel Cave chert artefacts

Except for three conchoidally initiated flakelets in layer 7-upper and one in layer 9-upper, flakes or flakelets produced from use have bending initiations.

K-S tests show that no types of use-wear in chert artefacts are found in any layer out of proportion to their overall representation in the deposit (*p* [H0] > 0.05).

Layer	artefact from tool working edge	use-wear on dorsal or bulbar face	use-wear on dorsal edge of platform	use-wear on other margins	whole piece produced from use	use-fracture: conchoidal initiation	use-fracture: bending initiation	use-fracture: feather term.	use-fracture: hinge or step term.	fresh edges	multiple use	Total chert
1												
2												
3												
5-upper					1	1						1
5-lower			2		3		2	1			1	156
6						1			1			33
7-upper	1	1	11	2	11	9	5	3	3	3	4	249
7-lower	2	4	15	14	22	14	16	11	7	24	6	850
8												18
9-upper			1	1	2	2	1	1			2	24
9-middle										1		27
9-lower, 10										2		8
Total	3	5	29	17	39	27	24	16	11	30	13	1366

Use and discard

Much the same use-wear types exist at Tunnel Cave as at Devil's Lair (Table 7.28). Quartz artefacts, not shown, provided no evidence of use-wear, probably because it is hard to detect on polycrystalline quartz surfaces. Table 7.28 shows no significant changes in the incidence of use-wear categories among chert artefacts (p [H0] > 0.05). However, the total number of incidences of use-wear is significantly high in layer 7-lower (n = 135, p [H0] < 0.05). Tool manufacture (see above) and tool use were more intensive relative to the amount of chert at the time of greatest chert abundance. Again, this intensive use of raw material is best explained by more intensive occupation.

No tools or flakes from former tool edges show evidence of tool rejuvenation or re-flaking. However, 13 artefacts suggest multiple uses (one in layer 5-lower, four in 7-upper, six in 7-lower, and two in 9-upper). These edges have overlapping flake scars with both bending- and conchoidal-initiated fractures (suggesting use on soft and hard materials, or scraping and cutting activities). Although the proportion of these artefacts does not increase at any one time (p [H0] > 0.05), allowing no inference about discard of artefacts and raw material treatment, they indicate that some chert tools were used intensively.

Summary

The trends at Tunnel Cave in raw materials, debitage types, knapping technique, artefact size, use-wear, and artefact re-use, parallel those at Devil's Lair. The observations in these areas suggest that at times of low sea-level, 20,000-13,000 BP, fossiliferous chert was in wide circulation among human groups. As long as sea-levels were low, people flaked more chert than quartz, suggesting that fossiliferous chert was a better material for flaking or use in tools. At these times when more chert was in circulation, people at Devil's Lair made fewer chert tools relative to the quantity of chert debitage, made larger flakes, and used bipolar technique on chert less often, all suggesting less intensive use of chert. In contrast, people commonly applied bipolar technique to quartz at all times. At both sites, throughout periods of falling and rising sea-level, there were no changes in the proportions of most debitage categories or shape of flakes, only in the size of debitage and an increased use of bipolar technique on chert (at Devil's Lair only) and quartz. Since bipolar technique is the common solution for flaking small cores, all these changes suggest no change in knapping techniques, only a change in the size of chert cores deemed worth flaking.

Devil's Lair provides no evidence that occupation intensity varied, but at Tunnel Cave, people occupied the site intensively in layer 7-lower time (17,600-16,500 BP). This intensive occupation is shown by the ratios of tools to debitage, and used artefacts to unused artefacts, which are the reverse of what one would expect at a time of chert abundance, and of what is inferred at Devil's Lair. These findings are now compared to preliminary studies at Witchcliffe Rock Shelter and Rainbow Cave (Cocks 1993, Coventry 1998), where in deposits dated less than 800 BP, people flaked quartz almost exclusively. The inferences made about quartz flaking at these sites help one understand the knapping and use of this perennially useful material, and hence allow a few comments about occupation intensity at these sites.

Witchcliffe Rock Shelter and Rainbow Cave

Interpretation of Witchcliffe Rock Shelter and Rainbow Cave stone artefact assemblages is based on analyses by Coventry (1998) and Cocks (1993). These investigations are respectively, an analysis of bipolar technique as an indication of the degree of hunter-gatherer mobility (or sedentism), and a taphonomic analysis of the Rainbow Cave deposit and quartz artefacts. Comments on the sites themselves are provided in the excavators' reports (J. Dortch 1996, Lilley 1993).

Post-depositional alteration

Weathering and burning are probably minor factors affecting the Witchcliffe Rock Shelter and Rainbow Cave quartz artefacts. Quartz is highly resistant to weathering and burning in campfires (Courty *et al.* 1989). Artefact breakage is still a potential factor that would increase the number of artefacts, but owing to the large quantity of debitage, and the difficulty of detecting breakage in irregularly-fractured polycrystalline quartz (Glover 1984, Hiscock 1982), neither Cocks nor Coventry attempted to distinguish broken quartz from knapped quartz. Given the relatively short periods of deposition at either site, and the observation that the fine layers show little evidence of heavy trampling (see Chapter 6), I presume that artefact breakage is a minor influence on the numbers of artefacts at the two sites.

Artefact movement is probably only a minor influence on the vertical distribution of artefacts. Most artefacts at Witchcliffe Rock Shelter are located in the occupation layers that lie above hearth F8 and the lowermost radiocarbon date for 800 BP (see Chapter 6). There is little evidence that trampling or other agents caused large numbers of artefacts to migrate below this feature. The well-defined and indurated hearth features above F8 suggest that vertical movement of artefacts would be limited to the layer that they are found in. A final test is provided by the trampling of modern visitors, which has not pushed the majority of the modern artefacts deeper than layer 1, less than 5 cm below the surface. At Rainbow Cave, Cocks (1993) suggests that trampling would have moved only a small proportion of artefacts in the unconsolidated sand of the deposit. The majority of Rainbow Cave artefacts are in stratigraphic association with occupation features (see Chapter 6).

The potential for trampling of deposit and hence artefacts at Witchcliffe Rock Shelter and Rainbow Cave suggests that there would have been slight vertical movement of most artefacts, and major vertical movement of a few artefacts. Some of these artefacts could have been edge-chipped or broken during their displacement, but extent of this damage would probably be in proportion to the movement. In conclusion, post-depositional factors probably do not greatly affect interpretations of raw material use and occupation intensity.

Table 7.29 Materials used for stone artefacts from square T20, Witchcliffe Rock Shelter

"Total artefacts" includes flaked stone, feldspar, and other exotic stone. Unprovenanced (unprov.) material is from section cleans.

Layer	Estimated 14C age, years BP Minimum	Maximum	Quartz pebbles	Sub-angular quartz	Flaked quartz	Other flaked stone	Feldspar	Other exotic stone	Total regarded as artefacts
1			141	95	307	2	2	2	313
2			12	5	9				9
3-upper	300	500	194	44	743		3	7	753
F3			36	12	60				60
F4			26	5	142				142
3-lower			34	15	307		2		309
F5			26	10	301			1	302
4-upper			53	25	414	1	1		416
F6			4	25	71	1			72
F7			5	4	11				11
4 middle			130	90	311	2	5		318
F8	500	860	38	7	117		1	4	122
4-lower			82	56	124			1	125
4-lower/5			84	19	104		3		107
5			83	84	34			1	35
unprov.			5	3	102	1	4		107
TOTAL			953	499	3157	7	21	16	3201

Raw materials

Flaked stone assemblages at both sites are mainly quartz (Tables 7.29, 7.30). At Witchcliffe Rock Shelter hundreds of rounded and sub-angular quartz pebbles are probably not artefacts. They were possibly formed in the underground stream bed inferred by Williamson (1979) to have exsurged from the hillside in the distant past at the height of Witchcliffe Cave, forming the cave and shelter. Williamson (1979) notes that granite-gneiss bedrock is located at more or less the same height in the hillside. Granite-gneiss is a possible source for quartz. The pebbles found in excavation are abundant in layer 5, which is probably a natural deposit pre-dating formation of the rock-shelter.

The quartz artefacts at Witchcliffe Rock Shelter and Rainbow Cave probably derive from nearby sources. Field survey by G. Coventry and myself found large, though somewhat incohesive, cobbles of vein quartz cropping out 400 m east of Witchcliffe Rock Shelter. They probably derive from veins in the granite-gneiss that we identified in road sections 1200 m east of the shelter. Similar sources exist near Rainbow Cave since it is located 1000 m from the shoreline, where granite-gneiss forms prominent headlands.

The quartz cobbles located by Coventry and myself contain mineral impurities and are composed of interlocking crystals. Glover (1984: 22) describes this type of material as quartzite, "polycrystalline rock of many origins ... [including] quartz vein material". Many artefacts at Witchcliffe Rock Shelter and Rainbow Cave seem to contain the same mineral impurities and have irregular surfaces, indicating different crystal planes.

Table 7.30 Numbers of quartz artefacts from pit D22, Rainbow Cave, after Cocks (1993)

Spit	Estimated 14C age, years BP Minimum	Maximum	Flaked quartz
1			9
2	250	430	125
3			213
4			243
5			713
6	690	890	280
7			142
8	740	920	91
9			24
10			28
11			6
12			10
13			1
14	4010	4290	0
15			3
16			7
Total			1895

Another type of quartz present in the late Holocene assemblages is rock crystal or "crystal quartz", that is, quartz deriving from a single crystal. This material was used for 18 out of 1083 artefacts (2%) that Coventry (1998: Table 6.3) analysed at Witchcliffe Rock Shelter and 138 out of 1574 (9%) that she analysed at Rainbow Cave. Moreover, layer 4-lower at Witchcliffe Rock Shelter contained one unflaked crystal weighing 8.7 g. Sources for rock crystal were not identified near either site. Rock crystal and other quartz may have been carried from locations suggested by Glover (1984), that is, "fault zones in Precambrian terrains", as found along the Darling Scarp, 80 km east of the sites (the Dunsborough fault is closer but is largely obscured by Tertiary sediments). Alternatively, rock crystal in the sites may derive from pebbles in the streams and beaches close to the two sites, and the rounded pebbles already present in the Witchcliffe Rock Shelter deposit might have included some made of crystal quartz, and could have been flaked to obtain very small fragments (layer 5, containing these pebbles, was exposed in a hole in the shelter floor).

In summary, large quantities of quartz, whether from veins ("quartzite") or in pure crystal form, could have been obtained from various locations close to the sites, and or even within Witchcliffe Rock Shelter. Rock crystal would perhaps not have been available in large pieces - the stream and beach pebbles I have seen are only a few centimetres long. Vein quartz is available as cobble-size chunks but may be harder to flake into large pieces as (a) the material is composed of interlocking crystals, so will not flake evenly in all directions (it is anisotropic), and (b), being incohesive, it tends to break into small pieces after a few blows with the hammerstone. These assertions are confirmed in Appendix 10, which describes a quartz flaking experiment, and they have implications for inferences about knapping techniques and occupation intensity.

Technology

Coventry (1998) identified 95 artefacts flaked by bipolar technique at Witchcliffe Rock Shelter and 88 bipolar artefacts at Rainbow Cave, respectively 9% and 6% of her sample from each site (Table 7.31). Since *c*.60% of the products of a single episode of bipolar flaking are unrecognisable as bipolar artefacts (Appendix 10), many more of the artefacts from both sites could have been produced by bipolar technique (a similar argument could be made at Devil's Lair and Tunnel Cave, since 13% of these sites' quartz artefacts is flaked by bipolar technique).

In these late Holocene sites people applied bipolar technique to small-sized quartz pebbles or chunks as a means of using the most suitable stone then available. The same technique had been applied to a more or less equal degree on quartz artefacts at Devil's Lair and Tunnel Cave (phi-squared values of less than 0.1, obtained from chi-squared tests, indicate that differences in the proportions of bipolar artefacts between any pair of sites are slight (Table 7.32). At Devil's Lair and Tunnel Cave, the presence of several large pieces of quartz flaked by direct percussion suggests that some large pieces of quartz were brought to those sites (although given the small size of quartz pieces at all sites, bipolar technique was probably always necessary).

Table 7.31 Bipolar artefacts at Witchcliffe Rock Shelter and Rainbow Cave

Data from Coventry (1998).

	Bipolar artefacts	Non-bipolar artefacts	Total
Witchcliffe Rock Shelter	95	988	1083
Rainbow Cave	88	1486	1574
Total	183	2474	2657

Table 7.32 Chi-squared tests of comparisons of bipolar and non-bipolar artefacts at each possible pair of sites (cf Appendix 8)

Values in cells indicate chi-squared statistic and phi-squared value (chi-squared divided by *n*) in brackets, and are shaded where the chi-squared statistic is greater than 3.841, which is the theoretically derived statistic for two samples drawn from identical populations at a confidence level of 0.05 and 1 degree of freedom. Phi-squared values indicate significant but small differences between sites. Data from Coventry (1998) and this chapter.

	Devil's Lair	Tunnel Cave	Witchcliffe Rock Shelter	Rainbow Cave
Devil's Lair	--------- ------	0.984 (n.a.)	**6.76 (0.004)**	**31.723 (0.015)**
Tunnel Cave	--------- ------	--------- --------	2.84 (n.a.)	**13.356 (0.008)**
Witchcliffe Rock Shelter	--------- ------	--------- --------	------------------ -------------	**10.12 (0.004)**
Rainbow Cave	--------- ------	--------- --------	------------------ -------------	------------ -----------

Table 7.33 Statistics comparing variation between layers in "average" weight of quartz artefacts at four sites (see text)

Data for Rainbow Cave from Cocks (1993: Table 6.2), data for other sites derived from Appendix 8. The first statistic in each column is the average weight of all quartz artefacts in the entire collection from each site; the "mean" is the mean of average quartz artefact weights from each layer.

Av. Wt	Devil's Lair. main excavation	Tunnel Cave. square G10	Witchcliffe R.S., square T20	Rainbow Cave, pit D22
All quartz	1.24 g	0.65 g	0.19 g	0.24 g
Min.	0.03	0.04	0.11	0.04
Max.	11.34	0.95	0.43	0.92
mean	1.73	0.30	0.18	0.23
st. dev.	3.16	0.33	0.08	0.21
variance	10.00	0.11	0.01	0.04

Although no measurements are available for all the Witchcliffe Rock Shelter and Rainbow Cave quartz artefacts, one can gain an idea of their small size from their average weight (calculated by dividing the weight of all quartz artefacts in a site or layer by their total number). Table 7.33 confirms that compared to similar data obtained from Devil's Lair and Tunnel Cave, Witchcliffe Rock Shelter and Rainbow Cave quartz artefacts are small and much less variable (compare the standard deviations and variance estimates). The people who occupied these sites probably flaked small quartz pebbles or chunks. This indication of uniformity is reinforced by Coventry's (1998) finding that the weight and length distributions of bipolar artefacts at Rainbow

Cave and Witchcliffe Rock Shelter are highly correlated (Table 7.34).

At Witchcliffe Rock Shelter and Rainbow Cave, no inference about occupation intensity can be drawn from the rate of tool production, as no attempt has been made to identify tools among the *c*.5000 quartz artefacts (as noted, the material is difficult to analyse even under the microscope). The apparent absence of tools at Rainbow Cave (Cocks, 1993, does not indicate whether she identified tools) and in the sample analysed from Witchcliffe Rock Shelter suggests that the small size of the artefacts and the material precludes easy identification of types such as geometric microliths.

Table 7.34 Length and weight classes of bipolar artefacts at Witchcliffe Rock Shelter and Rainbow Cave, from Coventry (1998: Table 6.2)

Significance test for Pearson's *r* from Startup and Whittaker (1982: 153).

Max weight (g)	Witchcliffe Rock Shelter	Rainbow Cave	Max length (mm)	Witchcliffe Rock Shelter	Rainbow Cave
0.2	50	31	2	0	0
0.4	11	14	4	0	0
0.6	11	8	6	2	4
0.8	0	10	8	14	5
1	4	6	10	18	19
1.2	4	5	12	17	12
1.4	0	3	14	11	12
1.6	1	2	16	12	14
1.8	2	3	18	10	9
2	1	1	20	5	7
2.2	1	1	22	2	5
2.4	1	1	24	1	0
2.6	1	0	26	1	1
2.8	0	1	28	1	0
3	0	0	30	0	0
3.2	1	0	32	0	0
3.4	0	0	34	0	0
3.6	0	0	36	0	0
3.8	0	1	38	0	0
4	0	0	40	1	0
4.2	0	0			
4.4	0	0			
4.6	1	1			
n	**89**	**88**	*n*	**89**	**88**
Pearson's *r*	**0.97**	Sig. test: 17.69	Pearson's *r*	**0.91**	Sig. test: 10.07
r^2	**0.94**	**Significant**	r^2	**0.83**	**Significant**

Use and discard

No attempt has been made to identify used edges on the chips and small fragments from Witchcliffe Rock Shelter or Rainbow Cave. A low-powered microscope examination of another Holocene quartz assemblage, from Nookanellup Rock Shelter on the Southern Ocean coast, suggests that quartz artefacts tend to show few traces of use-wear (Dortch and Kelly 1997). At this site, 14 observations of use-wear were made on 267 quartz artefacts (about 6% of the assemblage). The few signs of use-wear is in contrast to chert assemblages at Devil's Lair and Tunnel Cave, which included 172 instances (34%) and 190 instances (14%) respectively. Among the quartz assemblage at these sites, the 579 quartz artefacts from Devil's Lair returned 54 observations (9%), and the 163 quartz artefacts from Tunnel Cave returned one observation (0.6%). Thus one might expect to find relatively few signs of use-wear among the 4240 quartz artefacts from squares T20 and square 1 at Witchcliffe Rock Shelter, or the 1895 quartz artefacts from Rainbow Cave, especially as the small average weight of these artefacts implies that many are too small to use (Table 7.33).

Nonetheless, some of the sites' quartz pieces could have been used as tools, because their tough edges would suit use as finger-held implements or in composite tools such as *taap* knives (King 1827) or "quartz spears" (Roth 1903). These tools were used in the early 19th C (almost contemporary with occupation at the sites) and featured a row of sharp-edged quartz flakes or chips set into hardened gum. Given that the raw material appears to have been extremely abundant, one would expect that quartz pebbles were frequently broken or flaked as the need for fresh edges arose. The sheer number of artefacts deposited in relatively short time seems to provide the best indication that stone tool-using activities were frequent in the two sites.

Summary

Given the local abundance and small size of vein quartz and quartz pebbles, people probably would not have rationed quartz debitage or re-used tools at any particular time. The numbers of people occupying the site would have been a prominent factor in determining the number of artefacts at the site (although the number of artefacts does not indicate that more people used these sites than others, where different raw materials predominate). Since there are no significant changes in the numbers of artefacts throughout the period of human visits to either site, I suggest that the size or duration of human occupations could have been similar at each site throughout the period 800-400 BP.

The sources for the quartz in these recent sites were perhaps different from those used at Devil's Lair and Tunnel Cave. Witchcliffe Rock Shelter and Rainbow Cave are located within several hundred metres of quartz pebbles and cobbles, and the average size of the late Holocene quartz artefacts is even smaller than those at the Pleistocene sites. The heavy use of conveniently located but small or incohesive material suggests that human occupations of the sites were either small-scale or short, and based on very local resources.

Conclusion

To summarise, at Devil's Lair the period of greatest chert artefact use includes a period (*c.*18,000-13,000 BP) when relatively few tools were made from a given quantity of chert, fewer artefacts were used as tools, fewer chert cores were reduced by bipolar technique, and when the average size of chert artefacts was significantly larger. These trends all suggest that during the period of low sea-level and greatest chert availability, the raw material was less carefully rationed during occupations at the cave, away from the posited sources on the now-submerged coastal plain.

At Tunnel Cave, during this same period of low sea-level, chert artefacts were larger than other periods at the same site. This trend may have the same cause as inferred at Devil's Lair. However, even though chert may have been more widely available at this time, chert tools were more often made and more often used in the period represented by layer 7-lower (17,600-16,500 BP). This trend indicates intensive occupations involving sustained production and use of tools during the time that layer 7-lower accumulated.

These inferences and those made above for Witchcliffe Rock Shelter and Rainbow Cave suggest a progression in the types of occupations at cave and shelter sites, as follows:

Occupations at Devil's Lair and Tunnel Cave were usually either short or small-scale, perhaps made simply because the sites were attractive in bad weather (C.E. Dortch 1979a). The sites' human occupiers carried with them stone artefacts from near and distant sources, including large pieces of quartz. Later occupations at Devil's Lair may not have changed in size, length, or frequency, although more artefacts were deposited at times. At Tunnel Cave, several sustained occupations probably occurred around the time of the last glacial maximum. Besides evidence for more prolonged use of tools when raw material was abundant, there are indications of multiple firings of hearths at this time (see Chapter 6). Occupation at Tunnel Cave during the period 13,000-12,000 BP may have been similar to the pattern before the glacial maximum, since at both times there was a low level of tool use. Tunnel Cave provides little evidence that occupation intensity declined in response to major vegetation changes that probably took place from 12,000 to 8,000 BP (Chapters 8, 9; Balme *et al.* 1978), except that there are few artefacts dated to this period.

This trend towards small-scale or short occupations was seen again in the last millennium at Witchcliffe Rock Shelter and Rainbow Cave. Arguably these sites show an even greater reduction in the intensity of rock shelter occupations, since stone-knappers at these sites seem to have made almost exclusive use of raw materials found within a few hundred metres of the sites, as if they were making the most cursory use of local resources. The major inference about occupation intensity is that it was generally the same at all times, whenever people visited any of the four cave sites, except at Tunnel Cave soon after the glacial maximum. Whether or not vegetation was then very favourable to occupation of the Tunnel Cave locality, it is possible that factors such as high winds or cold temperatures may have made the cave particularly attractive. The decline in artefact numbers at Tunnel Cave from 12,000 to 8,000 BP is the only possible clue that there was a change in occupation intensity contemporaneous with known vegetational changes.

Chapter 8 Inferring palaeo-vegetation from faunal remains

This chapter discusses the faunal remains in the four archaeological deposits identified in Chapter 5. In Chapter 2, I argue that the animal bones in archaeological deposits can suggest what vegetation surrounded the sites at the time, on the basis of the environmental requirements or preferences of different animals. The presence or absence of animal species over the period of the deposits, and their relative frequencies in them, are therefore appraised here.

However, a problem in taking faunal remains from archaeological sites as palaeo-environmental indicators is that human and animal predators could have deposited the remains of prey animals disproportionately to their representation in local habitats. The ratios of prey to non-prey animal remains in a cave or shelter do not then necessarily reflect the animals' natural frequency. Nevertheless, it is still possible to make some inferences about the environment, taking into account what remains probably derive from human or other predator activity. Whether a result of human or animal predation or not, faunal remains still reflect, to a degree, local vegetation and climate (Balme *et al.* 1978).

Faunal analysts sometimes attempt to identify preferred human prey animals, for example, to infer human adaptations (Grayson 1984). However, evidence that some animals are more likely to be prey of humans, rather than of other predators, does not show that people hunted those animals in preference, since people potentially hunted almost any animal (see Chapter 4 and Appendix 1). It is only because other predators may hunt relatively few animals of certain species that human contributions sometimes bias the faunal record. The analyses below merely indicate the relative weights of humans and other predators' contributions. Human predation of apparently minor human prey, that is, animals also contributed by predators other than humans, could still have been significant for the people concerned.

Factors that form faunal samples

The faunal material at the four sites is made up of highly fragmented bones, and pieces of emu eggshell, aquatic mollusc shell, and crustacean exoskeleton. Bone fragments are the largest category and the one most likely to have been accumulated by several agents, which are identified below. They are also likely to have been modified after deposition. The potential interactions between bone accumulators and post-depositional processes are so complex that one has to consider them carefully before going on to find ways to extract information about palaeo-environments. Lyman (1994) and Grayson (1984) have outlined the copious effects of human and non-human carnivore predation, scavenging, fragmentation, and re-working of bone; "natural" deaths of potential prey animals in caves and rock shelters; re-working of deposits; humans burning sediments that contain bone and chewing bone; and selective modification of animal carcasses by humans and non-human carnivores.

Taphonomy, the study of these effects on organic remains after death, is often viewed as stages between the living assemblage (*i.e.*, ecological community) and the analysed assemblage (Grayson 1984, Lyman 1994, Ringrose 1993, Solomon 1990). In each stage components of the previous stage are lost by natural attrition or artificial selection, hence there are also death, depositional, fossil, and recovered assemblages. One must assume that all these attritive stages have affected faunal remains, but the relative *contributions* of non-human and human predators are probably most important to the study presented here. This is because most taphonomic factors probably operated equally in all the sites at all times, except at Devil's Lair, where, before *c*.45,000 BP, the cave entrance was small and admitted only small animals and raptors. If selective attrition acts the same way most of the time, changes in the fauna must be due to changes in the bone contribution. The exceptional, variable taphonomic factors are therefore the activities of human and non-human predators (I shall briefly consider other sources of bone also). In the discussion below I consider ways of assessing the relative roles of human and non-human predators on the basis of their distinctive faunal accumulations and bone-modifying behaviour.

Accumulation of bones by humans

Humans have probably accumulated bones in south-western Australian cave deposits since at least *c*.45,000 BP at Devil's Lair. Ethnographic records suggest that prey animals could have included almost all south-western Australian vertebrates and many invertebrates (Meagher 1974), and certainly animals that were hunted by other predators also. However, some faunal remains are contributed only by people: in cave sites, only people would probably have contributed a range of aquatic fauna, and carried in whole emu eggs to cook.

Terrestrial animal bones can be assumed to be probable human contributions if they have been cut during butchery (*e.g.*, in removing tendons from long bones). People are also known to make tools from bones, break bones to obtain marrow, chew them up into small pieces, or discard them into hearths. These processes are not as distinctively human as butchery marks, since a bone from any source could potentially be picked up to make a tool, other animals break bones (Johnson 1985), and any bone in a deposit can become incorporated into a hearth (Archer 1977, Balme 1980a, David 1990, Walters 1989), but bone tools and burnt bones, at least, suggest concurrent human activity.

It may be impractical to identify prey remains by their association with some types of humanly-modified bones. Research at Devil's Lair and Mammoth Cave suggests that flaked and spirally-fractured bones indicate human activity (Archer *et al.* 1980, C.E. Dortch 1979c), but these types of bones will not be considered here as such bones are also produced by non-human carnivores (Johnson 1985). Other studies suggest that there are definite bone artefacts in the deposits analysed here, in the form of jawbones with incisors removed, incised bones, and bone points (cf. Balme 1979, C.E. Dortch 1979a, J. Dortch 1996), but with the exception of macropod jawbones, modified bones are too scarce to infer prey animals other than macropods (Balme 1979).

The range of fragment sizes in bone assemblages is sometimes used to indicate the type of predator (Lyman 1994). Aboriginal people in food-stressed areas such as the Western Desert were observed to smash up chewed bones into minute pieces and eat the marrow and remaining scraps of meat along with many of the bone fragments (Gould 1996; J.E. Stanton, field-notes cited in Balme *et al.* 1978). The maximum size of the ingested fragments is 30 mm, but apparently averages 10-15 mm. Unconsumed fragments from pounded carcases were observed to be of similar size (Gould 1996: 83). However, animal predators also produce fragments of this size range (Walshe 1994a, b). Moreover, the ethnographic observations derive only from arid Australia, and there is no evidence from south-western Australia (cf. ethnographic records; Appendix 1) to indicate whether prehistoric food stress there would have been as great to motivate the same bone-chewing or pounding behaviour.

Another potential means of identifying humanly contributed bones is by assuming that they are those fragments that are too large to have been consumed by bone-ingesting carnivores such as *Sarcophilus* (Walshe 1994a). However, bones larger than the size consumed by carnivores might still derive from unconsumed meal debris of either human or non-human carnivores. I do not use bone fragment size as a means of identifying human prey animals.

One can infer human prey species by identifying the proportion of burnt specimens (cf. Balme 1980a) and the stratigraphic association of burnt bones and bones identified to species. Observations of hunter-gatherer peoples who discard bones of their prey into campfires, and build campfires on deposits containing discarded bones, suggest that burnt bones derive from hearth-building at the approximate time and place that remains of human prey are discarded (Gifford-Gonzalez 1989, Hammond 1933, Walters 1989, Yellen 1991). In this study, I assess at each site the associations between burnt bone fragments and identified specimens, and the proportions of specimens that are burnt. I identify burnt bones as bones that are charred (blackened and hardened) or calcined (white or blue in colour, often softened or weakened, and indicating higher temperature burning, as found in hearths, but not in brush-fires: Bellomo 1993, David 1990, Lyman 1994, Shipman *et al.* 1984).

If associations between identified specimens and burnt bones can indicate prey animals, so can associations between identified specimens and other products of human activity: stone artefacts, pieces of exotic stone such as ochre, and fragments of emu eggshell and aquatic mollusc shell can all be compared to specimen counts. At each site, I carry out test of association between all these classes of material.

Accumulation of bone by non-human carnivores

Here I assess the potential contributions of mammalian carnivores, and ways to identify them, in bone accumulations in Leeuwin-Naturaliste Region sites, mentioned by Balme *et al.* (1978), Merrilees (1979a) and Porter (1979). Carnivores contemporary with human occupation of the region that are known to bring bones of their prey to caves and rock shelters are predominately *Sarcophilus harissii* (Tasmanian Devil, now extinct in mainland Australia: Guiler 1970, 1982); *Canis familiaris dingo* (Dingo, present in Australia since *c.*3,500 BP: Gollan 1984); and owls (notably *Tyto* spp.: Pizzey and Doyle 1991). There is evidence that other predators such as *Dasyurus geoffroii* visited caves occasionally (Lundelius 1966). The bone-accumulating and cave-visiting behaviour of these animals and other mammalian carnivores, raptorial birds, and reptiles is detailed in Appendix 11.

The type of prey favoured by carnivores is not distinct from that of humans, partly because of the small size and vulnerability of most south-western Australian prey (Table 8.1, note column showing body weights). All of the extant carnivores have been observed to scavenge carcases of any animal, and can hunt live prey of any size up to a given maximum (Green 1967; Guiler 1970, 1982; Pizzey and Doyle 1991; Shepherd 1981; M. Smith 1982; Strahan 1995; Walshe 1994a, b; Whitehouse 1977). Table 8.1 shows which taxa could have been hunted by non-human carnivores. For extinct animals, there is limited historical evidence that *Thylacinus* (Thylacine, extinct in mainland Australia since *c.*3000 BP, but persisting into the 20th C in Tasmania: M. Smith 1982) hunted animals up to the size of kangaroos, while the prey of the larger *Thylacoleo* (Marsupial "lion", extinction date unknown) may have also included kangaroos, but no evidence of its prey is available.

Table 8.1 Probable capabilities of major predators suspected to have hunted vertebrates identified in Leeuwin-Naturaliste Region archaeological sites.
Prey weights from Strahan (1995).
✓ = predator could have hunted this animal
juv. = predator could have hunted juveniles only
? = predator's ability unknown but not discounted

Prey			Predator size								
			Small				Medium	Large			
Group	Species	Average prey weight	*Dasyurus geoffroii*	*Ninox novaeseelandiae*	*Tyto alba*	*T. novaehollandiae*	*Sarcophilus harrissii*	*Thylacinus cyanocephalus*	*Canis familiaris dingo*	*Thylacoleo carnifex*	*Homo sapiens*
Lizards	not distinguished	<0.5 kg	✓	✓	✓	✓	✓	✓	✓	?	✓
Murids	not distinguished	c. 0.1 kg	✓	✓	✓	✓	✓	✓	✓	?	✓
Dasyurids	not distinguished	<0.5 kg	✓	✓	✓	✓	✓	✓	✓	?	✓
Bandicoots	Isoodon obesulus	c. 0.5 kg				✓	✓	✓	✓	?	✓
	Perameles bougainville	c. 0.5 kg				✓	✓	✓	✓	?	✓
Possums	*Pseudocheirus peregrinus*	0.9 kg				juv.	✓	✓	✓	?	✓
	Trichosurus vulpecula	3.0 kg				juv.	✓	✓	✓	?	✓
	Potorous tridactylus	1.1 kg				?	✓	✓	✓	?	✓
Potoroids	*Bettongia penicillata*	1.3 kg				?	✓	✓	✓	?	✓
	B. lesueur	1.0 kg				?	✓	✓	✓	?	✓
	Setonix brachyurus	3.3 kg					?	✓	✓	?	✓
Small macropods	*Petrogale lateralis*	4.6 kg					juv.?	✓	✓	?	✓
	Macropus eugenii	6.0 kg					juv.?	✓	✓	?	✓
	M. irma	8.0 kg					juv.?	✓	✓	?	✓
Large macropod	*M. fuliginosus*	f: 27 kg						?	✓	?	✓
		m: 53 kg						?	?	?	✓

Non-human carnivore accumulations are difficult to distinguish from human accumulations. All of the extant non-human carnivores may have left bone-filled scats, meal debris, or regurgitated pellets in caves and rock shelters. Many carnivores probably scavenged other carnivores' meal debris, thereby removing the bone or re-depositing highly comminuted fragments of it in the site. All the carnivores, including humans, can ingest bone fragments up to 45 mm long (Walshe 1994a, b), and *Sarcophilus* and *Canis* particularly are known to leave their bone-filled scats in caves and rock shelters (Marshall and Cosgrove 1986, Douglas *et al.* 1968, Solomon and David 1990).

Since the non-human carnivores such as *Sarcophilus* and *Canis* could have scavenged remains of human meals, as well as carcases in or near sites, *all* bone fragments in Australian archaeological sites that are less than 45 mm long have potentially passed through the gut of a non-human carnivore (Walshe 1994a, cf. Schmitt and Juell 1994). Fragment size increases with the feeding animal's body size but no predator is associated with an exclusive size category of bone fragments.

Bone fragments longer than 45 mm are still, potentially, the unconsumed meal remains of any carnivore. Their characteristic meal debris is still not easy to identify, although *Canis* sometimes leave many fragmented bones uneaten in a range of sizes (Solomon and David 1990) while *Sarcophilus* probably eats the most fragments of any carnivore (Douglas *et al.* 1968, Guiler 1977). Both *Canis* and *Sarcophilus* leave gnawing and puncture marks, but if the bones are highly comminuted these marks will be lost. No characteristic imprint on bones is known for *Thylacinus*. Because of its small size, *Dasyurus* probably leaves more complete bones, but other scavengers could also easily obscure its imprint on bones. Bones from humans' meal debris may be cut, cracked, or broken. Bone breakage by humans may look like bone breakage by *Canis* (or possibly *Thylacinus*). Burnt bones are not necessarily from people's meals, but they do indicate human occupation at a time when bones are either at the surface or close to the surface of the deposit (Archer 1977, Balme 1980). Unconsumed meal debris of all animals, including humans, may bear gouge marks from teeth or claws and include spirally-fractured bones. While cut marks from stone artefacts sometimes indicate human use of a bone, and gnaw marks sometimes indicate a non-human scavenger (and its size), both marks may exist on the same bone.

Only bones from owls' meal debris are distinctive: they may be cracked or fissured, and tend to be whole if less than 45 mm long (Walshe 1994a, b). Bones deriving from owl predation are identified at Devil's Lair (Baird 1992, Balme *et al.* 1978) and Rainbow Cave (Jackson 1992). However, Balme *et al.* (1978) do not identify the owl contributions in that part of the Devil's Lair deposit dating to the period of *Sarcophilus* and human visits, presumably because identifying the owl contribution among hundreds of thousands of bone fragments would be extremely time-consuming. A similar problem exists at the other sites.

The implication of all these observations is that physical characteristics of bones are difficult to interpret as evidence for the source of bones in cave sites. However, Walshe (1994b) notes that only humans and owls bring portions of unconsumed food to caves and rock shelters. *Canis* and *Sarcophilus* are certainly bone-contributors, by virtue of their marking behaviour, but they do not often deposit unconsumed bones in caves. Since the largest owl, *T. novaehollandiae*, does not hunt animals heavier than two kilograms, Walshe argues that one need only identify bones longer than 45 mm that are also from animals heavier than two kilograms, to identify the human contribution.

The trouble with this approach is that the sites discussed here contain an overwhelming portion of identifiable bone fragments shorter than 45 mm (see below, and Balme *et al.* 1978: Figure 3), and scavengers have probably visited and comminuted most of the human deposits to small fragments (Balme *et al.* 1978). There is even evidence from Tunnel Cave that *Sarcophilus* was more active in hearth-containing layers, as if attracted to human food refuse (see below). The same behaviour is possible in *Canis*, especially pet animals or hunting dogs, at late Holocene sites (Hammond 1933, Hayden 1977). A method of inferring environmental records from identified bone fragments that could derive from any predator is discussed below. I first discuss other sources of potential confusion.

Other means of bone accumulation

Additional sources of bone are deaths of animals in caves. Cave-dwelling vertebrates like swallows (family *Hirundinidae*; Baird 1992) and bats could have occasionally died in caves, contributing their bones to the deposits below the nest or roost. However, with the possible exception of *Macrodermas gigas* (Ghost Bat), which is documented only at Devil's Lair, no permanently cave-dwelling vertebrate is very sensitive to vegetational conditions (Baird 1992, Balme *et al.* 1978), and none is analysed here.

Fast-moving animals could have sometimes fallen over the numerous cliffs concealed by vegetation in the Leeuwin-Naturaliste Region, and macropods, which occasionally seek shelter in caves (as indicated by droppings seen in the cave survey), might have died there from time to time. Bones that betray neither prey nor permanent residence in caves could be present without being identifiable as deriving from accident or disease. However, such contributions are probably effectively random over time, small compared to predator contributions, and hence would not confuse palaeo-environmental records.

Post-depositional alteration

Post-depositional processes that may confuse the potential palaeo-environmental records are either mechanical or chemical. Mechanical processes include movement and dispersal of bones by scavengers, such as those discussed above, fluvial action (Gifford 1981), and people sweeping or cleaning their occupation areas and removing large bone fragments (Hammond 1933, O'Connell 1987, Walters 1989). An additional action may be the digging of burrows or pits by animals or people. These last are evident in isolated areas in the deposits at Devil's Lair and Witchcliffe Rock Shelter, as discussed in Chapter 6.

Water-flow may have moved bones short distances, but most stratigraphic boundaries are intact and continuous, indicating no mixing from channel-cutting. The only evidence for channel-cutting is in layers 31-38 at Devil's Lair, which fall within a single analytical unit, Period I. All other layer boundaries at the four sites are sharp or mottled, neither case suggesting sub-surface disturbance. Water action in all the sites could have mixed sediments, and perhaps occasionally shifted bone fragments slightly, but it could not have introduced much material into underlying layers without destroying the layer boundaries identified above and used here to group bone assemblages.

Scavengers removing bones and people cleaning up camp-sites probably have influenced the composition of the faunal assemblage perhaps not too seriously, even though one cannot identify the extent of transport or lateral movement from test-excavations of limited area. Where layer boundaries are sharply-defined, bone dispersal is perhaps not serious, because it could not have been active enough to destroy the layer boundaries. Given the carnivore behaviour discussed above, carnivores could have removed many of the bones originally deposited, but such removal perhaps does not include teeth and jawbones, the skeletal parts identified here, since heads are least often scavenged (Marean 1991), and many of the identified teeth here are probably too small to have been noticed by people cleaning around the site, once they were incorporated into the surface deposit.

Chemical processes affecting bones include burning, weathering, and diagenesis. With the exception of calcined bone, which intense heat has weakened, and mussel-shell fragments, which are fragile to begin with, faunal remains at the sites are well-preserved. Almost all bone fragments are hard and resistant to breakage, as witnessed by their condition after post-excavation packaging and transport. During my laboratory analysis I noticed that bones in bags of sieve residue, which would have been pressed against rocks and other bones, rarely had fresh fractures. There has probably been some diagenesis of older bone fragments, manifested by the heavily mineralised appearance of bone from lower levels of the Devil's Lair deposit. In any case, chemical processes probably did not eliminate bones selectively. The chief observation in support of this is that the proportion of small animal bones, which are more delicate, is much the same throughout each deposit (see tables below).

In sum, post-depositional, largely attritional processes probably did not affect bone deposits as greatly as did the potential confusion created by different sources of bone. Of the latter, however, natural deaths of animals in caves are perhaps the least important.

Implications of taphonomic factors for palaeo-environmental interpretation

All of the above considerations imply that bones from human accumulations are difficult or impossible to distinguish from those left by other carnivores. The potential for one predator to deposit remains of one prey animal out of proportion to its representation in the local environment, and so confuse the palaeo-environmental record, cannot be easily assessed from physical criteria of the bone or from predator behaviour. Predators are shown above to leave their marks on the same bones, leave very similar marks in any case, and comminute all bones (regardless of origin) to barely identifiable fragments. Few or no terrestrial animals are exclusive to any category of predator. Several predators would have scavenged carcases of any animal, and then leave bone-filled scats in archaeological deposits.

Nevertheless, these assumptions are possible:

1. human occupation layers in caves and rock shelters do contain a proportion of human prey animals.

2. artefacts, pieces of exotic rock such as ochre or feldspar, burnt bones, emu eggshell fragments, and remains of fauna not hunted by non-human terrestrial carnivores, that is, fresh-water and marine molluscs and marine fish, were largely contributed by people. Stratigraphic associations between these items and faunal remains identified to species can therefore suggest which remains were more likely deposited by people.

3. a broadly similar range of non-human carnivores continued to visit caves during the long periods without human occupations, and during the shorter periods of human occupations. Carnivore contributions would therefore be expected in both human occupation and non-human occupation layers.

The major source of variation in taxa is therefore what is available in the environment, and what humans have contributed during relatively intense, concentrated occupations, compared to other animals' use of caves. By proceeding on this assumption, and provided that I identify probable human prey, I can compare the long term trends in the proportions of vertebrate taxa in different layers and periods. I identify human predation by identifying differences in the distribution of species in layers. Principal components analysis is used to indicate whether variation is largely explained by a single variable (such as changing environmental conditions) or more than one (such as changes in both environmental conditions and the relative importance of different predators).

Methods

Materials

Collection of faunal remains involved sieving excavated sediment through 3 and 5 mm mesh sieves in the field, and scrutinising and picking out faunal remains from the sieve residues in the field and in the laboratory. Some large bones and other faunal remains were identified and removed *in situ*. Faunal remains collected included bones, burnt bones, bone artefacts, emu eggshell, marine or fresh-water mollusc shell, and snail shells. The latter are from *Bothriembryon* spp., which live in dune sands like those found around the sites (Iredale 1939), that were probably washed into the deposits.

To remove large quantities of light plant material in surface layers, I wet-sieved sieve residues using a 1 mm mesh. Floating residue was removed and dried on absorbent paper. Tweezers, fluorescent lamps, and magnifying lenses were used to search through fine material from all layers. I retained most residues (some were discarded in the field when there were exceptionally good sorting conditions), even after laboratory sorting, in case of future analyses of fine charcoal or other material.

All faunal material, along with other archaeological and biotic material, was stored in resealable plastic bags in small boxes made of acid-free cardboard. The boxes contain faunal and other material from an entire spit, although some spits require two boxes.[1]

[1] All boxes of analysed material and bags of sieve residues are stored at the Western Australian Museum, Perth.

Table 8.2 Fauna recognised from laboratory sorting

Sorting class	Family	Genus & species	Common name (Nyoongar name)
Invertebrates:			
Molluscs			
terrestrial:	Bothriembryontidae	*Bothriembryon leeuwinensis*[1]	land snail
		Bothriembryon sayi[1]	
Aquatic:	Turbinidae	*Turbo torquatus*[2]	Turban shell
	Veranidae	*Katelysia scalarina*[2]	cockle shell
	Hyriidae	*Westralunio carteri*[3]	fresh-water mussel
	Amphineura	*Haliotis* spp[3]	abalone
Crustaceans	Parastacidae	*Cherax preissii* [4]	a type of koonak
		Cherax quinquecarinatus[4]	Jilgie
Vertebrates:			
Fish	Unidentified	species not distinguished	n.a.
Reptiles	Scincidae[5]	species not distinguished	skinks
Bats	mainly Vespertilionidae	species not distinguished	bats
Murids	Muridae [6]	species not distinguished	native mice and rats
Dasyurids	Dasyuridae[6]	small species not distinguished	small dasyurids
		Dasyurus geoffroii [6]	Western quoll (Chuditch)
Bandicoots	Peramelidae	*Isoodon obesulus*[6]	Southern Brown Bandicoot (Quenda)
		Perameles bougainville[6]	Western Barred Bandicoot
Possums	Petauridae	*Pseudocheirus peregrinus*[6]	Ring-tailed Possum (Nworra)
		Trichosurus vulpecula[6]	Brush-tail Possum
Potoroids (Rat-kangaroos)	Potoroidae	*Potorous tridactylus*[6]	Potoroo
		Bettongia penicillata[6]	Brush-tailed Bettong (Woylie)
		Bettongia lesueur[6]	Burrowing Bettong (Boodie)
Small macropods	Macropodidae	*Setonix brachyurus*[6]	Quokka
		Petrogale lateralis[6]	Black-footed Rock-wallaby
		Macropus eugenii [6]	Tammar
		Macropus irma[6]	Brush wallaby
Large macropod	"	*Macropus fuliginosus* [6]	Western Grey Kangaroo (Yongar)
Emu eggshell[7]	Dromaiidae	*Dromaius novaehollandiae*	Emu

[1] Identified by G. Kendrick, Palaeontology Department, Western Australian Museum, and by the author from Iredale (1939).
[2] Identified from Wells and Bryce (1986).
[3] Fragments of thin, nacreous mollusc shell, some identified by G. Kendrick. This is the only freshwater mollusc in south-western Australia.
[4] Genus and species identification by B. Knott, Department of Zoology, UWA.
[5] Some identified by Ken Aplin, Department of Terrestrial Vertebrates, WA Museum; others by J. Mead, University of Arizona.
[6] Identified by the author from comparative collections loaned from the WA Museum and UWA, and illustrations in Merrilees and Porter (1979).
[7] Identified by the author from the thickness of shell and exterior texture, which are unique to eggs of the emu in south-western Australia.

Sorting classes

Numerous classes of material recognised in preliminary sorting are summarised in Table 5.3. Faunal material was only one component of this material. Table 8.2 gives the taxonomic identifications of the bone material.

Given hundreds of sieve residues and many thousands of highly comminuted bone fragments to examine, I identified jaws and teeth only, which are distinctive skeletal parts. Most bones were so fragmented (by carnivore chewing, presumably) that they were impossible to identify to anatomical part. Any bias created by identifying only jaws and teeth may not be serious. The bias arises mainly because people and non-human carnivores may treat crania differently from other body parts. For example, some cultures esteem brains or tongue as food, while others view these parts as inedible. Crania might have been removed at the kill site, tossed aside at the consumption site, or heavily chewed beyond recognition (Capaldo 1998, Gould 1996). Two factors mitigate any negative influence:

1. Humans need not have removed any body parts to transport carcases to the sites. Apart from emus, whose bones are absent from the sites presented here, the extant Australian fauna, even adult male kangaroos, are small enough to be carried whole by an adult person (O'Connell and Marshall 1989). In Australia today, hunters often transport their prey whole to camps (Witter 1990). Crania may have been deposited in or near the sites regardless of their perceived value by prehistoric people.

2. Teeth and jaws are the most durable elements. If their deposition is unbiased, and given attrition of all the bones by carnivores, they can provide the best evidence for an animal's existence in a deposit.

To infer measures of bone fragmentation, potentially useful as a measure of carnivore comminution, I counted and weighed categories of unidentifiable and identifiable bone, defined as follows:

- *unidentifiable bone fragments*: lack end processes (epiphyses) or complete or partial cylindrical section, generally either a large, irregular scrap, or a tiny portion of a bone

- *potentially identifiable bone fragments*: whole bones, fragments with articulating ends or some other process permitting identification.

An additional class (*shaft fragments*), created for an aborted analysis, comprises unidentifiable fragments that are also bone cylinders or partial cylinders lacking epiphyses, and is treated here as unidentifiable bone.

I identified mammalian jaws and teeth by comparison with fossil material loaned by the Palaeontology Department, Western Australian Museum, and modern mammal skeletons loaned by the Zoology Department, UWA, and with the help of the *"Guide to the identification of teeth and some bones of native land mammals occurring in the extreme south west of Western Australia"* (Merrilees and Porter 1979). The guide provides photographs and descriptions of the distinct macroscopic features of the bones of all the species identified here. Where I felt that I lacked the expertise to identify specimens that were very similar in two or more species (such as molars from the potoroids *Bettongia lesueur* and *B. penicillata*, or the upper first incisors of small macropods), I made a general identification, as in "Potoroid, unidentified", or "Small macropod, unidentified". As species identifications are often more difficult to make with decreasing size of taxa, and given my lack of previous experience, I did not attempt to identify species of murids, bats, reptiles (mostly skinks), or the small dasyurids.

Estimation of taxonomic abundances

To estimate actual numbers of individuals, some analysts of vertebrate remains advocate using simply the number of identified specimens or "NISP" (Grayson 1984). As vertebrate individuals contain many bones, and some species' bones are more numerous, or more likely to be deposited, preserved, recovered, fragmented, or identified, according to various selective processes such as butchery or consumption of smaller bones, the NISP not only over-estimates the real number of individuals deposited, it varies from species to species (Horton 1984, Lyman 1994, Ringrose 1993).

Other estimates of abundance derive from the minimum number of individuals (MNI) that are needed to account for the specimens of a taxon in a stratum (Ringrose 1993). The basic method of MNI for vertebrates, first proposed in archaeology by White (1953, cited in Lyman 1994), is to identify skeletal elements for a the given taxon in the stratum and find the largest number for a single element from the left or right side. This is expressed as MNI= $max\ (L,R)$ where L and R denote the number of identified specimens of a given element and location. Variant expressions are $MNI = L + R - P$ and $MNI = (L + R)/2$, where P is the number of "paired specimens", i.e., pairs of left and right specimens that are potentially from the same individual. Paired specimens are sometimes examined to determine whether they are in fact from individuals of different age, size, or sex, and so represent more than one individual.

Methods derived from the MNI estimate - the Lincoln Index (LI), Minium Number of Elements (MNE), and Minimum Animal Units (MAU) - require many assumptions (Ringrose 1993), and the methods are not used here. The LI assumes that the two sides of the animal are statistically independent (unlikely for crania as both sides of the body part might be deposited together), and the MNE and MAU assume that different bone-accumulating and dispersing agents operated at different times (unlikely for reasons discussed above).

The major problems with the MNI estimate in all its forms are that certain types of bones may be highly fragmented, dispersed beyond the area of archaeological excavation, or mixed into adjacent layers (Horton 1984). This makes it difficult to determine whether specimens are from the same individual, and over time, the skeletal elements available for MNI estimation may decay selectively, according to circumstances of deposition or density of the bone. Some analysts (cited in Grayson 1984, Ringrose 1993) have attempted to identify the selective loss of bones by calculating the ratio NISP/MNI, but this ratio is known to be correlated with sample size (the larger the total sample, or NISP, the smaller the increase in MNI). However, by consistently identifying parts that are durable (*e.g.*, teeth), and comparing groups of similar-sized specimens, equally susceptible to movement, the problem of selective dispersal and fragmentation is reduced. It is still not easy to allow for selective removal of parts by scavengers, but teeth, at least, would have little attraction as food to most carnivores.

Moreover, MNIs increase with the number of the stratigraphic units used to divide the total sample (Grayson 1984). Combining MNIs from a finely-divided deposit will provide a higher overall estimate than calculating MNIs from the same deposit divided into a few large strata. However, age estimates and stratigraphic analysis show whether stratigraphically-organised samples are reliable. At Tunnel Cave and Witchcliffe Rock Shelter, I assume that the chrono-stratigraphic boundaries argued in Chapter 6 are reliable, and I therefore use the MNI.

Recording and excavation methods at the other sites constrain my choice of abundance estimate. At Devil's Lair, Balme *et al.* (1978) list MNI estimates by layer, and at Rainbow Cave, Jackson (1992) lists NISPs by spit. The former site is well-stratified, and the use of groups of layers divided by lithified bands (forming Periods) separates bones of different age reliably (see Chapter 6). Rainbow Cave was apparently not excavated stratigraphically, but by spits intersecting layers (at least, this is how the data are presented), which means that bones from the same taxa in adjacent spits cannot be assumed to be from different animals, and no MINI estimate is possible. Therefore, I use the published estimates for each site, as the most suitable for that site.

MNI estimates are calculated in various ways: here they are based on the symmetry of jaws about the midline, size variations between individuals, and stages of tooth development and wear in juveniles and adults. Every identified element has a potential opposite and matching pair. Within every layer or sub-layer (e.g., hearths, Tunnel Cave layers 7A through 7H), I attempted to find matching pairs by comparing all potential candidates. I eliminated jaw fragments and teeth as potential pairs if they included the same teeth on the same side and jaw (mandible or maxilla), had different wear patterns on the teeth, or indicated animals of different size or age. These principles allowed identification of more than one individual from teeth or jaw fragments even if they were from the same side and dentary. This MNI procedure is pair-matching, following the formula $MNI = L + R - P$, where L and R denote the number of non-matching specimens from both sides and P the number of pairs of opposing elements that could *not* be identified as being from different animals (Horton 1984, Ringrose 1993).

Assumptions

Differences in analytical approach at the four sites require me to make assumptions in four main areas:

Firstly, researchers at each site have estimated the number of animals using different methods. In contrast to Tunnel Cave and Witchcliffe Rock Shelter, where I identified species from jaws and teeth, species at Devil's Lair and Rainbow Cave were identified from jaws, teeth, scapulae, humeri, pelves, femora, and calcaneae. At Rainbow Cave, Jackson did not calculate MNIs. However, each deposit was analysed consistently, so I can compare periods within deposits. Hence I assume that variations in the estimates of abundance are due to changes at the site, either in the net predator contribution (*e.g.*, as might be expected when the Devil's Lair entrance widened, or when people occupied Tunnel Cave) or in vegetation.

Secondly, I combine MNI estimates for multiple Devil's Lair layers that are essentially the same radiocarbon age, or "Period". Normally, one should not combine MNIs from multiple layers that represent a single period, since there is a risk of exaggerating the apparent MNI by combining the layer MNIs. Examination of the most commonly-represented elements in the original data (not published, nor easily available) would probably suggest a lower number than the aggregate of relevant MNIs, in which MNIs from each layer are calculated separately and then combined. However, the "aggregate MNI" is perhaps not a greatly inflated estimate, since some periods may have lasted millennia, and seen the deposition of many more individuals than normally would be estimated from the specimens from that period.

Thirdly, the "aggregate MNI" over-estimate at Devil's Lair may be offset by a potential source of under-estimate, important where bones are highly dispersed (Grayson 1984, Hubbard 1993, Watson 1979). At Devil's Lair, bones were probably scattered and fragmented by scavengers, and individual prey animals probably contributed relatively few of their bones to the small area of the test-excavation (Balme *et al.* 1978: 57-59; see also quote from Chapter 6). I therefore assume that the use of aggregate MNIs and the probability of a high degree of bone dispersal and fragmentation mitigate each other.

Fourthly, researchers' expertise varies. As a relative novice to faunal identification, I may have failed to identify some bones at Tunnel Cave and Witchcliffe Rock Shelter that the Devil's Lair researchers would have identified. I may have also mis-identified a small proportion of bones, although I placed "difficult" bones in a general taxonomic group, so that this error would be the smaller. Balme *et al.* (1978) note that two researchers concurred on all bone identifications, while no palaeontologist has confirmed my bone identifications, *c.*7900 of them (1583 to species) at Tunnel Cave and 685 at Witchcliffe Rock Shelter (320 to species). However, many months' laboratory work on well-preserved material, work at other sites (Dortch and Kelly 1997), discussions with palaeontologists (A. Baynes, J. Long, Western Australian Museum), and my own checks of my earlier identifications, have mitigated this problem somewhat. Jackson (1992) states that he took similar care with the Rainbow Cave identifications. I therefore assume that unverified specimens from Tunnel Cave, Witchcliffe Rock Shelter, and Rainbow Cave are, on the whole, correctly identified.

In conclusion, these variations would introduce some error into comparisons of sites, but they should matter little for comparisons of periods within deposits, as carried out here.

Statistical methods

For each site I first characterise the variation in different species over time, using Principal Components Analysis (PCA) and analyses of diversity, and then proceed to identify biases caused by predators', especially humans', contributions to the bone samples. Statistical significance is assessed by several methods, predominately chi-squared (χ^2) test.

PCA condenses variation of multiple taxa in many samples into a few major vectors that account for a large percentage of the total variation. These vectors are then compared with known differences between layers, such as general environmental conditions at the time of their formation (see Chapter 3), and whether layers contain human occupational remains. Of several multi-variate analyses, PCA is used here because it is well-established (Pearson 1904, cited in Wright 1992; Shennan 1988) and because the vectors (components) are independent[2]. Components are ordered according to how much variation they account for, with the first principal component (PC1) accounting for the most variation.

The PCA results are shown as scattergrams with PC1 and PC2 as *x* and *y* axes. Variables (taxa) and objects (the layers or periods in which the taxa are counted) are points in these graphs, the position of each being determined by its component loading, or importance to each component (Wright 1992). Component loadings are calculated with the statistics program "MV-Arch" (Wright 1992) from a table of covariances (between each variable and object) calculated from square root transformations of the percentage frequencies of each taxon in each layer.

Chi-squared (χ^2) significance tests of the variations noted in PCA are based on the following contingency table:

taxon *x* in layer *a*	taxon *x* in all other layers besides *a*
all other taxa, not *x*, in layer *a*	all other taxa, not in *x*, in all other layers besides *a*

[2] Correspondence Analysis would also be appropriate for this data, except that it is designed to magnify small differences between variables and objects (as I found through preliminary comparison of the methods). Since the data may contain small inaccuracies, I use PCA for a more conservative interpretation.

As well as examining changes in abundance over time within a given taxon, I compare the faunal assemblages of different layers, as a means of identifying change in predator contributions. Differences between layers are assessed in terms of taxonomic diversity.

Diversity is a means of comparing groups of taxa used widely in archaeology and ecology (Leonard and Jones 1989; see also Appendix 12). Two concepts are employed: *richness* and *evenness*. Richness is the number of species in a sample; evenness is how evenly are species represented across the sample. Richness and evenness indicate whether samples are broadly similar to others, even if the species counts vary somewhat.

Assuming that non-human carnivores contributed bones throughout time, whereas humans contributed bones episodically, differences in diversity should be greatest between human occupation layers (such as hearth layers at Tunnel Cave and Witchcliffe Rock Shelter) and other layers. Human hunting strategies may also have resulted in a different range of taxa being deposited, but identifying this range with any precision is an undertaking beyond the scope of this study. I therefore base my comparisons on the probability that frequency and continuity of occupations by the various carnivores that occupied the sites were most marked between humans and all the others.

Nonetheless, some steps can be made towards identifying some of the human prey species. The first is to determine the proportion of burnt specimens in a taxon as an indication of animals that are deposited during occupations by people. This procedure is based on the premise that bones are either burnt because they were close to the surface where a hearth was built or because they were discarded into a hearth (Archer 1977, Balme 1980a). Balme (1980a) shows that burnt bone is virtually absent from sites never occupied by people (cf. Porter 1979), the weight of burnt bone is correlated from layer to layer with the number of artefacts, and more than 5% of the bones of the larger animals (*M. fuliginosus*, *M. eugenii*, *Petrogale*, *Setonix*, *B. lesueur*, *B. penicillata*, *Trichosurus*, and *Isoodon*) are burnt (Balme 1980a: Figure 3). Some larger animals' bones are rarely burnt (*Potorous* and *M. irma*), and all taxa are present in layers without burnt bone, demonstrating that the extremely frequent burning of some animals' bones is unrelated to selective destruction of other animals' bones through burning or other degradation. Bones of the larger animals, artefacts, and food remains such as fragments of emu eggshell usually appear together. Balme (1980a) therefore concludes that people most often hunted the animals listed above.

Walters (1989) questions Balme's assumption about how bones are burnt in occupation sites. Citing ethnographic research in central Australia, Walters states that people like to keep fires clean and discard bone away from campfires, nor do they build fires on bone-strewn surfaces. These practices are affirmed by other Australian ethnographic observations (Bindon 1986: 26-27), but unconfirmed in south-western Australia, where detailed comments are lacking[3]. Walter's comments may not apply to the sites discussed here. Bones are present in all layers, yet hearths were built, and the human occupiers of Devil's Lair and Tunnel Cave had no dogs to remove bones from the campsites. Possibly, sufficient intervals passed between occupations for people to be unconcerned with the bone content of the deposit beneath the campfire, allowing some bones to be burnt unintentionally.

Burnt bones suggest human activity also because Balme (1980a: Figure 2) shows that burnt bone and artefacts occur together in Devil's Lair layers without hearths. There is probably some lateral dispersal of burnt bone and artefacts from hearths not revealed by excavation, perhaps by scavengers. If vertical movement is confined within large stratigraphic units as used here, this dispersal would still result in correlations between burning and bones (among other things) deposited by people. Balme's method is therefore extended to the other sites, and to other materials.

Balme (1980a) mentions apparent correlations between the numbers of specimens of some taxa and the numbers of archaeological finds. I therefore test the rank-order correlations of identified bones and archaeological finds, including artefacts, burnt bones, and exotic stones (*e.g.*, ochre and feldspar) and food remains (*e.g.*, fragments of aquatic mollusc shell and emu eggshell). At the sites where they are represented, hearths (ranked by volume) and fish bone are included among the archaeological finds. The use of emu eggshell fragments as an indicator of human occupation rests on the assumptions that emus did not nest in caves and rock shelters, nor did non-human carnivores bring eggshell to caves (except possibly for small pieces in scats). Moreover, archaeological emu eggshell is frequently burnt, as if cooked on coals, a practice observed among south-western Australian Aboriginal people in the 19th C (Hammond 1933). In these analyses, I assess sample size dependency by calculating correlation between the total number of identified specimens (NISP) and the total amount of bone (cf. Grayson 1984).

Devil's Lair

Balme *et al.* (1978) and Baynes *et al.* (1975) detail the Devil's Lair faunal assemblage and its environmental implications (see Chapter 3). Merrilees (1979, 1984) discusses faunal succession further, and Lundelius (1983, 1989) examines associations of "disharmonious fauna", or taxa not now associated in the same environments. Balme (1979, 1980a, 1980b) identifies artificial biases in the representation of skeletal elements and burnt bones. Here, I identify environmental indicator species using new methods, and extend Balme's identifications of human prey species, by additional tests to hers.

PCA

Figure 8.1 shows Devil's Lair layers and taxa as points on a scattergram with Principal Components 1 and 2 as the axes. Variables (taxa) and objects (Periods) at the centre of the axes have contributed little to these components. PC1 extends from medium-size species *Trichosurus*, *Bettongia penicillata*, *B. lesueur*, and *Petrogale* (high positive values) to Period I at the extreme negative end. The only species that score negatively on PC1 are the only ones that are present in Period I (Table 8.4), so PC1 probably indicates the presence of species entering the site after Period I, that is, during the time when the former entrance is inferred to have been large enough to admit the entire Devil's Lair fauna, not just bandicoots, potoroos, and other potential owl prey. Thus PC1 probably indicates the size of the Devil's Lair entrance.

Setonix and *Perameles* lie at opposite ends of PC2. These species have opposite environmental requirements (thickets versus open vegetation), suggesting that PC2 reflects environment. Other species' scores are consistent with this interpretation. *B. lesueur* and *Petrogale* (positive on PC2) require open vegetation, and *Potorous* and *Pseudocheirus* (negative on PC2) inhabit closed vegetation. *Trichosurus*, *B. penicillata*, *Macropus eugenii*, *M. irma*, and *M. fuliginosus* are adapted to many environments and score close to zero on PC2.

Isoodon also ranges widely, so the reason for its high score on PC2 is unclear, though perhaps reflecting past preferences. *M. eugenii* requires thickets for shelter and open vegetation for forage. Its central position on PC2 suggests that it is less sensitive to changing environments than *Setonix* or *Petrogale*.

Consistent with the correlation of PC2 and environments, Periods VI through IX (representing the LGM and late glacial) are at the opposite end of PC2 from Periods II, III, XI, and XIII (representing the periods 34,600 to 25,000 BP and 13,800 to 12,300 BP, on either side of the glacial maximum). A low score for Period I is attributable to its lack of indicator fauna. The other periods, all low-scoring, represent transitions in the glacial cycle.

All of these observations support inferences of Balme *et al.* (1978) and hence application of PCA elsewhere. PCA also suggests that the first two components, probably the configuration of the cave entrance and local environmental conditions, explain as much as 81% of the variation in the identified fauna (Appendix 13).

---

3 One contemporary observation of 19th C Nyoongar in south-western Australia states that "such rubbish as fish bones were thrown on the fire to be destroyed" (Hammond 1933: 26), but treatment of other bone is not mentioned.

Table 8.3 **Distribution of fauna (MNIs) throughout the Devil's Lair deposit, arranged by taxonomic groups**

In Tables 8.3 and 8.4, small species not identified at other sites are aggregated into groups. As in Chapter 7, I exclude Period XIV because it potentially includes a mixture of modern and Pleistocene deposits. I combine the small samples from Periods XII and XIII.

Period	Birds	Lizards	Bats	Murids	Dasyurids	Bandicoots	Possums	Potoroids	Small macropods	Large macropod	All mammals	All vertebrates
XIV	7	3		9	2	3	3	2	1		20	31
XIII	1	1		1	1	2	4	5	5	1	19	21
XII	7	5	1	14	5	4	10	12	17	3	66	80
XI	11	3		18	14	4	12	13	12	3	80	95
X	34	95	1	130	69	47	53	70	49	10	445	576
IX	7	229	2	69	23	31	18	35	25	7	218	469
VIII	14	166	1	41	22	29	12	27	18	4	163	351
VII	15	62		18	17	25	23	41	30	7	163	250
VI	29	156	3	70	59	56	58	87	50	14	409	614
V	7	16	1	24	20	12	21	19	15	7	125	152
IV	12	27		16	20	13	26	23	20	4	125	171
III	20	30	1	104	42	17	50	49	43	8	332	393
II	56	73	11	630	206	75	35	55	130	24	1463	1616
I	44	31	11	304	101	59	19	20	12	2	840	931
Total	264	897	32	1448	601	377	344	458	427	94	4468	5750

Table 8.4 **Distribution of fauna (MNIs) throughout the Devil's Lair deposit, arranged by species**

Totals differ from Balme *et al.* (1978: Table 3) because of elimination of mixed units that cross phases, and because of minor counting errors in that table.

Period	*Pseudomys albocineurus*	*Pseudomys occidentalis*	*Pseudomys shortridgei*	*Pseuodmys praeconis*	*Notomys*	*Hydromys*	*Rattus*	*Phascogale*	*Antechinus*	*Sminthopsis*	*Dasyurus*	*Cercatetus*	*Tarsipes*	*Phascolarctos*	*Vombatus*	*Sthenurus brownei*	*Sthenurus occidentalis*	*Protemnodon*	*Zygomaturus*	*Sarcophilus*	*Thylacinus*	*Lagorchestes*
XIV	1						8		2													
XIII							1				1											
XII	1		1				12		2		3											
XI	3		1		1		13		9	3	2	4										
X	42		7		30		51	1	19	31	18	10	2							4		
IX	26		5	1	14	1	22	1	2	13	7	5	2							1		
VIII	14		3		7	2	15		1	12	9	6	1							2		
VII	7				5		6			6	11	1								1		
VI	34		4		4		28	1		34	24	8	1							3		
V	6		2				16	4	1	10	5	3	2							1		
IV	4						12	2	1	8	9	3										
III	27		5		3		69	1	6	23	12	12	4							2		
II	141	12	101		30	4	342	12	50	125	19	44	243	1		1		2		6		
I	66	1	47		1		189	10	22	61	8	24	271		1	4	3	2	1	2	2	2
Total	372	13	176	1	95	7	784	32	115	326	128	120	526	1	1	5	3	4	1	22	2	2

Period	*Isoodon*	*Perameles*	*Pseudocheirus*	*Trichosurus*	*Potorous*	*Bettongia penicillata*	*Bettongia lesueur*	*Setonix*	*Petrogale*	*Macropus eugenii*	*Macropus irma*	*Macropus fuliginosus*	Analysed sample
XIV	2	1	1	2	1	1		1					9
XIII	2		2	2	2	2	1	3	2			1	17
XII	4		5	5	4	7	1	11	4		2	3	46
XI	3	1	7	5	4	7	2	8	2	1	1	3	44
X	33	14	28	25	9	42	19	17	14	10	8	10	229
IX	22	9	10	8	3	23	9	6	13	3	3	7	116
VIII	18	11	5	7	2	14	11	5	9	1	3	4	90
VII	16	9	10	13	4	24	13	5	18	2	5	7	126
VI	38	18	29	29	9	53	25	14	27	5	4	14	265
V	10	2	13	8	4	11	4	4	3	2	6	7	74
IV	11	2	16	10	3	14	6	6	11	1	2	4	86
III	12	5	35	15	15	25	9	22	8	7	6	8	167
II	37	38	28	7	33	22		60	20	29	21	24	319
I	30	29	19		20			9		1	2	2	112
Total	238	139	208	136	113	245	100	171	131	62	63	94	1700

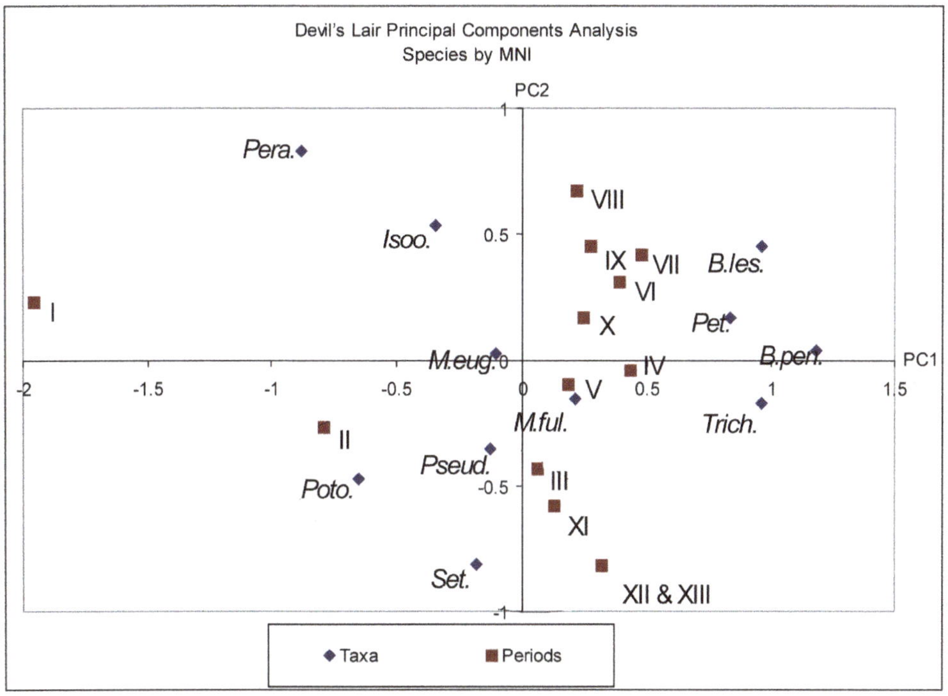

Figure 8.1 PCA scattergram of Devil's Lair taxa and Periods

Table 8.5 Results of chi-squared tests comparing changes over time in Devil's Lair taxa

Null Hypothesis or H0 (variation in each taxon in each Period is no more that would be expected from random variation in identical populations) is rejected if p (H0) < 0.05. H = significant difference is higher than would be expected from H0; L = significant difference is lower than expected from H0; other cells - cannot reject H0. Sources of bone in Period I probably differ markedly from subsequent Periods, and it is excluded from the analysis.

Period	*Isoodon*	*Perameles*	*Pseudocheirus*	*Trichosurus*	*Potorous*	*B. penicillata*	*B. lesueur*	*Setonix*	*Petrogale*	*M. eugenii*	*M. irma*	*M. fuliginosus*
XII & XIII		L						H				
XI												
X												
IX	H											
VIII	H	H					H					
VII							H	L	H			
VI						H	H	L			L	
V												
IV			H									
III	L	L	H									
II		H		L	H	L	L	H	H	H	H	

Figure 8.2 Proportions of Devil's Lair species in total MNI

Chi-squared tests of the frequencies of species in each layer suggest that several species vary independently over time (Table 8.5). Considering these results and Figure 8.2 together, *Pseudocheirus*, *Potorous*, and *Setonix* are exceptionally abundant in relatively humid Periods II, III, XII and XIII, consistent with their preference for closed vegetation. *Perameles, B. penicillata, B. lesueur*, and *Petrogale* are most abundant in the relatively arid Periods VI, VII, VIII, and IX, which is also consistent with their present day ecology. *M. fuliginosus* and *Trichosurus* are no better represented in any Period, perhaps as a result of these animals' adaptability (Strahan 1995). *Trichosurus* is rare in Period II, perhaps because of a change in predator.

The abundance of *Perameles* in Period II, when other species suggest closed vegetation, is perhaps due to its vulnerability to owls, which were still making a significant contribution in Period II. The larger taxa in this period could have been contributed by a range of predators, and perhaps more faithfully indicate their abundance in the environment. In the pre-glacial Period III, when the relative importance of the owl contribution is suddenly much reduced (*i.e.*, above layer 18: Balme *et al.* 1978), *Perameles* becomes lower than expected. Hence the significant changes in the proportions of taxa at Devil's Lair are all consistent with vegetational changes proposed by Balme *et al.* (1978).

Diversity

Measures of richness are potentially misleading at Devil's Lair, as they are negatively correlated with sample size (Table 8.6). Spearman's r_s values between *dl* and layer MNI, and between *R* and layer MNI, are -0.85 ($p < 0.001$) and -0.68 ($p < 0.014$). On the other hand, measures of evenness are uncorrelated with layer MNI (for $1/\lambda$, Spearman's $r_s = 0.47, p > 0.124$; for J, $r_s = 0.44, p > 0.150$). The proportion of the most abundant taxon, $n_{i\ max}/N$, and MNI are also uncorrelated ($r_s = 0.14, p > 0.65$).

In the terminology of Frankel and Holdaway (1993), most periods are moderately even. Period I is less even than all other phases, probably because of its distinctive contribution from owls. Period II is very even than most others: Balme *et al.* (1978) note that both owls and other predators contributed bones in Period II, the only period with heavy contributions from all these predators, and the one providing the largest sample.

Differences in taxonomic structure do not seem attributable to potential human influences. All periods but Period I might include bone contributions from humans, but two-sided Smirnov tests show that the distribution of taxa in Period I is dissimilar only to Period XI ($p < 0.05$). Human contributions may have been a small addition to the total bone accumulation by many predators, suggesting that palaeo-environmental indications are unbiased by human contributions. I investigate this suggestion further by identifying correlations between taxa and cultural material in the uppermost part of the deposit that was accessible to human occupiers, represented by Periods II to XIII.

Table 8.6 Devil's Lair faunal diversity estimates

$n_{i\ max}$ = number of the most numerous taxon in the sample.

Period	Sample data					Richness estimates				Evenness estimates			
	MNI	rank	S	$n_{i\ max}/N$	rank	*dl*	rank	*R*	rank	$1/\lambda$	rank	J	rank
XII & XIII	63	11	9	0.222	2	1.931	8	1.134	5	7.891	11	0.953	2
XI	44	12	11	0.182	11	2.643	1	1.658	1	8.381	8	0.927	9
X	229	3	11	0.183	10	1.840	9	0.727	9	9.390	3	0.948	3
IX	116	6	11	0.198	6	2.104	5	1.021	6	8.353	9	0.929	8
VIII	90	8	11	0.200	4	2.222	4	1.160	4	8.411	7	0.914	10
VII	126	5	11	0.190	7	2.068	6	0.980	7	8.975	4	0.932	6
VI	265	2	11	0.200	4	1.792	10	0.676	10	8.658	6	0.937	4
V	74	10	11	0.176	12	2.323	2	1.279	2	9.641	1	0.936	5
IV	86	9	11	0.186	9	2.245	3	1.186	3	8.254	10	0.912	11
III	167	4	11	0.210	3	1.954	7	0.851	8	8.685	5	0.932	6
II	319	1	10	0.188	8	1.561	11	0.560	12	9.568	2	0.960	1
I	112	7	7	0.268	1	1.272	12	0.661	11	4.847	12	0.817	12

Human contributions to the bone sample

In this analysis the upper part of the Devil's Lair deposit (Periods II-XIII) is of most interest. The lower part (Period I) has a large proportion of small mammals (murids, bats, and dasyurids), mammals over lizards, and unbroken over broken bones. This pattern suggests the accumulation of bones by owls, as well as a few deaths of small mammals entering the cave voluntarily (Balme *et al.* 1978).

In the upper part of the deposit, broken bones are attributed to both Tasmanian devils and people; the former could have also scavenged meal remnants of the latter. Balme *et al.* (1978) do not distinguish the human contribution on the basis of prey body size, since *Sarcophilus* could have hunted many species. The dual contribution is confirmed by comparisons of the quantities of archaeological and faunal remains (Table 8.7). The former are mostly not correlated with the total MNI (Table 8.9; the exception is emu eggshell).

Table 8.7 Quantities of archaeological and faunal material at Devil's Lair, standardised against volume of sediment excavated for each period

		Archaeological remains			Faunal remains
Period	Volume sediment (*l*)	Flaked stone artefacts (number)	Other stone artefacts (number)	Emu eggshell fragments (weight in g)	Total MNI
XII, XIII	240	<0.001		0.019	0.263
XI	60	0.400		0.160	0.733
X	755	0.895	0.017	0.032	0.303
IX	385	0.151		0.002	0.301
VIII	480	0.330	0.025	0.000	0.188
VII	385	3.312	0.021	0.181	0.327
VI	1635	2.331	0.011	0.128	0.162
V	340	0.180		0.017	0.218
IV	725	0.059	0.003		0.119
III	470	0.041		0.007	0.355
II	2140	1.780	0.014		0.149
I	2300	0.030			0.049

Table 8.8 Spearman's rank order correlations (r_s) between classes of archaeological and faunal material at Devil's Lair, as outlined in Table 8.7

The probability of the Null Hypothesis *p* [H0] is calculated from the T-statistic, which in turn is calculated from Spearman's r_s using a standard formula (Appendix 12; Startup and Whittaker 1982). Where *p* is equal to or greater than 0.05, r_s is deemed non-significant; where *p* is less than 0.05 results are deemed significant and are shaded in the table.

Category	Estimate	Flaked stone (no.)	Unflaked exotic stone (no.)	Emu eggshell (wt)	Total MNI
Flaked stone	r_s	---------	**0.65**	0.55	0.18
	T-stat	---------	2.682	2.077	0.561
	p	---------	**0.023**	0.065	0.587
Unflaked,	r_s	---------	---------	0.43	0.23
	T-stat	---------	---------	1.52	0.762
	p	---------	---------	0.159	0.464
Emu eggshell	r_s	---------	---------	---------	**0.71**
	T-stat	---------	---------	---------	3.186
	p	---------	---------	---------	**0.010**
Total MNI	r_s	---------	---------	---------	---------
	T-stat	---------	---------	---------	---------
	p	---------	---------	---------	---------

Table 8.9 Spearman's rank-order correlations (r_s) between categories of archaeological material and taxa at Devil's Lair.

The probability of the Null Hypothesis *p* is calculated from the T-statistic shown above, which in turn is calculated from Spearman's r_s using a standard formula (Appendix 12; Startup and Whittaker 1982). Where *p* is equal to or greater than 0.05, r_s is deemed non-significant; where *p* values are less than 0.05, results are deemed significant and are shaded in the table. Mammal species are calculated from MNIs; lizards (species unidentified) from NISPs.

Category	Estimate	*Isoodon*	*Perameles*	*Pseudocheirus*	*Trichosurus*	*Potorous*	*Bettongia penicillata*	*B. lesueur*	*Setonix*	*Petrogale*	*Macropus eugenii*	*M. irma*	*M. fuliginosus*	Lizards
Flaked stone (number)	r_s	0.31	0.55	-0.04	0.245	-0.077	0.287	0.500		0.231	0.399	0.259	0.178	0.476
	T-stat	1.023	2.077	0.144	0.798	0.244	0.946	1.826		0.750	1.374	0.847	0.573	1.709
	p	0.331	0.065	0.888	0.4433	0.8122	0.3663	0.0979	1.000	0.471	0.199	0.417	0.579	0.118
Unflaked, exotic stone (number)	r_s	0.35	**0.59**	0.021	0.255	0.045	0.241	0.503	0.136	0.402	0.220	0.227	0.084	0.556
	T-stat	1.194	2.337	0.066	0.835	0.144	0.786	1.843	0.435	1.389	0.714	0.738	0.266	2.115
	p	0.260	**0.042**	0.948	0.4233	0.8884	0.4500	0.0952	0.673	0.195	0.492	0.478	0.795	0.061
Emu eggshell (weight in g)	r_s	0.56	0.20	**0.61**	**0.85**	0.423	**0.80**	**0.75**	0.437	**0.61**	0.318	0.556	**0.63**	0.381
	T-stat	2.154	0.631	2.424	5.095	1.477	4.227	3.567	1.537	2.402	1.061	2.115	2.561	1.304
	p	0.057	0.542	**0.036**	**0.001**	0.1706	**0.002**	**0.005**	0.155	**0.037**	0.313	0.061	**0.028**	0.222
Total MNI	r_s	**0.80**	0.25	**0.84**	**0.94**	**0.63**	**0.92**	**0.83**	0.74	**0.72**	0.629	0.741	**0.81**	0.476
	T-stat	4.176	0.810	4.949	9.052	2.561	7.589	4.682	3.492	3.284	2.561	3.492	4.332	1.709
	p	**0.002**	0.4366	**0.001**	**0.0000**	**0.0283**	**2×10^{-5}**	**0.001**	**0.006**	**0.008**	**0.028**	**0.006**	**0.002**	0.118

The weak correlation between archaeological remains and identified bones extends to comparisons of stone artefacts and individual taxa as well (Table 8.9). Since there are few correlations between taxa and artefacts, the amount of faunal material deposited from non-human sources is probably relatively large. Only emu eggshell, almost certainly deposited by people, may indicate that some species are human prey. The significant correlations between it and many taxa are probably not a sample size effect, as there is no rank-order correlation between any of the rank-order correlations and the total number of any species throughout the deposit (smallest $p > 0.443$). The correlations between emu eggshell weight and taxa affirm Balme's (1980) identifications of human prey. As in Balme's analysis, the small samples provided by *M. irma* and *M. eugenii* are not correlated with human activity. In contrast, the absence of correlations between emu eggshell weight and large samples of *Potorous* and *Isoodon* suggest that other carnivores preyed on these animals as well.

Major human prey animals probably included all the animals discussed here except for *Isoodon*, *Pseudocheirus*, *Potorous*, and *Perameles*. However, as argued above, if human contributions of bone were small relative to the total amount deposited, then even the inclusion of human prey would not unduly bias palaeo-environmental interpretation. The agreement between human prey and non-human prey as environmental indicators is illustrated in Figure 8.3.

In Figure 8.3, the strongest environmental indicator species (xeric species are *Perameles* and *Petrogale*; mesic species are *Potorous* and *Setonix*) appear to have responded to glacial aridity and post-glacial humidity regardless of whether there were human influences on their representation in the deposit (Periods VII and VIII include the last glacial maximum, and Period XIII indicates the terminal Pleistocene) These inferences are supported by the record of Devil's Lair lizards, which peak at the LGM (Balme *et al.* 1978: Figure 11), and are not major human prey (*p* H0, that they are not rank-order correlated with archaeological material > 0.05).

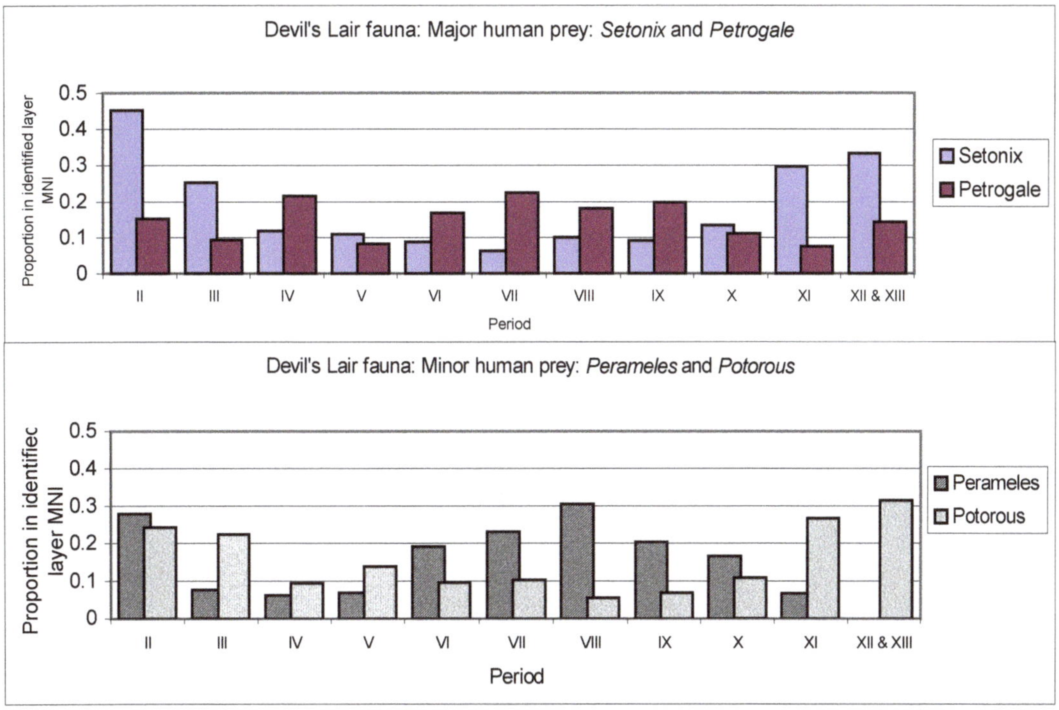

Figure 8.3 Proportions of environmental indicator species in respective populations of major and minor human prey

Summary

Remains of species representing closed or open vegetation seem to have been deposited in proportion to the species' distribution in the environment, regardless of their source. The availability of suitable habitat is probably a major influence on species numbers in the deposit. The high proportions of lizards, *Perameles*, and *Petrogale* suggest large areas of open vegetation, in Periods VII, VIII, and IX (18,400- 13,100 BP).

Today, open vegetation exists within a few hundred metres of Devil's Lair (F.G. Smith 1973), but the total area of such habitat is evidently insufficient to support *Perameles*, *B. lesueur*, and *Petrogale* – none of these taxa are extant locally. A large expanse of open coastal vegetation also grows west of the Leeuwin Ridge crest, *c.*3 km west of Devil's Lair (F.G. Smith 1973), and due to wind conditions, the vegetation west of the ridge crest was probably always open, even at the end of the Pleistocene, equivalent to Periods XII-XIII (Balme *et al.* 1978). Pleistocene animals would have only found more areas of open vegetation than exist today if vegetation east of the crest was also open. Since rainfall is the major determinant of vegetation composition and structure (holding soil and topography constant; cf. Beard 1981), the Devil's Lair faunal records indicate that rainfall was lower, and that vegetation around the site was more open in the Late Pleistocene. Tunnel Cave, the next site analysed, extends the same type of record from the terminal Pleistocene to Holocene.

Tunnel Cave

Tables 8.10 and 8.11 show the taxa identified at Tunnel Cave, by NISP and MNI.

Table 8.10 Distribution of fauna at Tunnel Cave, arranged by taxonomic group (NISP data, cf. Appendix 14)

Type:	Invertebrates		Bird egg	Autochthonous vertebrates		Allochthonous vertebrates								
Layer	Land-snail	Aquatic mollusc shell	Emu egg-shell	Birds	Bats	Lizards	Murids	Dasyurids	Bandicoots	Possums	Potoroids	Macropods	Human (decid. tooth)	Total alloch. vert.
1	872	1		3	1	4	79	20	28	14	30	34		209
2	1150			7	4	5	131	53	11	17	20	55		292
3	469			10	1	1	57	18	4	4	3	21		108
5-upper	213		1	41	6	9	125	86	33	54	69	119	1	496
5-lower	275	17	72	11	1	128	248	72	121	50	130	95		844
6	249	13	25	15	2	130	428	118	171	81	165	128		1221
7-upper	175	24	127	3	6	66	244	78	120	64	77	108		757
7-lower	178	84	180	1	16	106	257	113	140	100	179	197		1092
8	303	12	9		1	80	80	27	54	37	61	47		386
9-upper	391	14	48		1	60	130	38	80	39	100	67	1	515
9-middle	542	7	11		1	56	161	40	40	31	37	58		423
9-lower, 10	2132	7	18	5	8	167	669	130	137	140	181	161		1585
Total	6949	179	491	96	48	812	2609	793	939	631	1052	1090	2	7928

Table 8.11 Distribution of identified species at Tunnel Cave (NISP and MNI data, cf. Appendix 14)

Layer NISP	*Isoodon*	*Perameles*	*Pseudocheirus*	*Trichosurus*	*Potorous*	*Bettongia penicillata*	*B. lesueur*	*Setonix*	*Petrogale*	*Macropus eugenii*	*M. irma*	*M. fuliginosus*	Total
1	28		10	4	5	10	4	27		4		2	94
2	7		8	9	4	4	7	28	1	11	2	2	83
3	2		3	1		1		8	2	8	1		26
5-upper	33		44	10	21	23	5	24	28	24	5	7	224
5-lower	101	20	26	24	9	40	21	18	36	8	6	8	317
6	117	53	34	47	8	42	44	7	49	19	4	18	442
7-upper	85	16	42	20	3	34	10	11	49	7	5	9	291
7-lower	94	24	66	35	7	66	30	9	98	20	4	24	477
8	29	23	19	17	6	17	11	5	24	2		2	155
9-upper	53	22	17	21	9	25	10	2	32	7		7	205
9-middle	27	6	15	15	2	6	2	7	36			2	118
9-lower, 10	66	35	104	33	18	42	13	6	62	19		15	413
TOTAL	642	199	388	236	92	310	157	152	417	129	27	96	2845
Layer MNI													
1	3		2	1	2	3	1	3		1		1	17
2	2		2	2	1	1	2	4	1	2	1	1	19
3	1		2	1		1		3	1	2			11
5-upper	2		7	2	4	7	1	5	5	5	3	1	42
5-lower	11	4	4	5	4	7	4	5	7	2	2	2	57
6	9	7	4	4	2	9	6	2	9	3	1	3	59
7-upper	13	7	8	6	2	12	5	3	8	4	1	4	73
7-lower	17	9	11	9	4	20	9	3	17	6	3	9	117
8	4	5	3	3	3	3	2	2	6	2		1	34
9-upper	8	4	5	3	3	5	3		8	2		2	43
9-middle	5	1	3	2		1	1	1	7			1	22
9-lower, 10	13	8	15	6	7	8	5	3	14	4		6	89
TOTAL	88	45	66	44	32	77	39	34	83	33	11	31	583

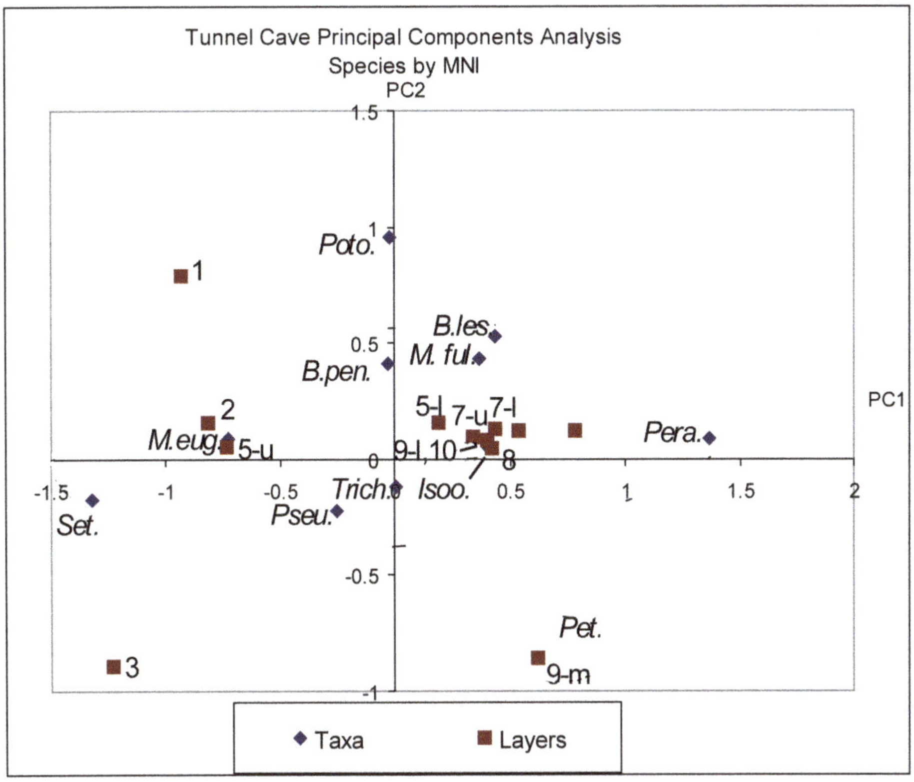

Figure 8.4 **PCA scattergram of PC1 and PC2 for Tunnel Cave layers and species**

PCA

PCA of Tunnel Cave species' MNI shows that most species and layers lie at the centre of the first two components (Figure 8.4). However, *Perameles* and *Setonix* lie at opposite ends of PC1, suggesting that this component summarises the range of vegetational conditions suitable to these species. On PC2 most species score near zero, except for *Potorous* (moderately positive) and *Petrogale* (moderately negative). These species indicate closed and open vegetation, but perhaps the reason for their opposite behaviour on PC2 relates to their predator. At Devil's Lair, *Petrogale* was probable human prey; *Potorous* was not.

The positions of layers 1, 2, and 3 at the extreme negative end of PC1, and other layers on its positive axis, is consistent with the age of layers and environmental changes inferred at Devil's Lair. Layers 9-middle, 3, and 1, with extreme scores on PC2, contain relatively little evidence for human occupation, suggesting that PC2 indicates the effect of predation patterns.

PC1 and PC2 describe 74% of the variation in 12 identified species (Appendix 13), so the four species with extreme scores are probably major influences on the total variation, as they are at Devil's Lair. Considering the results of chi-squared tests (Table 8.12) and Figures 8.4 and 8.5 together, the main difference between Devil's Lair and Tunnel Cave may be caused by a somewhat smaller sample at the latter site, especially in Holocene layers 1, 2 and 3 (combined to reduce small sample effect). This difference emphasises the need to test for non-environmental influences on taxonomic abundance. In particular, human predation may have had more influence at Tunnel Cave.

Table 8.12 Results of chi-squared tests comparing changes over time in Tunnel Cave taxa

H0 (variation in each taxon in each Period is no more than expected from random variation in identical populations) is rejected if p (H0) < 0.05. H = difference is higher than would be expected from H0; L = significant difference is lower than expected from H0; other cells - cannot reject H0.

Layer	*Isoodon*	*Perameles*	*Pseudocheirus*	*Trichosurus*	*Potorous*	*B. penicillata*	*B. lesueur*	*Setonix*	*Petrogale*	*M. eugenii*	*M. irma*	*M. fuliginosus*
1, 2, 3		L						H	L			
5-upper											H	
5-lower												
6												
7-upper												
7-lower												
8												
9-upper												
9-middle									H			
9-lower & 10												

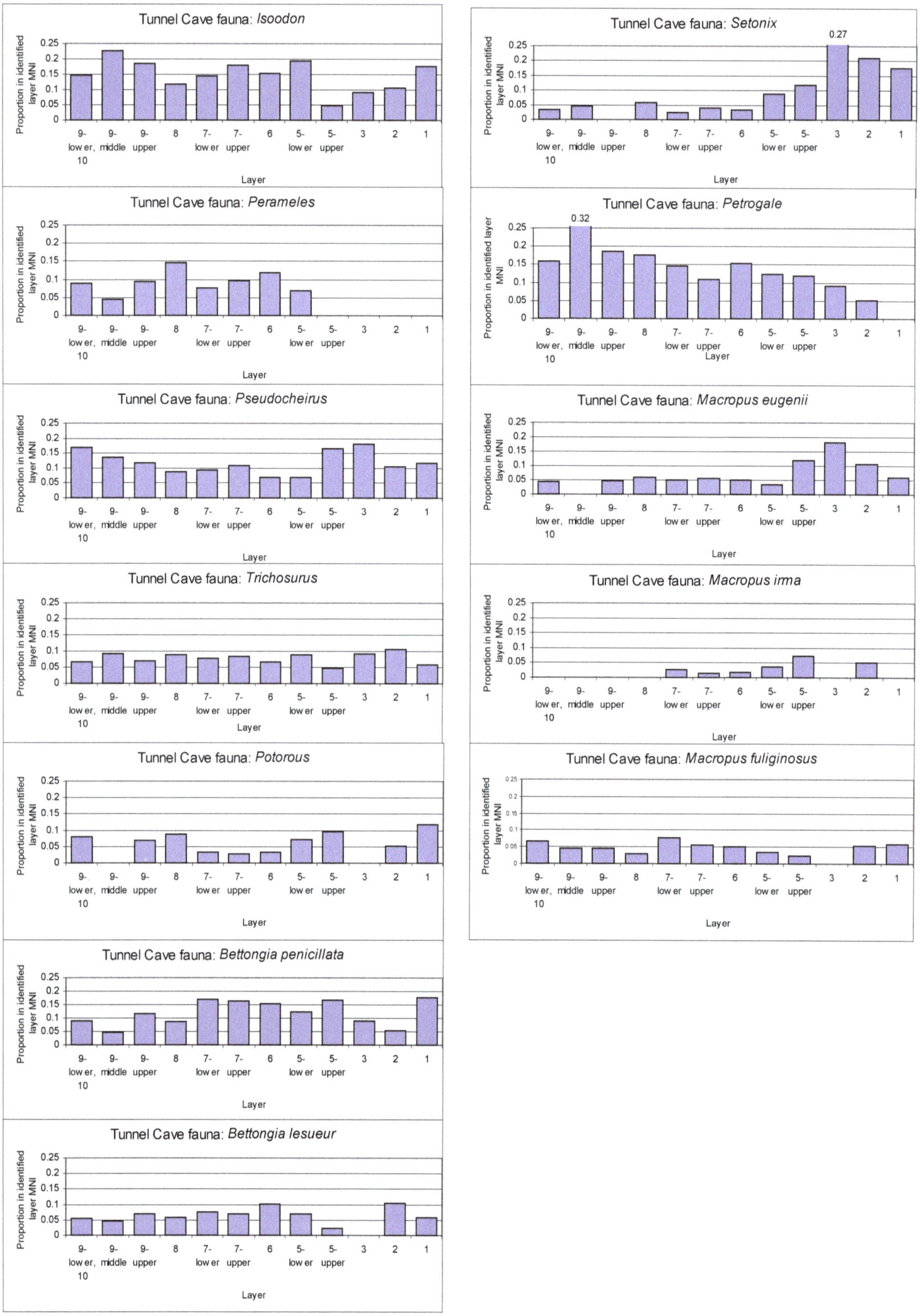

Figure 8.5 Proportions of Tunnel Cave species in layer MNI

Diversity

Unlike Devil's Lair, measures of richness (Table 8.13) are negatively correlated with sample size. Spearman's r_s values between *d*I and layer MNI, and between *R* and layer MNI, are -0.84 ($p < 0.001$) and -0.97 ($p < 1 \times 10^{-6}$). The proportion of the most abundant taxon, $n_{i\ max}/N$, is negatively correlated with layer MNI ($r_s = -0.61$, $p < 0.038$). The richness estimates seem to over-compensate for apparent richness, caused by very large samples like layer 7-lower that contain many more specimens than other layers. Comparing layers with similar layer MNI, richness values are fairly close.

The reciprocal of Simpson's index is correlated with layer MNI also (Spearman's $r_s = 0.65$, $p < 0.022$), but J is not ($r_s = 0.08$, $p > 0.81$). Since $1/\lambda$ is not correlated with J ($r_s = 0.54$, p > 0.07) one could reasonably reject the former estimate and accept J. J ranges even less widely than at Devil's Lair, suggesting that layers are more or less equally even. Layer 9-middle is the least even: it is one of the few layers dominated by one taxon (*Petrogale*; see PCA, above).

Sample size may influence evenness. Two-sided Smirnov tests confirm that small-sample layers 1, 2, 3, and large-sample layers 7-lower and 9-lower & 10, differ from the largest number of other layers ($p < 0.01$).

Interpretations of evenness estimates are therefore dubious as each of these layers contains a broadly similar range of taxa, with only one or two absences in any. The layers with the most bone and largest MNIs are those with the most evidence for human occupation. One must therefore ask whether human contributions of bones could have affected potential palaeo-environmental interpretations.

Table 8.13 Tunnel Cave faunal diversity estimates
$n_{i\ max}$ = number of the most numerous taxon in the sample.

Layer	Sample data					Richness estimates				Evenness estimates			
	MNI	rank	S	$n_{i\ max}/N$	rank	dI	rank	R	rank	$1/\lambda$	rank	J	rank
1	17	11	9	0.176	8	2.824	3	2.183	2	7.410	10	0.949	5
2	18	10	11	0.222	3	3.114	1	2.357	1	8.100	7	0.954	2
3	11	12	11	0.273	2	2.502	5	2.111	3	5.762	11	0.948	6
5-upper	39	7	11	0.179	7	2.457	8	1.601	6	7.643	9	0.915	11
5-lower	55	5	11	0.200	4	2.495	6	1.483	8	8.871	3	0.948	6
6	58	4	11	0.155	12	2.463	7	1.444	9	8.715	5	0.945	10
7-upper	72	3	11	0.181	6	2.338	10	1.296	10	8.698	6	0.947	9
7-lower	114	1	11	0.175	10	2.111	12	1.030	12	8.757	4	0.951	3
8	34	8	11	0.176	8	2.836	2	1.886	5	9.175	1	0.951	3
9-upper	43	6	11	0.186	5	2.393	9	1.525	7	8.074	8	0.948	6
9-middle	22	9	10	0.318	1	2.588	4	1.919	4	5.261	12	0.864	12
9-lower, 10	89	2	7	0.169	11	2.228	11	1.166	11	8.910	2	0.960	1

Human and other contributors to the bone sample

Predator activity at Tunnel Cave is indicated by archaeological remains, indicating humans, and coprolites, apparently from *Sarcophilus*. The lack of bones from predators is probably not significant, considering their scarcity in much larger Australian faunal assemblages (cf. Hope *et al.* 1977, Balme *et al.* 1978: Table 3).

Table 8.14 shows the standardised quantities of different classes of archaeological and non-archaeological materials; Table 8.15 shows Spearman's r_s values for correlations between them. As at Devil's Lair, the weight of burnt bone and artefact numbers are strongly correlated; unlike that site, other classes of material are also strongly correlated throughout the deposit. These correlations hold whether Pearson's *r* (as used by Balme 1980a) or Spearman's r_s (as preferred here) is calculated.

The table suggests that each of the indicators of human occupation – flaked stone, unflaked exotic stone, emu eggshell, aquatic mollusc shell, burnt bone, burnt white (calcined) bone, and hearth volume – is a good indication of human occupation, and that humans left similar proportions of these items whenever they visited the site. Layer (total) MNI is correlated with this archaeological evidence, but the total amount of bone deposited is not. The quantity of flaked stone and emu eggshell is correlated with identified MNI, but only emu eggshell is moderately correlated with the total quantity of bone deposited (possibly because both these classes include remains of people's meals). These results suggest that many of the species making up the total MNI in each layer were human prey species, and that non-human carnivores contributed a larger proportion of the bone that was not identified to species.

Bone quantities are highly correlated with one another, suggesting that there is little or no variation in the relative proportions of unidentifiable or potentially identifiable bone. Hence there is little variation in the level of bone comminution, as probably carried out by *Sarcophilus*. The role of this animal can be assessed further thanks to the presence of its coprolites.

Table 8.14 Quantities of archaeological and faunal material at Tunnel Cave, standardised against weight of sediment excavated in each layer

Flaked stone includes fossiliferous chert, quartz, and calcrete artefacts; unflaked exotic stone includes fragments of ochre and feldspar; burnt bone includes black (charred) bone and whole (calcined) bone; burnt white bone is the latter category on its own; potentially identifiable and unidentifiable bone fragments are discussed in the text. I use bone fragment numbers as using their combined weights would introduce a bias from the influence of large animals' bones. Coprolite indicates weight of fragmentary coprolite identified as that of *Sarcophilus harissii*. Empty cells indicate no material present.

Layer	Weight sediment (kg)	Flaked stone artefacts (number)	Unflaked exotic stone artefacts (number)	Emu eggshell fragments (weight in g)	Aquatic mollusc shell fragments (weight in g)	Burnt bone fragments (number)	Burnt white bone fragments (number)	Estimated hearth volume (l)	Total MNI	Potentially identifiable bone fragments (number)	Unidentifiable bone fragments (number)	Total bone fragments (number)	Coprolite (weight in g)
1	198.5	0.010			0.0001	0.065	0.035	3.0	0.086	4.146	9.995	14.141	
2	301.2	0.003	0.003			0.027	0.003		0.063	5.305	14.366	19.671	0.033
3	163	0.006				0.196	0.166		0.067	5.350	10.706	16.055	
5-upper	119	0.025		0.001		0.042	0.017		0.353	27.681	103.597	131.277	0.018
5-lower	450	0.551	0.007	0.028	0.004	5.924	1.920	14.7	0.127	14.140	48.422	63.738	0.029
6	327	0.128	0.006	0.012	0.003	2.538	0.034		0.180	21.636	78.667	100.303	0.156
7-upper	242	1.095	0.372	0.067	0.007	14.165	3.372	5.4	0.302	17.430	84.463	101.893	0.585
7-lower	408	2.108	0.189	0.046	0.020	32.478	17.201	136.25	0.287	12.988	74.554	95.328	0.116
8	235.5	0.089	0.004	0.005	0.001	1.648	0.348		0.144	7.911	37.919	45.830	0.068
9-upper	243.5	0.148	0.029	0.026	0.016	4.583	0.550	20.9	0.177	9.474	51.585	61.060	0.040
9-middle	251.5	0.135	0.008	0.004	0.003	2.783	0.294		0.087	8.270	38.835	47.105	0.140
9-lower, 10	755.5	0.026	0.013	0.003	0.002	1.355	0.466	6.25	0.118	11.463	56.019	67.481	0.051

Table 8.15 Spearman's rank order correlations (r_s) between classes of archaeological and faunal material at Tunnel Cave

The probability of the Null Hypothesis p is calculated from the T-statistic, which in turn is calculated from Spearman's r_s using a standard formula (Appendix 12; Startup and Whittaker 1982). Where p is equal to or greater than 0.05, r_s is deemed non-significant; where p values are less than 0.05, results are deemed significant and are shaded in the table. The column headings are based on the same means of quantification (weight or number) as the row headings.

Category	Estimate	Flaked stone	Unfl'd exotic stone	Emu egg-shell	Aquatic mollusc shell	Burnt bone	Burnt white bone	Est. hearth vol.	Total MNI	Potentially identifiable bone	Unidentifiable bone	Total bone	Coprolite
Flaked stone (number)	r_s	---------	0.86	0.95	0.95	0.97	0.85	0.66	0.63	0.55	0.54	0.55	0.65
	T-stat	---------	5.251	10.127	9.376	13.088	5.021	2.811	2.561	2.058	2.021	2.058	2.682
	p	---------	**4×10^{-4}**	**1×10^{-6}**	**3×10^{-6}**	**1×10^{-7}**	**0.001**	**0.0185**	**0.028**	0.067	0.071	0.067	**0.023**
Unflaked, exotic stone (number)	r_s	---------	---------	0.86	0.92	0.86	0.84	0.61	0.50	0.44	0.55	0.53	0.74
	T-stat	---------	---------	5.333	7.589	5.251	4.949	2.446	1.826	1.537	2.077	1.966	3.492
	p	---------	---------	**3×10^{-4}**	**2×10^{-5}**	**4×10^{-4}**	**0.001**	**0.0345**	0.098	0.155	0.065	0.078	**0.006**
Emu eggshell (weight in g)	r_s	---------	---------	---------	0.92	0.94	0.83	0.71	0.67	0.60	0.59	0.60	0.70
	T-stat	---------	---------	---------	7.589	8.758	4.682	3.186	2.892	2.359	2.316	2.359	3.094
	p	---------	---------	---------	**2×10^{-5}**	**1×10^{-5}**	**0.001**	**0.0097**	**0.016**	**0.040**	**0.043**	**0.040**	**0.011**
Aquatic mollusc shell (weight in g)	r_s	---------	---------	---------	---------	0.95	0.85	0.82	0.53	0.44	0.48	0.47	0.64
	T-stat	---------	---------	---------	---------	9.376	5.095	4.500	1.966	1.567	1.726	1.661	2.657
	p	---------	---------	---------	---------	**3×10^{-6}**	**5×10^{-4}**	**0.0011**	0.078	0.148	0.115	0.128	**0.024**
Burnt bone fragments (number)	r_s	---------	---------	---------	---------	---------	0.89	0.70	0.49	0.42	0.40	0.41	0.63
	T-stat	---------	---------	---------	---------	---------	6.110	3.064	1.775	1.462	1.374	1.403	2.585
	p	---------	---------	---------	---------	---------	**1×10^{-4}**	**0.0120**	0.106	0.175	0.199	0.191	**0.027**
Burnt white bone fragments (no.)	r_s	---------	---------	---------	---------	---------	---------	0.73	0.37	0.24	0.27	0.27	0.42
	T-stat	---------	---------	---------	---------	---------	---------	3.385	1.262	0.798	0.872	0.896	1.447
	p	---------	---------	---------	---------	---------	---------	**0.0069**	0.236	0.443	0.404	0.391	0.179
Estimated hearth volume (l)	r_s	---------	---------	---------	---------	---------	---------	---------	0.465	0.360	0.381	0.374	0.350
	T-stat	---------	---------	---------	---------	---------	---------	---------	1.661	1.221	1.304	1.276	1.180
	p	---------	---------	---------	---------	---------	---------	---------	0.128	0.250	0.222	0.231	0.265
Total MNI	r_s	---------	---------	---------	---------	---------	---------	---------	------	0.85	0.90	0.88	0.47
	T-stat	---------	---------	---------	---------	---------	---------	---------	------	5.172	6.349	5.892	1.661
	p	---------	---------	---------	---------	---------	---------	---------	------	**4×10^{-4}**	**1×10^{-4}**	**2×10^{-4}**	0.128
Potentially identifiable bones (no.)	r_s	---------	---------	---------	---------	---------	---------	---------	------	---------	0.94	0.97	0.43
	T-stat	---------	---------	---------	---------	---------	---------	---------	------	---------	9.052	11.642	1.506
	p	---------	---------	---------	---------	---------	---------	---------	------	---------	**4×10^{-6}**	**4×10^{-7}**	0.163
Unidentifiable bone fragments (no.)	r_s	---------	---------	---------	---------	---------	---------	---------	------	---------	---------	0.99	0.53
	T-stat	---------	---------	---------	---------	---------	---------	---------	------	---------	---------	26.599	1.966
	p	---------	---------	---------	---------	---------	---------	---------	------	---------	---------	**1×10^{-10}**	0.078
Total bone fragments (number)	r_s	---------	---------	---------	---------	---------	---------	---------	------	---------	---------	---------	0.51
	T-stat	---------	---------	---------	---------	---------	---------	---------	------	---------	---------	---------	1.895
	p	---------	---------	---------	---------	---------	---------	---------	------	---------	---------	---------	0.087
Coprolite (weight in g)	r_s	---------	---------	---------	---------	---------	---------	---------	------	---------	---------	---------	---------
	T-stat	---------	---------	---------	---------	---------	---------	---------	------	---------	---------	---------	---------
	p	---------	---------	---------	---------	---------	---------	---------	------	---------	---------	---------	---------

Throughout the deposit, coprolite weight is moderately correlated with all indicators of human occupation except hearth volume and burnt-white bone, but it is not correlated with overall quantity of bone. One explanation for these correlations with *human* activity may be that coprolite material is not from devils, but from humans. This possibility cannot be rejected out of hand but shape, size, and bone content of the coprolites all conform to scats of present-day *Sarcophilus* (Brendan Marshall, La Trobe University, pers. comm.). An alternative explanation, advanced by Balme *et al.* (1978) for Devil's Lair, is that devils were attracted to the site when humans visited it, and hence left more scats there at these times. Coprolites would not indicate *Sarcophilus* prey, only its tendency to scavenge human leavings. The search for human prey animals therefore returns to the methods used above at Devil's Lair.

As Balme (1980a) found at Devil's Lair, burnt bones at Tunnel Cave are strongly associated with archaeological remains (and also hearth volume), and the bones of larger animals are more often burnt, as shown by a graph (Figure 8.6), similar to one for Devil's Lair (Balme 1980a: Figure 3).

Figure 8.6 does not show the only way of analysing burnt specimens. Some bones are burnt merely because they happen to be present in a surface deposit that is burnt in a hearth. All bones in the surface deposit are equally likely to be burnt. In that case, the most commonly represented taxon in the burnt assemblage *alone* is the taxon whose bones are discarded most often just before the burning episode. Balme's measure, on the other hand, includes both these types of surface bones and those burnt by other means, most likely those discarded into the hearth during its life (ethnographic records of campfire hygiene notwithstanding). Comparison of Balme's measure (proportion of species that is burnt) and that measuring burning of the deposit (proportion of burnt bones in species x) helps distinguish stages of site formation. The deposit-burning measure would include more bones deposited by agents other than people. Figure 8.7 shows the proportions of each species' burnt specimens in the total number of burnt specimens.

The two figures seem to show roughly similar orderings of the most commonly burnt taxa, but a test of the rank-ordered values indicates that the apparent correlation is not significant (p of Spearman's $r_s > 0.236$; cf. Startup and Whittaker 1982: 153). The two measures of burnt bones are perhaps unrelated, and may be investigated further. The second graph suggests that among burnt specimens, the most common species are *Petrogale*, *Isoodon*, and *Trichosurus*. Either graph indicates the species whose remains are common in hearth layers but Figure 8.7 indicates the proportions of species present before the hearth was built. Comparison of the two graphs indicates the species whose remains are added to hearths during burning. The most-often burnt specimens in Figure 8.6, that are not so often burnt in Figure 8.7 (because they are not so common in the deposit *before burning*), are probably human prey. Thus *B. lesueur* and macropods larger than *Petrogale* were probably human prey; *Petrogale*, *Isoodon*, and *Trichosurus* were possibly human prey too, but were hunted by other carnivores; and all the other species were perhaps mainly hunted by non-human carnivores.

One way of confirming this tentative inference would be to test the frequency of burnt specimens, but significance testing (using Fisher's Exact Probability Test) indicate that in most layers the burnt sample of most species is too small to show that any are significantly burnt more often ($p < 0.05$).

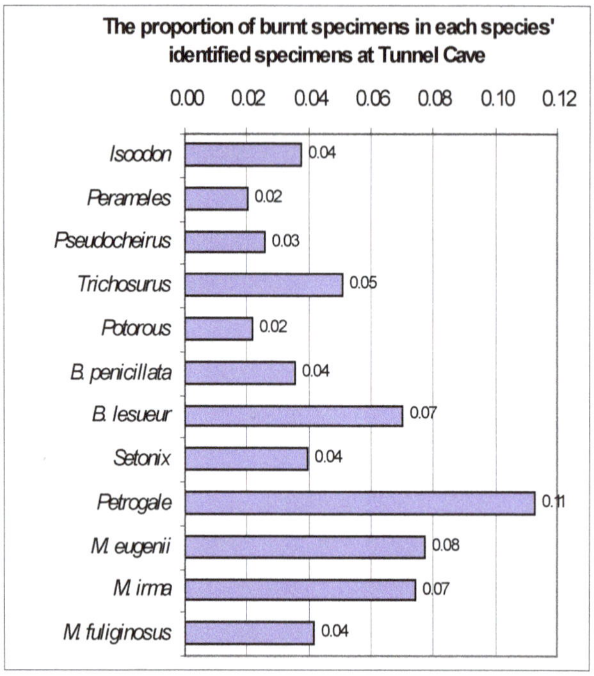

Figure 8.6 The proportion of burnt specimens in each species' identified specimens at Tunnel Cave

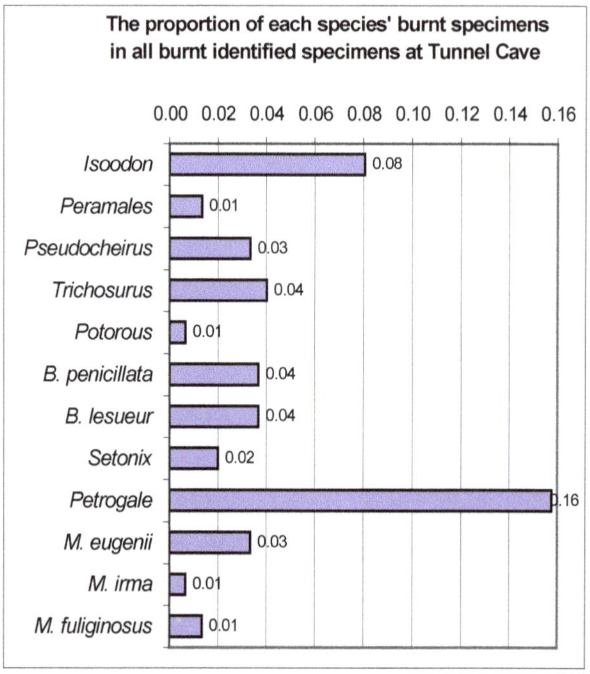

Figure 8.7 The proportion of each species' burnt specimens in all burnt identified specimens at Tunnel Cave

Table 8.16 Spearman's rank-order correlations (r_s) between categories of archaeological material and taxa at Tunnel Cave.

The probability of the Null Hypothesis p is calculated from the T-statistic, which in turn is calculated from Spearman's r_s using a standard formula (Appendix 12; Startup and Whittaker 1982). Where p is equal to or greater than 0.05, r_s is deemed non-significant; where p values are less than 0.05, results are deemed significant and are shaded in the table. Mammal species are calculated from MNIs; lizards (species unidentified) from NISPs.

Category	Estimate	*Isoodon*	*Perameles*	*Pseudocheirus*	*Trichosurus*	*Potorous*	*Bettongia penicillata*	*B. lesueur*	*Setonix*	*Petrogale*	*Macropus eugenii*	*M. irma*	*M. fuliginosus*	Lizards
Flaked stone (number)	r_s	0.94	0.78	0.38	0.71	0.14	0.55	0.75	-0.51	0.62	0.10	0.45	0.64	0.72
	T-stat	9.052	3.893	1.290	3.218	0.435	2.058	3.567	1.877	2.514	0.311	1.598	2.609	3.284
	p	**4×10^{-6}**	**0.003**	0.226	**0.009**	0.673	0.067	**0.005**	0.090	**0.031**	0.762	0.141	**0.026**	**0.008**
Unflaked, exotic stone (number)	r_s	0.90	0.81	0.48	0.61	0.10	0.39	0.65	-0.51	0.58	0.15	0.43	0.62	0.52
	T-stat	6.476	4.443	1.726	2.446	0.311	1.332	2.733	1.860	2.234	0.481	1.491	2.492	1.930
	p	**1×10^{-4}**	**0.001**	0.115	**0.034**	0.762	0.213	**0.021**	0.093	**0.050**	0.641	0.167	**0.032**	0.082
Emu eggshell (weight in g)	r_s	0.94	0.88	0.42	0.77	0.20	0.61	0.86	-0.30	0.60	0.28	0.55	0.67	0.79
	T-stat	8.758	5.999	1.477	3.850	0.655	2.446	5.418	1.010	2.359	0.934	2.096	2.892	4.126
	p	**1×10^{-5}**	**1×10^{-4}**	0.171	**0.003**	0.527	**0.034**	**3×10^{-4}**	0.336	**0.040**	0.372	0.063	**0.016**	**0.002**
Aquatic mollusc shell (weight in g)	r_s	0.93	0.79	0.37	0.61	0.16	0.47	0.73	-0.49	0.57	0.09	0.47	0.63	0.63
	T-stat	8.237	4.030	1.248	2.446	0.515	1.693	3.385	1.759	2.193	0.300	1.677	2.538	2.538
	p	**1×10^{-5}**	**0.002**	0.240	**0.034**	0.618	0.121	**0.007**	0.109	0.053	0.770	0.124	**0.029**	**0.029**
Burnt bone fragments (number)	r_s	0.90	0.78	0.29	0.60	0.00	0.42	0.68	-0.52	0.50	0.06	0.37	0.52	0.68
	T-stat	6.611	3.893	0.972	2.380	0.011	1.462	2.919	1.948	1.809	0.177	1.248	1.948	2.919
	p	**1×10^{-4}**	**0.003**	0.354	**0.039**	0.991	0.175	**0.015**	0.080	0.101	0.863	0.240	0.080	**0.015**
Burnt white bone fragments (number)	r_s	0.74	0.69	0.38	0.50	0.16	0.32	0.53	-0.45	0.36	0.03	0.34	0.43	0.47
	T-stat	3.492	3.034	1.290	1.843	0.504	1.074	1.984	1.583	1.234	0.111	1.140	1.491	1.677
	p	**0.006**	**0.013**	0.226	0.095	0.625	0.308	0.075	0.145	0.245	0.914	0.281	0.167	0.124
Estimated hearth volume (l)	r_s	0.73	0.66	0.66	0.83	0.227	0.80	0.73	-0.133	0.81	0.55	0.68	0.81	0.52
	T-stat	3.351	2.758	2.758	4.746	0.738	4.279	3.420	0.424	4.387	2.058	2.947	4.387	1.948
	p	**0.007**	**0.020**	**0.020**	**0.001**	0.4775	**0.002**	**0.007**	0.6806	**0.001**	0.067	**0.015**	**0.001**	0.080
Total MNI	r_s	0.69	0.69	0.80	0.92	0.53	0.93	0.76	-0.10	0.88	0.66	0.62	0.86	0.58
	T-stat	2.976	2.976	4.176	7.224	2.002	8.006	3.724	0.311	5.892	2.811	2.492	5.333	2.254
	p	**0.014**	**0.014**	**0.002**	**3×10^{-5}**	0.073	**1×10^{-5}**	**0.004**	0.762	**2×10^{-4}**	**0.018**	**0.032**	**3×10^{-4}**	**0.048**
Potentially identifiable bone (number)	r_s	0.61	0.55	0.57	0.71	0.22	0.78	0.62	-0.11	0.73	0.49	0.64	0.72	0.48
	T-stat	2.424	2.058	2.174	3.218	0.714	3.984	2.514	0.356	3.420	1.775	2.633	3.284	1.742
	p	**0.036**	0.067	0.055	**0.009**	0.491	**0.003**	**0.031**	0.729	**0.007**	0.106	**0.025**	**0.008**	0.112
Unidentifiable bone fragments (number)	r_s	0.66	0.62	0.72	0.79	0.28	0.79	0.66	-0.19	0.84	0.58	0.62	0.82	0.55
	T-stat	2.758	2.469	3.284	4.078	0.934	4.078	2.811	0.608	4.879	2.254	2.492	4.500	2.058
	p	**0.020**	**0.033**	**0.008**	**0.002**	0.372	**0.002**	**0.018**	0.557	**0.001**	**0.048**	**0.032**	**0.001**	0.067
Total bone fragments (number)	r_s	0.64	0.60	0.67	0.78	0.26	0.78	0.66	-0.15	0.80	0.55	0.65	0.80	0.46
	T-stat	2.657	2.380	2.864	3.893	0.835	3.984	2.758	0.469	4.279	2.096	2.682	4.176	1.645
	p	**0.024**	**0.039**	**0.017**	**0.003**	0.423	**0.003**	**0.020**	0.649	**0.002**	0.063	**0.023**	**0.002**	0.131
Coprolite (weight in g)	r_s	0.75	0.83	0.29	0.59	-0.14	0.25	0.58	-0.52	0.52	0.17	0.27	0.50	0.69
	T-stat	3.605	4.682	0.959	2.316	0.447	0.810	2.234	1.930	1.930	0.550	0.896	1.826	3.005
	p	**0.005**	**0.001**	0.360	**0.043**	0.665	0.437	**0.050**	0.082	0.082	0.594	0.391	0.098	**0.013**

The other way is to test rank-order correlations between taxa and archaeological remains, as I do above for Devil's Lair. The results of this method are given in Table 8.16. Overall quantity of bone and total MNI are correlated with some taxa, suggesting sample size dependency. However, as at Devil's Lair, there is no rank-order correlation between the correlations given in the table and the total number of specimens in each species (p in all cases > 0.145). Sample size, and the activity of all potential bone accumulators, may be responsible for overall bone abundance, but sample size does not account for all the variation observed.

Species correlated with archaeological remains are *Isoodon*, *Perameles*, *Trichosurus*, *B. lesueur*, *Petrogale*, and *M. fuliginosus*. These species I regard as major human prey species at Tunnel Cave, and all but *Perameles* have been tentatively identified as potential human prey in the preceding Tunnel Cave analyses and in the Devil's Lair analyses. I argued that *Isoodon* and *Perameles* were unlikely human prey at Devil's Lair. It is probably impossible to determine whether the bandicoot species were regular human prey, as the proportion deposited by people may have varied between the sites, perhaps according to their local availability. With the possible exception of *B. penicillata* (correlated with emu eggshell), the remaining species are uncorrelated with archaeological material, and are unlikely human prey. Contrary to the burning analysis, the macropods *M. eugenii* and *M. irma* are not correlated with archaeological material, but this could be because of their increased representation in layer 5-upper, where artefacts are less numerous.

If coprolites indicate visitations of Tasmanian devils, and are also correlated with human visits to Tunnel Cave, then correlations between them and various taxa indicate scavenged animals, including some that were remains of human meals. *Isoodon*, *Perameles*, *Trichosurus*, and *B. penicillata*, all small to medium-size animals within the hunting capability of *Sarcophilus*, are correlated with coprolite fragments. Specimens of the larger animals tend not to be correlated with coprolite fragments, although the correlations are moderately positive and almost pass significance tests for *Petrogale* and *M. fuliginosus*. From these results, it seems that *Sarcophilus* deposited both remains of carcases scavenged from human campsites and its own prey.

The two sources of bone (*Sarcophilus* and humans) are confirmed by the correlations between most taxa and bone quantity, shown in all taxa but *Potorous* and *Setonix*. Further support for the two sources, mentioned above, is that bone quantity is not correlated with most archaeological remains.

This analysis of correlations suggests that the species likely to be human prey (as well as the prey of other predators) include the three (*Perameles*, *B. lesueur,* and *Petrogale*) that indicate more open vegetation than present. But as their remains are highly correlated with archaeological remains, their representation in the deposit could equally be the result of human hunting as much as environmental trends. As these animals are also correlated with layer MNI and with bone quantity, their representation is probably a result of human predation and other factors, including non-human predator activity and changes in vegetational conditions. This ambiguity would present problems for palaeo-environmental interpretation, but the species associated with closed vegetation (*Potorous*, *Setonix*, and *M. eugenii*) show a strong opposing trend, and are not associated with either sample size or human activity.

Of additional interest is the distribution of lizard specimens, which show a similar trend to that seen at Devil's Lair (Figure 8.8, cf. Balme *et al.* 1978: Figure 12). However, unlike the situation at Devil's Lair, they are correlated with various archaeological remains (Table 8.19). At Tunnel Cave, human predation may have been partly responsible for the deposition of lizard bones. At the same time, their peak during the formation of layer 8 (dated to the LGM), which is bereft of archaeological remains, confirms the view that the environment was more arid at that time.

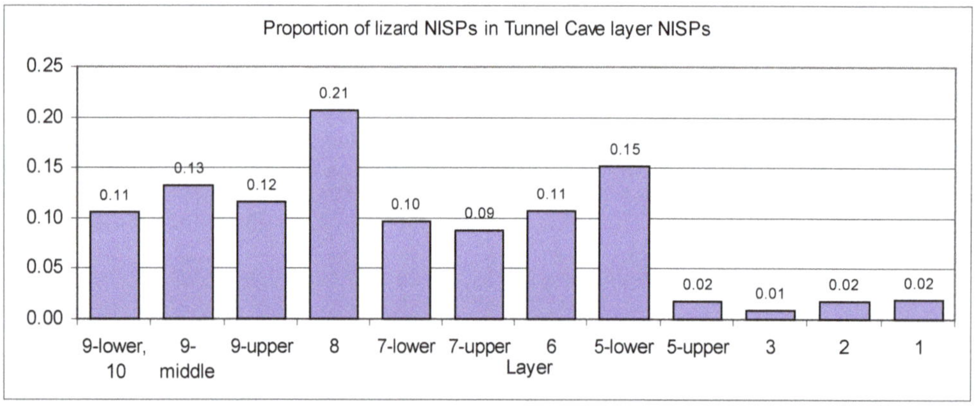

Figure 8.8 Proportion of lizard NISPs in Tunnel Cave layer NISPs

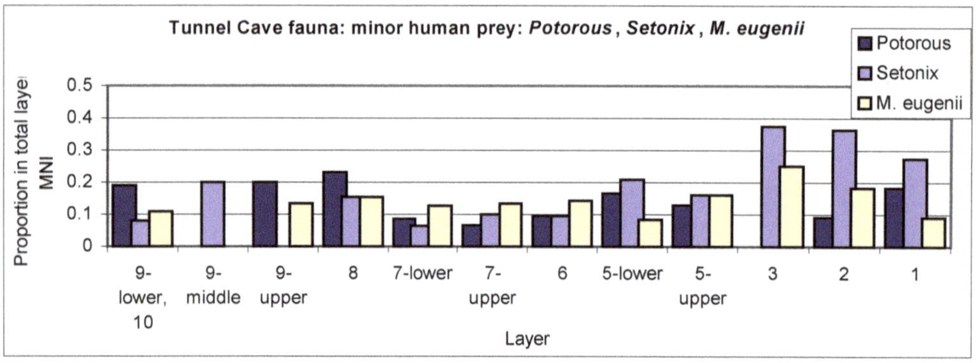

Figure 8.9 Proportions of environmental indicator species among the minor human prey animals at Tunnel Cave

Summary

At Tunnel Cave, from 22,000-1,400 BP, marked changes in the proportions of closed vegetation-adapted mammals were the opposite of those among open vegetation mammals and lizards. During and after the last glacial, the Tunnel Cave locality was suitable for *Petrogale*, *B. lesueur*, *Perameles*, and lizards. These groups dwindled or disappeared altogether in Holocene. The numbers of *Setonix*, and to a lesser extent *Potorous*, increase from layer 5-upper time and later (*after* the end of major human occupation). *Macropus eugenii* also requires dense habitat, and may increase after the Pleistocene.

To some extent these trends were biased by heavy contributions of human prey species. The latter are identified from their association with artefacts, exotic faunal remains, hearths and burnt bone as *M. fuliginosus*, *Petrogale*, *B. lesueur*, *Trichosurus*, *Isoodon*, and *Perameles*. Considering only those environmental indicator species that were minor human prey, graphed in Figure 8.9, the proportions of *Setonix* are much higher after layer 5-upper, that is, after 8,000 BP (Figure 8.9). The proportions of *Potorous* conform less well to its predicted environmental preference, since layer 8, dated to the LGM, is now its most "successful" period. This anomaly may be due to the small sample in layer 8 , made even smaller by the removal of human prey species. Proportions of the thicket-dwelling *M. eugenii* support the trend suggested by *Setonix*.

This analysis has necessarily simplified several variations within fairly long time periods. The recent deposits at Witchcliffe Rock Shelter and Rainbow Cave offer the potential to test the assumed vegetation associations of various species and their attractiveness (or not) as human prey. This potential derives from the existence of well-stratified and varied archaeological remains and the likelihood that vegetation a few centuries ago was little different from that around the sites today (McNicol 1999).

Witchcliffe Rock Shelter

Taxa identified at Witchcliffe Rock Shelter are given in Tables 8.17 and 8.18, below. *Perameles* and *Bettongia lesueur* are absent. Although some open habitat suitable for these animals might exist near the site today (F.G. Smith 1973), they probably had become extinct in the Leeuwin-Naturaliste Region as the total extent of open vegetation declined (Merrilees 1979, 1984). *Petrogale*, in decline after the terminal Pleistocene, is represented here by three specimens at the base of the deposit. Its absence from the upper part of the deposit could indicate that open vegetation around Witchcliffe Rock Shelter gave way to closed vegetation, *c.*800 BP.

Petrogale's presence at the base of the Witchcliffe Rock Shelter deposit is interesting because the depositional sequences at Tunnel Cave (above) and Skull Cave (Porter 1979) suggest that it became extinct in the southern part of the Leeuwin-Naturaliste Region by 7,000 BP, yet the Witchcliffe Rock Shelter deposit suggests that it was extant only 10 km further north until 900-700 BP. Gould (1863) reported *Petrogale* living near Toodyay, which is 400 km north of the Leeuwin-Naturaliste Region (Figure 2.1). There are no other known historical or archaeological records of *Petrogale* for regions south-west of Toodyay.

Table 8.17 Number of identified specimens (NISPs) for all taxa at Witchcliffe Rock Shelter (cf. Appendix 14)

Type:	Invertebrates			Bird egg	Autochthonous vertebrates		Allochthonous vertebrates								
Layer	Land-snail	Aquatic mollusc shell	Crustacean	Emu egg-shell	Birds	Bats	Fish	Lizards	Murids	Dasy-urids	Bandi-coots	Possums	Poto-roids	Macro-pods	Total alloch. vert.
1	55	14	1	45		8	9	2	55	1	24	24	10	37	162
2		1		2		1			15		4		1	4	24
3 upper	37	27		112	18	1	17		31	4	10	16	34	37	149
F3	1	3		12	1	1		1	4	1	2	3	1	8	20
F4	1			6		16			2	1	4	1		4	12
3 lower	7	4	3	37	1		2	1	28	2	4	6	4	10	57
F5	9	3		26	2		3	2	9	2	6	1	5	12	41
4 upper	20	3		36	3		5	1	11		8	9	14	30	78
F6				7			1					1	4	1	7
F7	1			3											0
4 middle	25	14		58	8		6	14	53	3	9	34	31	49	199
F8	9	3	1	5	1		5	1	7		3	1	1	5	23
4 lower	37	2		6		6	4	4	7	3	9	8	12	30	77
4 lower & 5	34	2		40	3	2	1	5	11		4	11	9	26	67
5	11			3		1					1		1	1	3
Total	247	76	5	398	37	36	53	31	233	17	88	115	127	254	919

Table 8.18 NISPs and MNIs of mammal species identified at Witchcliffe Rock Shelter (cf. Appendix 14)

	Layer NISP									
	Isoodon	*Pseudocheirus*	*Trichosurus*	*Potorous*	*Bettongia penicillata*	*Setonix*	*Petrogale*	*Macropus eugenii*	*M. fuliginosus*	Total
1	14	15	2	2	3	15			2	53
2	1									1
3 upper	6	13	2	1	9	15		1	2	49
F3		2			1	3			1	7
F4										0
3 lower	2	1	2		2	1		2	1	11
F5					1	3				4
4 upper	6	5	4	3	7	14		8	3	50
F6		1			2			1		4
F7										0
4 middle	9	22	12	5	8	16	5	17	5	99
F8	3	1				3				7
4 lower	9	3	5	3	4	15	1	10	2	52
4 lower & 5	4	8	2	2	5	8	1	11	2	43
5	1				1					2
Total	55	71	29	16	43	93	7	50	18	382
	Layer MNI									
1	2	3	1	1	2	2			1	12
2	1		1							2
3 upper	1	3	1	1	1	4		1	1	13
F3		1			1	2			1	5
F4										0
3 lower	1		1		1	1		1	1	6
F5					1	1				2
4 upper	1	2	1	1	3	3		2	2	15
F6		1			1			1		3
F7										0
4 middle	2	3	3	2	2	4	1	4	2	23
F8	1	1				2				4
4 lower	1	1	2	2	1	3	1	2	2	15
4 lower & 5	1	1		1	1	2	1	2		9
5		1			1	1				3
Total	11	17	10	8	15	25	3	13	10	112

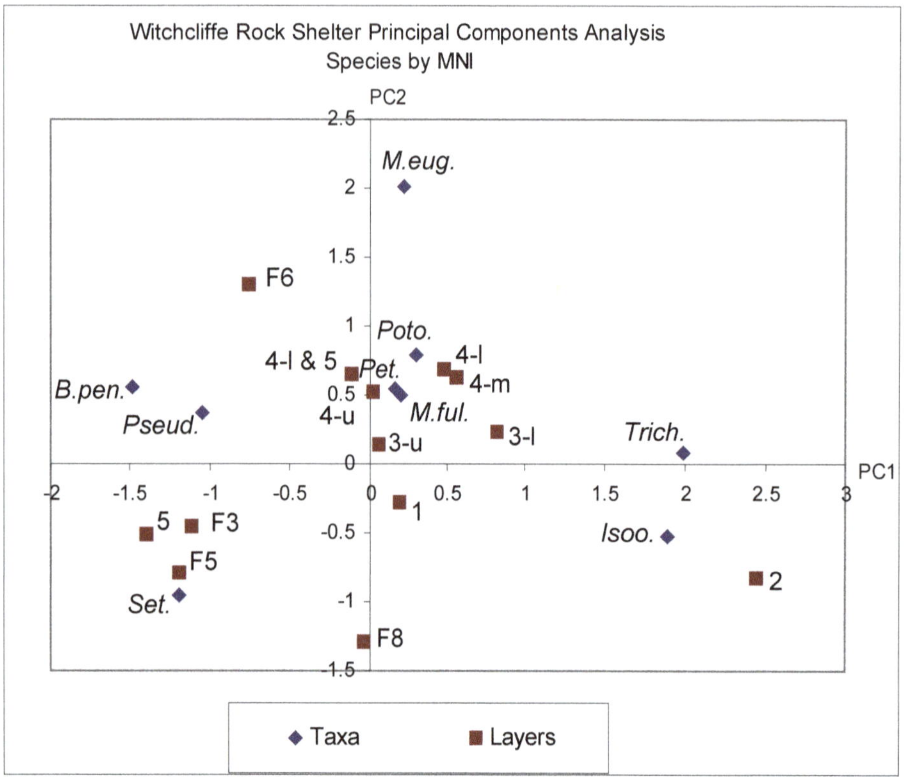

Figure 8.10 PCA scattergram of PC1 and PC2 for Witchcliffe Rock Shelter layers and species.

PCA

PCA of Witchcliffe Rock Shelter fauna gave a scattergram in which most taxa are clustered at the centre of PC1 and PC2 (Figure 8.10; hearths F4 and F7 are not shown as they contain no identified specimens). The components describe only some of the total variation (58%: Appendix 13), and it is unclear what they represent. Potential human prey species identified above are scattered widely across the graph.

A further complication is that the small sample is unsuitable for diversity analyses, which require at least 25 individuals per sample (Cruz-Uribe 1988). This analysis therefore turns to identification of human prey species.

Human contributions to the bone sample

The most likely and obvious human contributions are remains of aquatic fauna (fish bone, otoliths, and scales, fragments of crustacean exoskeleton, and fragments of marine and freshwater mollusc shell); all these species are absent or rare in the Tunnel Cave and Devil's Lair deposits. The obvious explanation for the additional species is the location of Witchcliffe Rock Shelter 200 m from Devil's Pool and Boodjidup Brook, and 1.6 km from the coast.

The activity of non-human carnivores is inferred from the highly fragmented bone (fragmentation indices in the layers at Tunnel Cave and Witchcliffe Rock Shelter are similar, ranging from 11 to 18 fragments per gram of bone at Tunnel Cave, and from 4 to 23 fragments per gram at Witchcliffe Rock Shelter; average rates are 17.4 and 14.4 respectively). However, there are no coprolites to indicate the type of non-human carnivore. *Sarcophilus*, the main scat-depositor identified at Tunnel Cave, was possibly extinct by the time of deposition at Witchcliffe Rock Shelter. There seems to be no reason why coprolites would not have been preserved. The Witchcliffe Rock Shelter and Tunnel Cave deposits are both highly alkaline (pH 9-10). At Witchcliffe Rock Shelter, dingoes could have comminuted the bone to a similar degree as *Sarcophilus* probably did at Tunnel Cave. If dingoes left scats containing fragmented bone, they have not been preserved.

Table 8.19 shows the distribution of categories of archaeological and faunal material, standardised against the weight of deposit excavated. Correlations between these categories, calculated as at Tunnel Cave, are shown in Table 8.20. Most Witchcliffe Rock Shelter layers apart from layer 5 contain some archaeological material. Both archaeological and non-archaeological materials show rank-order correlation with at least one (other) class of archaeological material. Rank-order correlations between definite indicators of human activity (artefacts, burnt bones), and between these indicators and definite human food remains (emu eggshell, aquatic mollusc shell, fish bones), are high. Not all archaeological categories are correlated with overall bone quantity, so many bones probably derive from non-human sources. Correlation between identifiable and unidentifiable bone is high, suggesting little variation in their ratio, and as at Tunnel Cave, a constant high degree of bone comminution.

Table 8.19 Quantities of archaeological and faunal material at Witchcliffe Rock Shelter, standardised against weight of sediment excavated in each layer

Flaked stone includes fossiliferous chert, quartz, and calcrete artefacts; unflaked exotic stone includes fragments of ochre and feldspar; burnt bone includes black (charred) bone and whole (calcined) bone; burnt white bone is the latter category on its own; potentially identifiable and unidentifiable bone fragments are discussed in the text. I use bone fragment numbers as using their combined weights would introduce a bias from the influence of large animals' bones. Coprolite indicates weight of fragmentary coprolite identified as that of *Sarcophilus harissii*. Empty cells indicate no material present.

Layer	Weight sediment (kg)	Flaked stone artefacts (number)	Unflaked exotic stone artefacts (number)	Emu eggshell fragments (weight in g)	Aquatic mollusc shell fragments (weight in g)	Fish & crustacean (number)	Burnt bone fragments (number)	Burnt white bone fragments (number)	Total MNI	Potentially identifiable bone fragments (number)	Unidentifiable bone fragments (number)	Total bone fragments (number)
1	35.0	8.77	0.14	0.08	0.04	0.57	4.11	2.00	0.34	6.89	55.89	62.77
2	1.5	6.00		0.07	0.85		4.00	2.00	1.33	16.00	36.00	52.00
3 upper	65.5	11.34	0.29	0.08	0.08	0.64	11.13	4.89	0.20	8.64	38.82	47.47
F3	11.5	5.22	0.09	0.14	0.00		17.83	6.78	0.44	3.48	33.30	36.78
F4	29.0	4.90	0.03	0.01			4.83	2.66		1.62	6.17	7.79
3 lower	16.0	19.19	0.25	0.07	0.01	0.38	35.94	20.94	0.38	14.88	95.88	110.75
F5	33.5	8.99	0.09	0.04	0.02	0.18	27.18	16.23	0.06	5.06	48.42	53.48
4 upper	62.0	6.68	0.16	0.06	0.00	0.11	18.61	10.23	0.24	17.57	75.36	92.92
F6	5.5	12.91	0.55	0.06		0.18	70.55	22.00	0.55	26.91	77.46	104.36
F7	2.0	5.50		0.05			38.00	18.00		4.50	45.50	50.00
4 middle	93.5	3.33	0.14	0.05	0.01	0.12	15.78	9.59	0.25	25.56	102.20	127.77
F8	18.5	6.32	0.43	0.01	0.00	0.32	18.05	8.00	0.22	22.05	62.92	84.97
4 lower	72.0	1.72	0.07	0.01	0.00	0.07	4.17	1.64	0.21	12.01	57.36	69.38
4 lower/5	154	0.68	0.03	0.02	0.00	0.01	0.58	0.24	0.06	2.62	11.98	14.60
5	49.5	0.69	0.02	0.00			0.47	0.18	0.06	2.04	8.16	10.20

Table 8.20 Spearman's rank order correlations (r_s) between classes of archaeological and faunal material at Witchcliffe Rock Shelter, as outlined in Table 8.22

The probability of the Null Hypothesis p is calculated from the T-statistic, which in turn is calculated from Spearman's r_s using a standard formula (Appendix 12; Startup and Whittaker 1982). Where p is equal to or greater than 0.05, r_s is deemed non-significant; where p values are less than 0.05, results are deemed significant and are shaded in the table. The column headings are based on the same means of quantification (weight or number) as the row headings.

Category	Estimate	Flaked stone	Unfl'd exotic stone	Emu eggshell	Aquatic mollusc shell	Fish bone	Burnt bone	Burnt white bone	Total MNI	Potentially identifiable bone	Unident-ifiable bone	Total bone
Flaked stone (number)	r_s	---------	0.72	0.58	0.49	0.76	0.65	0.71	0.43	0.47	0.50	0.48
	T-stat	---------	3.699	2.558	2.023	4.273	3.055	3.624	1.702	1.927	2.062	1.984
	p	---------	**0.003**	**0.024**	0.064	**0.001**	**0.009**	**0.003**	0.113	0.076	0.060	0.069
Unflaked, exotic stone (number)	r_s	---------	---------	0.30	0.31	0.85	0.53	0.54	0.41	0.65	0.68	0.64
	T-stat	---------	---------	1.141	1.171	5.774	2.235	2.331	1.641	3.099	3.315	2.983
	p	---------	---------	0.274	0.263	**1×10^{-4}**	**0.044**	**0.037**	0.125	**0.008**	**0.006**	**0.011**
Emu eggshell (weight in g)	r_s	---------	---------	---------	0.64	0.45	0.24	0.34	0.61	0.20	0.13	0.14
	T-stat	---------	---------	---------	2.969	1.799	0.903	1.308	2.742	0.722	0.467	0.494
	p	---------	---------	---------	**0.011**	0.095	0.383	0.213	**0.017**	0.483	0.648	0.630
Aquatic mollusc shell (weight in g)	r_s	---------	---------	---------	---------	0.63	0.01	0.14	0.52	0.36	0.27	0.31
	T-stat	---------	---------	---------	---------	2.941	0.026	0.514	2.172	1.410	1.002	1.179
	p	---------	---------	---------	---------	**0.011**	0.980	0.616	**0.049**	0.182	0.334	0.260
Fish bone (number)	r_s	---------	---------	---------	---------	---------	0.40	0.45	0.38	0.54	0.61	0.58
	T-stat	---------	---------	---------	---------	---------	1.574	1.826	1.499	2.331	2.807	2.558
	p	---------	---------	---------	---------	---------	0.140	0.091	0.158	**0.037**	**0.015**	**0.024**
Burnt bone fragments (number)	r_s	---------	---------	---------	---------	---------	---------	0.97	0.16	0.44	0.59	0.54
	T-stat	---------	---------	---------	---------	---------	---------	14.297	0.594	1.763	2.654	2.309
	p	---------	---------	---------	---------	---------	---------	**2×10^{-9}**	0.563	0.101	**0.020**	**0.038**
Burnt white bone fragments (number)	r_s	---------	---------	---------	---------	---------	---------	---------	0.28	0.52	0.65	0.62
	T-stat	---------	---------	---------	---------	---------	---------	---------	1.031	2.193	3.099	2.820
	p	---------	---------	---------	---------	---------	---------	---------	0.321	**0.047**	**0.008**	**0.014**
Total MNI	r_s	---------	---------	---------	---------	---------	---------	---------	---------	0.64	0.48	0.56
	T-stat	---------	---------	---------	---------	---------	---------	---------	---------	3.040	1.956	2.408
	p	---------	---------	---------	---------	---------	---------	---------	---------	**0.009**	0.072	**0.032**
Potentially identifiable bone (number)	r_s	---------	---------	---------	---------	---------	---------	---------	---------	---------	0.86	0.89
	T-stat	---------	---------	---------	---------	---------	---------	---------	---------	---------	6.048	7.148
	p	---------	---------	---------	---------	---------	---------	---------	---------	---------	**1×10^{-5}**	**7×10^{-6}**
Unidentifiable bone fragments (number)	r_s	---------	---------	---------	---------	---------	---------	---------	---------	---------	---------	0.99
	T-stat	---------	---------	---------	---------	---------	---------	---------	---------	---------	---------	24.432
	p	---------	---------	---------	---------	---------	---------	---------	---------	---------	---------	**3×10^{-12}**
Total bone fragments (number)	r_s	---------	---------	---------	---------	---------	---------	---------	---------	---------	---------	---------
	T-stat	---------	---------	---------	---------	---------	---------	---------	---------	---------	---------	---------
	p	---------	---------	---------	---------	---------	---------	---------	---------	---------	---------	---------

Searching for human influences on indicator species, Fisher Tests indicate that only three species are more often burnt than would be expected under the Null Hypothesis, that all specimens are equally burnt. Each one is more often burnt in only layer: *Isoodon* in layer 4-middle ($p < 0.017$), *Trichosurus* in layer F7 ($p < 0.016$), and *Potorous* in layer 4-upper ($p < 0.007$).

Correlations between taxa and archaeological remains suggest that all taxa apart from *Petrogale* (a small sample) are correlated with one or more classes of archaeological material (Table 8.20). *Potorous* is associated with only one type of material, fish bone, which is in turn correlated with most types of material, and most other taxa also. The apparent association between fish-bone and other material is perhaps because fish-bone is abundant where bone is abundant. Moreover, if people hunted *Potorous* often, they would probably have deposited other exotic material with its bones. *Potorous* was probably not regular human prey.

The correlations between taxa and classes of archaeological material suggest taxa hunted by humans. These taxa are all those listed in Table 8.21, except for *Potorous* and *Petrogale*. Only *B. penicillata*, *Setonix*, and *M. eugenii* are associated with more than two classes of strictly archaeological material. These species are not the most often burnt, but are probably still the most likely human prey species, given that the analysis of burnt bone found so few cases of any species being more significantly burnt than any other.

Table 8.21 Spearman's rank-order correlations (r_s) between categories of archaeological material (row headings) and taxa (column headings) at Witchcliffe Rock Shelter.

The probability of the Null Hypothesis p is calculated from the T-statistic, which in turn is calculated from Spearman's r_s using a standard formula (Appendix 12; Startup and Whittaker 1982). Where p is equal to or greater than 0.05, r_s is deemed non-significant; where p values are less than 0.05, results are deemed significant and are shaded in the table. Mammal species are calculated from MNIs; lizards (species unidentified) from NISPs.

Category	Estimate	*Isoodon*	*Pseudocheirus*	*Trichosurus*	*Potorous*	*Bettongia penicillata*	*Setonix*	*Petrogale*	*Macropus eugenii*	*M. fuliginosus*	Lizards
Flaked stone (number)	r_s	0.45	0.38	0.35	0.19	0.56	0.39	-0.15	0.46	0.36	0.11
	T-stat	1.826	1.491	1.347	0.681	2.419	1.507	0.534	1.890	1.410	0.395
	p	0.091	0.160	0.201	0.508	**0.031**	0.156	0.603	0.081	0.182	0.699
Unflaked, exotic stone (number)	r_s	0.41	0.69	0.27	0.34	0.59	0.62	-0.03	0.59	0.42	0.36
	T-stat	1.607	3.447	1.024	1.293	2.642	2.846	0.110	2.667	1.667	1.410
	p	0.132	**0.004**	0.324	0.219	**0.020**	**0.014**	0.914	**0.019**	0.119	0.182
Emu eggshell (weight in g)	r_s	0.44	0.48	0.43	0.26	0.55	0.44	-0.12	0.33	0.53	0.23
	T-stat	1.790	1.965	1.728	0.988	2.397	1.745	0.441	1.254	2.245	0.840
	p	0.097	0.071	0.108	0.341	**0.032**	0.104	0.666	0.232	**0.043**	0.416
Aquatic mollusc shell (weight in g)	r_s	0.76	0.33	0.70	0.50	0.32	0.54	0.15	0.35	0.50	0.44
	T-stat	4.202	1.239	3.499	2.102	1.209	2.331	0.534	1.363	2.102	1.754
	p	**0.001**	0.237	**0.004**	0.056	0.248	**0.037**	0.603	0.196	0.056	0.103
Fish bone (number)	r_s	0.66	0.61	0.56	0.60	0.55	0.66	0.25	0.64	0.55	0.46
	T-stat	3.128	2.807	2.442	2.704	2.374	3.174	0.945	3.026	2.374	1.844
	p	**0.008**	**0.015**	**0.030**	**0.018**	**0.034**	**0.007**	0.362	**0.010**	**0.034**	0.088
Burnt bone fragments (number)	r_s	0.12	0.29	0.16	0.10	0.48	0.25	-0.09	0.51	0.34	0.32
	T-stat	0.434	1.104	0.587	0.362	1.946	0.917	0.336	2.162	1.316	1.216
	p	0.671	0.289	0.567	0.723	0.074	0.376	0.742	**0.050**	0.211	0.245
Burnt white bone fragments (number)	r_s	0.24	0.28	0.27	0.13	0.52	0.23	-0.07	0.56	0.36	0.33
	T-stat	0.903	1.053	1.024	0.474	2.193	0.868	0.265	2.453	1.387	1.270
	p	0.383	0.311	0.324	0.643	**0.047**	0.401	0.795	**0.029**	0.189	0.226
Total MNI	r_s	0.64	0.58	0.60	0.32	0.63	0.42	-0.02	0.52	0.54	0.43
	T-stat	2.997	2.570	2.717	1.201	2.955	1.684	0.058	2.172	2.320	1.710
	p	**0.010**	**0.023**	**0.018**	0.251	**0.011**	0.116	0.955	**0.049**	**0.037**	0.111
Potentially identifiable bone (number)	r_s	0.62	0.47	0.54	0.39	0.34	0.30	0.11	0.68	0.34	0.46
	T-stat	2.860	1.918	2.298	1.507	1.293	1.141	0.402	3.364	1.308	1.862
	p	**0.013**	0.077	**0.039**	0.156	0.219	0.274	0.694	**0.005**	0.213	0.085
Unidentifiable bone fragments (number)	r_s	0.53	0.38	0.49	0.40	0.50	0.40	0.01	0.68	0.43	0.51
	T-stat	2.235	1.459	2.003	1.590	2.082	1.574	0.052	3.299	1.710	2.112
	p	**0.044**	0.168	0.066	0.136	0.058	0.140	0.960	**0.006**	0.111	0.055
Total bone fragments (number)	r_s	0.58	0.35	0.53	0.40	0.49	0.37	0.01	0.67	0.43	0.51
	T-stat	2.570	1.363	2.266	1.557	2.023	1.442	0.052	3.235	1.693	2.112
	p	**0.023**	0.196	**0.041**	0.143	0.064	0.173	0.960	**0.007**	0.114	0.055

Summary

The proportions of fauna at Witchcliffe Rock Shelter differ little from those in the uppermost layer 1 at Tunnel Cave, of similar age, and the range of taxa differs little from that identified in zoological surveys of lower south-western Australia (How *et al.* 1987). At Tunnel Cave three taxa were identified as environmental indicators unbiased by human prey preferences: *Potorous*, *Setonix*, and *Macropus eugenii*. At Witchcliffe Rock Shelter, the last two are probable human prey, and they are therefore possibly represented out of proportion to their representation in the local environment. However, the low numbers of *Potorous*, identified above as infrequently hunted by humans, and even scarcer in Pleistocene layers at Tunnel Cave and Devil's Lair, confirms the increased spread of closed vegetation at Tunnel Cave in the late Holocene, and the probable value of this species as an environmental indicator. Suitable habitat for *Potorous* and *Setonix* at Witchcliffe Rock Shelter probably exists in the dense vegetation around Devil's Pool, 200 m below the site.

Rainbow Cave

Jackson (1992) revised the taxonomic identifications of specimens made by Adie *et al.* (1990), which are those presented in Lilley (1993: Table 2)[4]. Jackson's (1992) listing is shown here (Tables 8.22, 8.23).

[4] The article by Lilley (1993) was in press before Jackson revised the identifications, and so omitted the revised identifications (Jackson, pers. comm.).

Table 8.22 Distribution of fauna in square D22, Rainbow Cave, arranged by taxonomic group (NISP data)

Spit	Birds	Fish	Frogs	Lizards	Murids	Dasyurids	Bandicoots	Possums	Potoroids	Small macropods	Large macropod	All mammals	All vertebrates
1	1				2	1		2				5	6
2	8	21	2	1	7	1	7	9	1	6	4	35	67
3	3	13			1	2	10	4	2	2	3	24	40
4	3	4			2	2	2	4	1		1	12	19
5		15		1	2		3		1	3	5	14	30
6		7		1	1		3	2		1		7	15
7	12	1		1	5		8	1		1	2	18	32
8	3	1			4	3	1	2	2	2		14	18
9	2	1		2	2	3						5	10
10	2	2			3	3	2	2	1	1		11	15
11	2				2				6			8	10
12		1			3	1						4	5
13					3		1		2	1		7	7
14								1				1	1
15	1	1			1		1	2		2		6	8
16						2	1					3	3
Total	37	67	2	6	38	18	39	27	16	19	15	174	286

Table 8.23 Distribution of fauna in square D22, Rainbow Cave, arranged by species

Spit	*Pseudomys albocineurus*	*Pseudomys shortridgei*	*Pseudomys praeconis*	*Rattus tunneyii*	*Rattus fuscipes*	*Cercatetus concinnus*	*Tarsipes spencerae*	*Sminthopsis murina*	*Antechinus flavipes*	*Dasyurus geoffroii*	*Isoodon obesulus*	*Pseudocheirus peregrinus*	*Trichosurus vulpecula*	*Bettongia penicillata*	*Setonix brachyurus*	*Petrogale* sp.	*Macropus eugenii*	*Macropus irma*	*Macropus fuliginosus*	Analysed sample
1		1		1				1				1	1							2
2	1			3	3				1		7	4	5	1	5		1		4	27
3				1					1	1	10	1	3	2	1	1			3	21
4		1			1	1			1	1	2	2	1	1					1	7
5	1			1							3			1	2	1			5	12
6				1							3	1	1		1					6
7		1	1	3			1				8	1			1			2		12
8	1	1		1	1				2	1	1	2		2		2				7
9		1			1				3											
10		1		1	1				3		2	1		1	1					5
11		1		1										6						6
12	2	1						1												
13		1		2							1			2	1					4
14													1							1
15				1							1	2			2					5
16								2			1									1
Total	5	9	1	16	7	1	1	4	11	3	39	15	12	16	14	4	1	2	13	116

The preference for Jackson's listing is based on examination of Rainbow Cave macropods. To confirm the Rainbow Cave record of *Petrogale*, I examined all the teeth of *Setonix*, *Petrogale*, and *Macropus eugenii* from excavation square D22, concurred with all of Jackson's identifications, and rejected several of Adie *et al.* (1990). These species' teeth have potentially confusing morphological similarities (Merrilees and Porter 1979). As Jackson (1992) clearly avoided mis-identifying these specimens, his identifications of other groups of morphologically similar taxa are probably reliable.

Jackson bases his analysis on the premise that most bones at Rainbow Cave are attributable to owl predation, and the small quantity remaining to human predation. Citing characteristics of owl-pellet bone (Marshall 1986, Dodson and Wexler 1979), Jackson finds that most of the bones in test-pit D22 are fragmented in a way consistent with their consumption by owls. The few bones of medium- and large-size mammals are the only bones definitely not consumed by owls. These bones are found only in the upper, artefact-bearing spits of pit D22 (see Table 8.26), and not in pit K8, which has few fish remains or artefacts. Because all the medium- and large-size animals are historically recorded as prey of Aboriginal people, Jackson concludes that people contributed these animals' bones. The only exceptions to this body size definition of human prey are *Potorous*, of medium size and absent from pit D22, and *Isoodon*, of small size and possibly hunted by both owls and humans. However, Jackson acknowledges that because his analysis concentrates on animals that are within the size range of owl prey, exclusive human predation of the larger animals is not demonstrated.

Some attempt can be made to identify the respective contributions of humans and other predators of medium to large-size prey. I compare faunal and artefact data arranged by excavation unit, or spit, assuming that spits (marked by depth intervals) correspond approximately to layers. Lilley (1993) indicate a broad division of the deposit at *c*.40 cm depth into an upper, cultural part (layers above "dark sand", including ashy and charcoal-rich deposits), and a lower, non-cultural part ("dark sand"). These upper and lower parts of the deposit might be expected to show the greatest difference in faunal remains and hence which remains are left by people.

Square D22 contains no specimens of *Perameles* and *B. lesueur*, the open-vegetation species that are also lacking from the Holocene deposits at Witchcliffe Rock Shelter and Tunnel Cave. *Petrogale* is another open-vegetation species, scarce in Holocene deposits, and now locally extinct. Its presence at Rainbow Cave affirms the record from Witchcliffe Rock Shelter. The stratigraphic position of the two uppermost *Petrogale* specimens, between Wk 1875 and Wk 1876, suggests an age of 250-900 BP.

Species preferring thick vegetation, whose remains are abundant in the Holocene deposits at Witchcliffe Rock Shelter and Tunnel Cave, include *Potorous tridactylus*, *Setonix brachyurus*, and *Macropus eugenii*. Of these animals, only *Setonix* is abundant at Rainbow Cave, and *Potorous* is absent from square D22 (it is present in moderate quantities in square K8, which contains relatively few artefacts). The absence or scarcity of *Potorous* and *M. eugenii* from square D22 is consistent with the proposition that people infrequently hunted these animals. Analyses at Devil's Lair, Tunnel Cave, and Witchcliffe Rock Shelter show that *Potorous* especially was not favoured by people (above, and Balme 1980).

PCA

Removing spits with very small samples (1, 9, 12, 13, 16), PCA of the Rainbow Cave fauna gives the scattergram shown in Figure 8.11. PCA describes 68% of the variation in the Rainbow Cave faunal distribution (Appendix 13).

PC1 seems to describe abundance of small and medium-size taxa in both upper and lower parts of the deposit. *Isoodon*, *Pseudocheirus*, and *Setonix* are located at the positive end of PC1 and *B. penicillata* (mostly rare) is located at its negative end. On PC2, *Pseudocheirus* is moderately negative, and *Trichosurus* and *M. fuliginosus* are at the positive end. As these last two species are concentrated in spits 2, 3, 4, and 5, also positive on PC2 (and neutral on PC1) it seems that PC2 indicates species' tendency to be concentrated in these few spits, which happen to be richer in archaeological remains. Possibly, *Trichosurus* and *M. fuliginosus* are favoured human prey species, while *Pseudocheirus* is not. PC1 and PC2 do not describe variation well in *Petrogale*, *M. eugenii*, and *M. irma*; these species score low on both components.

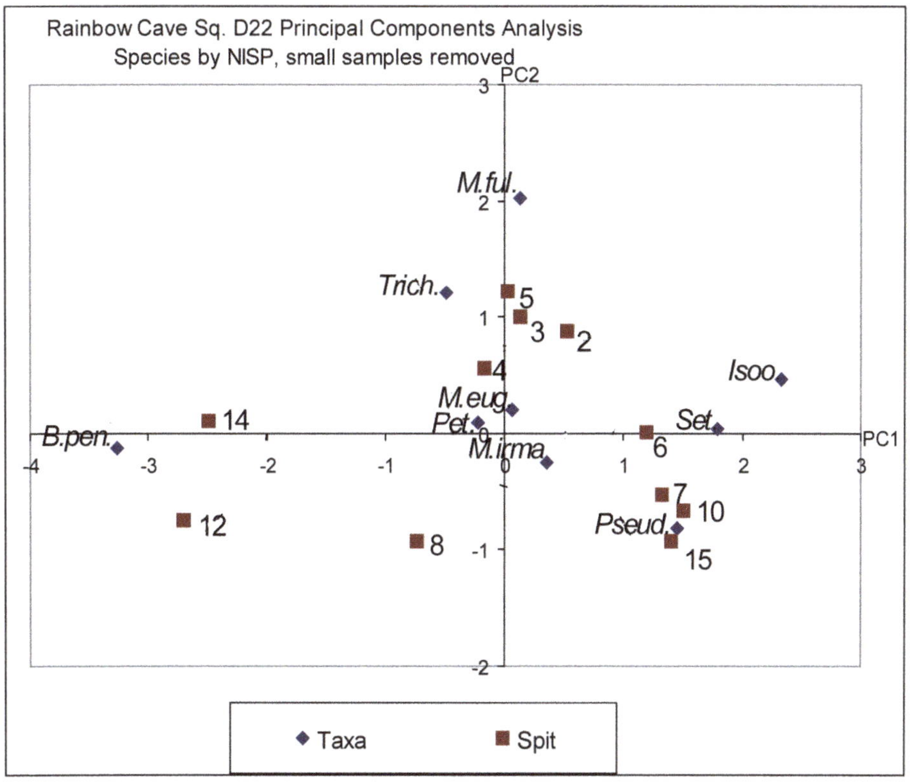

Figure 8.11 **PCA scattergram of PC1 and PC2 calculated for Rainbow Cave spits and species**

Small sample sizes make significance testing of the PCA results difficult. Several of the units can be combined, because they are not stratigraphically distinct. I combine the spits 1 and 2; spits 9 and 10; spits 12, 13, and 14; and spits 15 and 16. Numbers of *Petrogale*, *M. eugenii*, and *M. irma* are too few for chi-squared tests of their distributions.

Chi-squared tests show that specimen counts differ significantly from the value expected for random variation, in the following species and spits ($p < 0.05$):
Isoodon: higher than expected in spit 7
Trichosurus: higher than expected in spits 1 & 2
B. penicillata: higher than expected in spit 11
M. fuliginosus: higher than expected in spit 5

Pseudocheirus and *Setonix* show no significant variation throughout the deposit. Local vegetation formations (including scrub, thicket, and low woodland) would have provided ideal habitat for these animals. Their specimens are abundant throughout the deposit and various predators could have deposited their bones. The number of specimens deposited may vary little despite human contributions in the upper part of the deposit.

Similar considerations apply to *Isoodon* and *Trichosurus*, which are also small animals vulnerable to several predators, and known to inhabit the type of vegetation that surrounds the site (Strahan 1995). They are significantly better represented in the upper part of the deposit containing more artefacts and may have been preferred by human hunters to other animals. The same interpretation is proposed for *M. fuliginosus*, an animal that would be difficult for non-human carnivores to hunt, ranges through nearby habitats, and whose bones are confined entirely to the upper part of the deposit.

The significant result for *B. penicillata* in spit 11 is obviously related to the presence of six specimens in that spit, but there is no obvious reason why it contains so many specimens. Remains of this animal are otherwise evenly distributed throughout the deposit. It may be that a relatively undisturbed carcase in this part of the deposit contributed several bones from the same animal.

As at Witchcliffe Rock Shelter, diversity measures for Rainbow Cave layers are unwarranted because of the small samples in each layer.

Human contributions to the bone sample

Unlike the other sites, there are no records of emu eggshell fragments or of the burnt identified specimens. However, other classes of archaeological material allow comparison of taxonomic and archaeological distributions throughout the deposit (Table 8.24). The exotic or humanly transported fauna are marine. Although Rainbow Cave is located at roughly the same distance from Margaret River, Boodjidup Brook, and the coast, it probably has better access to marine resources, as dune ridges separate it from the rivers. The deposit contains fish bones, otoliths, and scales, but no fresh-water mussel shells are recorded (Adie *et al.* 1990, Jackson 1992).

Coprolites are not mentioned in the Rainbow Cave reports, but activity by non-human carnivores is indicated by the high degree of bone fragmentation. By weight, the unidentified and therefore highly fragmented bones make up 61% of the bone from square D22. The equivalent figure for Tunnel Cave and Witchcliffe Rock Shelter, where only cranial bones were identified, is 72% at each site, so a similar degree of bone fragmentation is implied.

Correlations between archaeological categories, calculated as at other sites above, are shown in Table 8.25. Because unidentifiable bones were not counted, and for consistency within the assemblage, I calculate weights for all categories showing bone quantity (cf. Jackson 1992, Tables A1, A2).

Most categories show strong correlation with the other categories of archaeological material. Rank-order correlations between definite indicators of human activity (artefacts and burnt bones) are high; correlations between these indicators and specialised human food remains (fish bones and marine shell) are high; and correlations between archaeological categories and total MNI and overall bone quantity are also high. Correlation between identifiable and unidentifiable bone weight suggests little variation in their ratio, hence, a consistently high degree of comminution.

Despite correlations between archaeological remains and bone quantity, few of the rank-order correlations between archaeological items and faunal specimens are significant, according to the test applied to the other sites, above. Only *Isoodon* and *Trichosurus* specimens are rank-order correlated with any class of archaeological material (the weight of marine mollusc shell; p in both cases < 0.036). This result is not enough to infer human prey animals. None of the previous analyses shows which specimens are burnt, so the method of Balme (1980a) is inapplicable. In any case, a potential problem with this type of analysis is that the spits may have intersected many layer boundaries in the upper part of the deposit. Material from occupation and non-occupation layers could therefore be mixed together, negating the use of rank-order comparisons of small stratigraphic units.

The entire upper part of the deposit might be compared with the lower part, as the only stratigraphic boundary that can be easily inferred from the data, but chi-squared tests have already indicated that no apparent increase in taxonomic frequencies across this boundary is significant, probably because of small sample sizes (thus even *M. fuliginosus* failed the significance tests, above).

Table 8.24 Quantities of archaeological and faunal material at Rainbow Cave (raw figures), from Adie *et al.* (1990), Jackson (1992), and Lilley (1993)

Flaked stone includes quartz and silcrete artefacts; burnt bone includes black (charred) bone and whole (calcined) bone; potentially identifiable and unidentifiable bone fragments are discussed in the text. Empty cells indicate no material present. Quantities are not standardised against weight of deposit as spits are of identical size.

Spit	Volume sediment (l)	Flaked stone artefacts (number)	Fish & crustacean (number)	Marine shell (weight)	Burnt bone fragments (weight)	Total MNI	Potentially identifiable bone fragments (weight)	Unidentifiable bone fragments (weight)	Total bone fragments (weight)
1	50	9			0.13	2	3.2	13.09	16.29
2	50	125	21	6.15	34.03	27	57.18	95.16	152.34
3	50	213	13	7.56	28.63	21	46.85	39.43	86.28
4	50	243	4	0.81	12.65	7	12.243	39.437	51.68
5	50	713	15	1.22	31.18	12	62.37	59.4	121.77
6	50	280	7		15.76	6	14.63	37.52	52.15
7	50	142	1	0.37	12.53	12	26.5	38.06	64.56
8	50	91	1	0.24	6.45	7	16.08	41	57.08
9	50	24	1		2.28		5.25	8.79	14.04
10	50	28	2		5.61	5	10.76	16.37	27.13
11	50	6			3.04	6	1.99	10.99	12.98
12	50	10	1		0.59		2.45	3.99	6.44
13	50	1				4	4.08	3.45	7.53
14	50	0			1.49	1	1.67	0.51	2.18
15	50	3	1		11.11	5	9.57	19.15	28.72
16	50	7			0.26	1	0.55	2.91	3.46
Total	800	1895	67		165.74	116	275.373	429.257	704.63

Table 8.25 Spearman's rank order correlations (r_s) between classes of archaeological and faunal material at Rainbow Cave

The probability of the Null Hypothesis *p* is calculated from the T-statistic shown above, which in turn is calculated from Spearman's r_s using a standard formula (Appendix 12; Startup and Whittaker 1982). Where *p* is equal to or greater than 0.05, r_s is deemed non-significant; where *p* values are less than 0.05, results are deemed significant and are shaded in the table. The column headings are based on the same means of quantification (weight or number) as the row headings.

Category	Estimate	Flaked stone	Fish bone	Marine shell	Burnt bone	Total MNI	Potentially identifiable bone	Unidentifiable bone	Total bone
Flaked stone (number)	r_s	---------	0.80	0.57	0.84	0.63	0.73	0.71	0.75
	T-stat	---------	4.064	2.096	4.577	2.445	3.179	3.017	3.355
	p	---------	**0.003**	0.066	**0.001**	**0.037**	**0.011**	**0.015**	**0.008**
Fish bone (number)	r_s	---------	---------	0.77	0.94	0.82	0.91	0.83	0.92
	T-stat	---------	---------	3.600	8.335	4.342	6.547	4.418	6.953
	p	---------	---------	**0.006**	**2×10^{-5}**	**0.002**	**0.000**	**0.002**	**4×10^{-4}**
Marine shell (weight)	r_s	---------	---------	---------	0.78	0.84	0.78	0.73	0.78
	T-stat	---------	---------	---------	3.762	4.662	3.762	3.179	3.762
	p	---------	---------	---------	**0.004**	**0.001**	**0.004**	**0.011**	**0.004**
Burnt bone fragments (weight in g)	r_s	---------	---------	---------	---------	0.69	0.89	0.80	0.88
	T-stat	---------	---------	---------	---------	2.867	5.743	4.064	5.483
	p	---------	---------	---------	---------	**0.019**	**1×10^{-4}**	**0.003**	**1×10^{-4}**
Total MNI	r_s	---------	---------	---------	---------	---------	0.85	0.79	0.89
	T-stat	---------	---------	---------	---------	---------	4.841	3.819	5.743
	p	---------	---------	---------	---------	---------	**0.001**	**0.0041**	**0.0003**
Potentially identifiable bone (number)	r_s	---------	---------	---------	---------	---------	---------	0.86	0.99
	T-stat	---------	---------	---------	---------	---------	---------	5.140	22.097
	p	---------	---------	---------	---------	---------	---------	**6×10^{-4}**	**4×10^{-9}**
Unidentifiable bone fragments (number)	r_s	---------	---------	---------	---------	---------	---------	---------	0.87
	T-stat	---------	---------	---------	---------	---------	---------	---------	5.363
	p	---------	---------	---------	---------	---------	---------	---------	**5×10^{-4}**
Total bone fragments (number)	r_s	---------	---------	---------	---------	---------	---------	---------	---------
	T-stat	---------	---------	---------	---------	---------	---------	---------	---------
	p	---------	---------	---------	---------	---------	---------	---------	---------

This apparent contradiction between well-correlated classes of archaeological material on one hand, and few apparent correlations between those classes and faunal specimens on the other, might have several explanations. The effect may be a result of some bones and artefacts filtering down through the deposit, as proposed by Lilley

(1993), or it may indicate that there are many different sources for the faunal specimens, with the human contribution outweighed by contributions from other animals, as suggested by Jackson (1992). However, neither of these alternatives explains the high correlation between artefact numbers and total bone quantity (composed largely of unidentified bones). This correlation alone suggests that people were contributing large quantities of animal bones to the Rainbow Cave deposit, notwithstanding possible carnivore contributions. At the same time, the identified sample is too small to permit identification of humanly-contributed species.

Possibly, the identified sample from square D22 is too small to indicate correlations between taxa and classes of archaeological material, or significant changes within a taxon from one part of the deposit to another. An alternative explanation might be that the weight of unidentified animal bone fragments, which is correlated with artefacts, includes few of the terrestrial species discussed here. Bones of marine mammals, not yet identified, could greatly outweigh the marsupial component. A third explanation could be that unknown carnivores comminuted the humanly-contributed bones to such a degree that the vast majority of specimens could not be recognised in the various faunal analyses. None of these possibilities can be confirmed at present.

Summary

Among mammalian remains at Rainbow Cave, the favoured human prey species cannot be identified, but analysis of a larger sample may confirm the tentative predictions made so far, to show that *Isoodon*, *Trichosurus* and *M. fuliginosus* are particularly well-correlated with classes of archaeological material.

As Jackson (1992) points out, given present day vegetation, all of the Rainbow Cave fauna would have found suitable habitat close to the site. Unlike Witchcliffe Rock Shelter, no marsupial species is unduly abundant because of human predation. Perhaps the terrestrial animals at Rainbow Cave were a minor resource compared to potential marine foods. The above analysis supports the use of species such as *Potorous*, least likely to be human prey, as environmental indicators, and confirms the suggestion from Tunnel Cave that *Setonix*, if not the focus of human hunting, also serves as a palaeo-environmental indicator.

Discussion

My aims in this chapter were to identify the contributions of humans and other carnivores to bone deposits, as a means of allowing for bias in faunal records, which can then be used to identify changes in vegetation structure.

At Devil's Lair and Tunnel Cave, people seem to have accumulated bones of *Macropus fuliginosus*, *Petrogale*, and possibly other macropodids. At Tunnel Cave, these animals and also *Trichosurus* appear to have been human prey. At Witchcliffe Rock Shelter, the same animals (minus those which became locally extinct, *Petrogale* and *Bettongia lesueur*) plus *Setonix* were human prey. At Rainbow Cave, one cannot be certain of the human prey but it may have included all the species mentioned in the analyses above.

Excluding the relevant prey species from each site, which may have an unduly high representation because people occupied the sites, the Devil's Lair and Tunnel Cave deposits show changes from open to closed vegetation after the LGM. The Witchcliffe Rock Shelter and Rainbow Cave deposits indicate that archaeological abundance of fauna is in proportion to the extent of their habitat around the sites.

In the Pleistocene sites, the LGM is associated with fauna suited to an open vegetation structure, more like woodland or open-forest than the present-day tall open-forest. This fauna probably included relatively large populations of species now rare or locally extinct (Merrilees 1984). All the species living near these sites today were present then too, but species requiring closed vegetation for habitat were previously less common. The Devil's Lair and Tunnel Cave analyses affirm the statements of Balme *et al.* (1978) and palaeontological interpretations for south-western Australia in the terminal Pleistocene.

At about 10,000 years BP, and for two or more millennia afterwards, there were relatively large changes in the fauna. Numbers of some species suited to open vegetation declined sharply. Notable declines are seen in *Perameles*, *Bettongia lesueur*, and *Petrogale*. At Tunnel Cave, which includes the region's only deposits currently dated to 10,000-8000 BP, these declines may be a result of declines in human predation, but also at Tunnel Cave, species suited to dense vegetation (*Potorous*, *Setonix*, and *Macropus eugenii*) increased rapidly. One cannot easily envisage predation having opposite effects on different species of similar size and vulnerability to predators.

By 1,000 BP, as human occupation began at Witchcliffe Rock Shelter and Rainbow Cave, the fossil fauna were almost entirely consistent with the fauna thought to have existed in the Leeuwin-Naturaliste Region just before European settlement (see Chapter 2). The only species that may be added to this fauna is *Petrogale*, whose presence in late Holocene deposits in far south-western Australia was not previously suspected (Merrilees 1979). The presence of *Petrogale* is possible evidence for a small population of these animals within a few kilometres of these sites.

These findings confirm faunal succession and vegetational changes argued by Balme *et al.* (1978), but they stem from only one line of evidence, and do not indicate whether humans or climate affected vegetation. Human occupiers at Devil's Lair and Tunnel Cave apparently did not strongly bias the deposition of faunal remains, but they may have actually influenced vegetation according to the frequency of their firing it. Any prolonged human absence from Tunnel Cave and Devil's Lair may have allowed different vegetation to emerge, independently of climatic conditions. These possibilities are discussed again in Chapter 10.

This study has shown that for most of the period of human occupation in the Leeuwin-Naturaliste Region cave and shelter sites, vegetation was generally open. People also used three of the four sites in the last millennium, when vegetation was relatively closed. Additional evidence for local site environments will clarify their possible significance for human populations. This evidence is provided by the burnt remnants of the plants that predominated at the archaeological sites. Chapter 9, below, analyses charcoal fragments as palaeontological evidence of canopy trees and vegetational conditions.

Chapter 9 Palaeo-environmental interpretations from identified charcoal fragments

The aim of this chapter is to infer environmental changes from plant remains found in Leeuwin-Naturaliste Region cave and rock shelter floor deposits. These remains and faunal remains provide the bulk of regional evidence for Pleistocene-Holocene vegetation change (Balme et al. 1978, Burke 1997, Merrilees 1984). While faunal remains mainly indicate whether palaeo-vegetation structure was closed or open, floral remains indicate dominant plants, which may be equally important to hunter-gatherer use of and influence on vegetation. Obtaining direct evidence from charcoal fragments helps address this question.

Plant remains as palaeo-vegetational records

For use in palaeo-environmental interpretations, plant remains must have been deposited approximately at the time of the plants' growth, they must be common throughout a deposit, and in Australia, because a few families have long dominated Australian vegetation (Groves 1994), they should be distinctive enough allow species or genus identification. The type of plant material that satisfies these requirements here is charcoal, which derives from trees and shrubs. The taxa considered here are mostly Australian hardwoods. Plant remains other than charcoal at Devil's Lair and Tunnel Cave are presently unsuitable for analysis.

Preliminary examination of four Tunnel Cave soil samples revealed no phytoliths (D. Bowdery, Department of Anthropology and Archaeology, The Australian National University, pers. comm.). More exhaustive analyses would probably indicate otherwise, but the time involved, and the current lack of comparative material for south-western Australia, precludes phytolith analysis for now. Meanwhile, pollen analysts state that the Tunnel Cave soil samples that they examined contain no well-preserved pollen, and little can be identified (J. Newsome, Biology Department, Murdoch University, and J. Dodson, Geography Department, UWA; pers. comms). Coprolites, however, contain abundant well-preserved pollen grains and the potential for identifying palaeo-vegetation is good (M. MacPhail, Australian National University, pers. comm.). At the time of writing, the analysis of coprolite pollen is continuing.

Numerous carbonised seeds are preserved in upper layers at Tunnel Cave but few appear in lower layers. These seeds are extremely fragile, and only one or two types appear to survive at all. Nonetheless, Tunnel Cave seeds would merit future analyses, including assessment of preservational biases and collection of reference material.

The remaining abundant plant material at Tunnel Cave is charcoal. Charcoal fragments are common in all layers, although there appears to be a decline in charcoal weight downwards through the deposit. Figure 9.1 gives the weight of charcoal recovered per kilogram of sediment, by layer (Layer 1 is omitted as much of the large quantity of charcoal it contains was not collected). Layers containing hearths tend to have more charcoal than adjacent "non-hearth" layers. Thus the graph indicates that charcoal is abundant in the deposit, more so in hearth layers, and it degrades over time. At the base of the deposit, reasonable quantities remain for taxonomic identification. Pieces from layer 10, measuring 2 or 3 mm long and weighing *c.*0.01 g, could be identified if necessary.

No detailed records from pollen grains, seeds, or phytoliths have been reported at Devil's Lair, and again, the most well-documented plant material in that deposit is charcoal. Burke (1997) identified the species of 144 fragments of Devil's Lair charcoal and inferred palaeo-vegetational change at that site (see Chapter 3). In the Results, below, his findings are assessed with a new statistical treatment. This treatment is the same as employed for Tunnel Cave, discussed below.

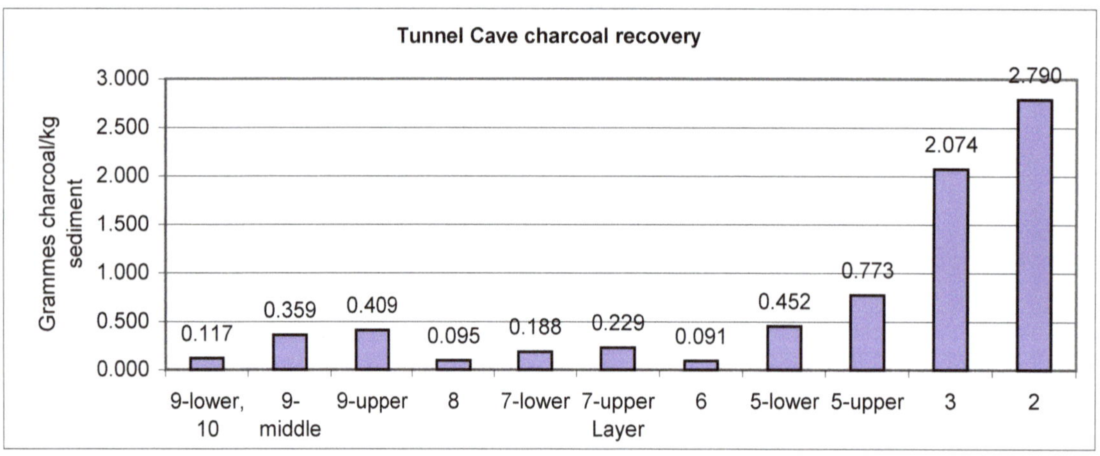

Figure 9.1 Weight of charcoal fragments throughout the Tunnel Cave deposit

Palaeo-vegetational records have not been inferred from Witchcliffe Rock Shelter and Rainbow Cave, as these records are relatively short. However, McNicol (1999; see Chapter 3) suggests that an intensification of Aboriginal burning and its later cessation caused changes in floristic composition at Devil's Pool, 200 m from the former site, at 800 BP (further changes at 100 BP are linked to European settlement rather than climatic change). These changes and the apparent disappearance of *Petrogale* at Witchcliffe Rock Shelter and Rainbow Cave at about the same time suggest structural changes in vegetation took place also.

Methods

Archaeological applications of taxonomic identifications of charcoal fragments date back 60 years (Salisbury and Jane 1940, Godwin and Tansely 1941, cited in Smart and Hoffman 1988). Identifying prehistoric woods from archaeological or fossil charcoal fragments that retain characteristic cell structures is now a widely used method of inferring the composition of prehistoric vegetation (cf. Cartwright and Parkington 1997; Dolby 1995; Donoghue 1989; February 1992, 1994; Figueiral 1995; Hope 1998 and refs; Hopkins *et al.* 1993; Leney and Casteel 1975; Smart and Hoffman 1988; M.A. Smith *et al.* 1995; Thompson 1994, Willcox 1974). However, palaeo-environmental interpretation of charcoal identifications has always been controversial and charcoal analysts therefore use several statistical methods. These methods are reviewed below. A summary of identification and sample collection methods precedes this review.

Taxonomic identification of wood or charcoal is time-consuming, but methodologically simple. The cell structure (anatomy) of wood is characteristic to genera and often to species. Characteristic anatomical features include size, density, distribution pattern, and shape of cell types (Jane 1970). The analyst requires only a flat, correctly oriented surface on the specimen. Since the level of magnification needed to identify it to genus or species is only ×50 to ×500, identification is often feasible with an ordinary light microscope.

Charcoal, however, is fragile, brittle, and extremely porous. Flat sections cannot be obtained with wood preparation techniques such as thin-sectioning, polishing, or impregnating samples with resin which on hardening is polished or sliced (Cartwright and Parkington 1997, Jane 1970, Leney and Casteel 1975). Charcoal surfaces are usually irregular, so analysts need microscopes with a good depth of field, as given by light microscopes at low magnification (c. × 50 or less), or Scanning Electron Microscopes (SEMs) at almost any magnification needed. A requirement of charcoal samples is that they are cut or split into three orthogonal planes, obtained relative to the growth pattern of the trunk or stem. Each section reveals distinctive cell anatomy. The transverse section (TS), perpendicular to the stem, reveals the longitudinal cells in section. In hardwoods, these cells include vessels (pores), fibres, and parenchyma cells. The radial longitudinal section (RLS), parallel to both the stem and its radius, reveals the ray cells in side view. The tangential longitudinal section (TLS, parallel to the stem and perpendicular to the RLS, shows ray cells in section and a side view of fibres and vessels. High magnification (×500) reveals features of vessels such as pits, which sometimes distinguish genera (Hope 1998), but these were not required here.

In the current study of hardwoods, abundant comparative material derives from studies made by the Commonwealth Science and Industrial Research Organisation (CSIRO) in response to commercial interest in timber. The CSIRO has published identification keys for hardwoods at magnifications ranging from ×10 upwards and thin sections of many species (Dadswell 1972, Dadswell and Eckersley 1935, Ilic 1991). In the present study, I also examined thin slides kept in the Forestry Department at the Australian National University (ANU).

Using all of these sources obviates the need to collect many reference samples of trees. However, species can show local variation, according to soils or rainfall regimes (February 1994, February *et al.* 1995), so charcoal analysts should also collect local reference material (Cartwright and Parkington 1997). Moreover, burning distorts, splits, and shrinks wood cells (Donoghue 1989, Prior and Gasson 1993). These changes are not well documented in the CSIRO literature, and Australian charcoal analysts should assess the effect of burning on identification procedures (cf. Rossen and Olsen 1985).

Accordingly, S. Burke and I, as researchers in the same region, collected 50 dry wood samples from 22 major tree species in the Leeuwin-Naturaliste Region and at Devil's Lair and Tunnel Cave, and burnt them in an oven to simulate archaeological charcoal. We assessed the transformation of burnt wood cells and applied it to our respective identifications. Our reference collection procedure is described in Appendix 15, and our wood and charcoal comparative collection is stored in the Centre for Archaeology, UWA. SEM micrographs of reference charcoal specimens are stored on compact disc at the Centre for Archaeology and at the Department of Archaeology and Natural History at the ANU. Charcoal fragments viewed under high-vacuum SEMs are stored at the Department of Archaeology and Natural History at the ANU.

Examination of the charcoal reference collection showed that burning does not seriously affect cell shape, distribution patterns, and density, but it does shrink vessels (the cells with the most airspace) by at least 25% for Jarrah and Karri. This shrinkage is close to the 30% volumetric shrinkage of eucalypt wood blocks observed by Donoghue (1989). The reference charcoal samples also suggested a crucial difference between Karri and Jarrah charcoal, which can be difficult to distinguish if vessels are small (as in juvenile wood from either tree). The CSIRO wood key (Ilic 1993) states that Karri fibre cells tend to have thick walls compared to Jarrah, a distinction preserved in charcoal samples. Perhaps because of these thick walls, burnt Karri fibre cells are fused together, so that very little of the fibre cell lumen (the space inside the cell walls) is visible. The TS of fused fibre cells can appear glassy, even to the naked eye. In contrast, burning seems to make Jarrah fibre cells even more thin-walled, so that fibre cells in TS look more like a loose grid pattern than a mass of cells. No research has yet been done to show why burning should cause a marked glassy appearance in Karri charcoal. However, Prior and Gasson (1993) suggest that in some circumstances, fibre cells in different species respond to heat differently according to the plasticity and thickness of the cell walls. These characteristics are among those that distinguish Karri from Jarrah wood (Dadswell 1972, Ilic 1996).

Allowing for these differences in archaeological collections permits identification of many more archaeological charcoal fragments than would have been feasible with the wood keys alone.

Sample selection

Sample selection is important because charcoal identification is time-consuming and only a small sample can be analysed (cf. Cartwright and Parkington 1997, February 1992). However, the number of identified specimens needed is roughly proportional to the number of taxa (Smart and Hoffman 1988). During analysis of an assemblage, one can determine that a sample is adequate when progressive identifications reveal few or no new taxa: "the point of diminishing returns". This point can be illustrated, if necessary, by plotting the cumulative number of taxa discovered against the number of fragments identified, and showing the levelling-off point on the cumulative growth curve.

In the study of Tunnel Cave charcoal, I selected seven periods for analysis (Table 9.1), and randomly selected charcoal from the spit collection bags. Six periods indicate different environmental conditions from present (see Chapter 3), and one indicates the Late Holocene ("modern") period, as a means of indicating the contributions from recent plants at the site (assumed to be similar to the present-day ones there). The layer 5-upper sample is split into two sub-samples (early Holocene 1 and 2), on the basis of radiocarbon dates in that layer. Further details of sample selection and preparation are given in Appendix 15.

Table 9.1 Charcoal fragments selected from the Tunnel Cave deposit

Layer and sample groups' 14C ages are indicated by estimates given in Chapter Six.
The "Early Holocene" sample of 65 fragments includes 50 fragments identified using a light microscope and submitted for ^{14}C dating.

Sq.	Spit	Old layer	New layer	^{14}C Age (years BP)	Group ^{14}C age	Sample name	Specimens examined	Group total
G10	2	1	1	<1,400	modern to 1,400	modern	27	30
G10	8	2	2	1,400-4,300	modern to 1,400	modern	3	
G10	10	3	3	c. 4,300	4,300-6,900	mid Holocene	10	20
G10	13	3	3	4,300-6,900	4,300-6,900	mid Holocene	10	
G10	15	5A	5-upper	c. 6,900	6,900-8,300	early Holocene 2	10	25
G10	16	5A	5-upper	c. 8,300	6,900-8,300	early Holocene 2	15	
G10	17	5A	5-upper	9,800-9,900	9,800-9,900	early Holocene 1	65**	65
G10	18	5A	5-lower	12,900	12,400-13,000	post-glacial	12	65
G10	19	5A	5-lower	12,800-12,900	12,400-13,000	post-glacial	11	
G10	20	5A	5-lower	12,800	12,400-13,000	post-glacial	10	
G10	28	5B	5-lower	12,400	12,400-13,000	post-glacial	32	
G10	92	7H	7-lower	17,100	17,100-20,000	full glacial	5	39
G10	96	8	8	17,100-20,000	17,100-20,000	full glacial	30	
G10	100	8	8	17,100-20,000	17,100-20,000	full glacial	4	
G10	107	9A	9-upper	19,300	18,000-20,000	pre-glacial	10	30
G10	108	9A	9-upper	19,300	18,000-20,000	pre-glacial	10	
G10	110	9A	9-upper	19,300	18,000-20,000	pre-glacial	10	
						Total	**274**	

Sample preparation

To obtain the TS, TLS, and RLS I split the specimens with chisel-ended pins and scalpel blades sharpened on both sides. Almost all specimens gave at least one TS, TLS, and RLS. Very small or fragile samples only provided two sections, but in most of these cases, the two sections still permitted identification.

All sections were viewed under a low-powered stereoscopic microscope before further preparation, to check for broken or smeared surfaces, and to confirm that the section was orthogonal, as oblique sections are of little use.

For electron microscopy I glued the split fragments to aluminium SEM stubs (aluminium discs about 1-2 cm diameter) with nail varnish, with the section for viewing uppermost, and splutter-coated them with gold to improve conductivity.

For transmitted light microscopy, with 50 specimens from spit 17, I did not prepare samples other than fixing them with poster putty onto plastic lids, so that the correct side faced upwards. The identifications would not have been as definite, but the specimens were exceptionally good quality, and my experience of Tunnel Cave charcoal taxa had improved by the time I used the light microscope.

Identification

I used the SEM facilities at the Electron Microscopy Unit (EMU) at the Australian National University. Most of the time I used a Hitachi S2250N, adequate for low-magnification work (×5-200) but I also used higher resolution machines, a Cambridge S360 and a Jeol 6400. Every section was photographed at ×50 or ×100 magnification. I took photographs or, in later stages of analysis, made digital images, which I stored on compact disc.

For light microscopy I examined samples at ×10, ×20, ×50, and where possible, ×100 magnification with an Olympus BH-2 darkfield transmitted light microscope in the Centre for Archaeology, UWA. The highest magnification gave adequate depth of field only on exceptionally flat specimens.

I compared archaeological samples with wood thin-sections held by the Department of Forestry at the Australian National University, photographs (Ilic 1991), and the computer-based CSIRO Wood Identification Key (version 3.2, Ilic 1996). The CSIRO computer key for eucalypts relies on the operator identifying up to 25 characteristics in wood specimens, including the Australian state of origin (Ilic 1993). Many of these characteristics are preserved in archaeological charcoal, and in this study the state can be assumed, as all the trees considered here are endemic to south-western Australia (Beard 1981, cf Table 2.1).

The computer program eliminates taxa that are not known to have the variations shown by the specimen in question. For example, in a Western Australian eucalypt, the presence of vessels with a tangential width greater than 250 μm (0.25 mm) shows that the specimen is Jarrah, Karri, or Marri. If the vessels are also arranged in lines that cross the grain obliquely, the specimen is Jarrah or Karri; and if the fibres have thick walls, or if many rays are more than 15 cells high, then it is Karri.

Cell structures in Australian hardwoods are well-described (Dadswell and Eckersley 1935, Dadswell 1972), and Jane (1970) provides a general guide to wood anatomy. I considered the following anatomical characteristics: vessel arrangement, vessel size, vessel density, grain straightness, presence of tyloses and kino, ray height, proportion of multiseriate rays, height of multiseriate portions of heterogenous rays, fibre cell wall thickness, parenchyma density and parenchyma distribution. Appendix 15 summarises the expression of these characteristics in the major south-western Australian hardwood species.

Interpretation

Palaeo-environmental interpretations may be complicated where some species produce more charcoal than others, or the species are differentially preserved or deposited. Depositional conditions and characteristics of different woods can affect charcoal preservation and ease of identification and bias the charcoal assemblages. Considered here are some the following potential biases of wood and charcoal assemblages:

1. Under the same burning conditions, some woods produce up to 50% more charcoal than others, relative to the weight of the sample before burning (Shackleton and Prins 1993). Quantitative comparisons of charcoal taxa *within layers* may be misleading, unless adequate work has been done to determine the charcoal yields of the various identified species. This work has not been done in south-western Australia.

2. Smart and Hoffman (1988) show that shrubs are less likely than trees to be preserved as charcoal. This bias indicates a positive correlation between the fire temperature and the thickness of the stem that the fire converts to ash, or put another way, the tendency for small wood to burn to ash during a bush-fire (Shackleton and Prins 1992, Thompson 1994).

3. Some woods resist burning, as suggested by the CSIRO "ash-test" for distinguishing Karri from Jarrah in the field (both woods can have a rosy colour). Burning a match-sized splinter from the wood suspected to be one of the two converts Karri to ash and Jarrah to charcoal.

However, oven tests show that moderate heat or short-lived intense heating does not convert larger pieces of Karri wood to ash (Burke 1997). Only very hot or prolonged fires would bias against the representation of Karri, and perhaps only if it is splintered. However, this bias is another reason to analyse charcoal from natural sources, as bushfires typically do not produce the temperatures experienced in the centre of hearths, nor do they last as long on a small area (David 1990).

4. A possible bias of south-western Australian charcoal records is susceptibility of different woods to biological decay. Anecdotal evidence from builders, timber-workers and pest-controllers states that termites prefer to eat Karri and Marri over Jarrah, which is regarded as extremely hard and durable. Research into timber durability confirms that Jarrah resists fungal and insect attack far longer than Karri and Marri. Marri timber lying on the forest floor lasts up to nine years, Karri lasts up to 15 years, and Jarrah lasts up to 24 years (Thorton *et al.* 1996). Species that were more susceptible to decay would produce less charcoal, since their wood is less common on the forest floor when a bush fire occurs.

However, all these biases are probably more or less constant for each species throughout the periods discussed here. At one site, where topography and soil structure may change little during several millennia, the taxa that are abundant at one period and rare at another probably reflect vegetational change at that site.

The same principle suggests that depositional biases (indicated by field observations) may be treated in a similar manner. At Tunnel Cave, charcoal fragments probably derive from trees growing near the site. Charcoal fragments lying on the ground, probably deriving from burnt logs, may have been deposited by water run-off. Charred logs are located near the site today, and one above the site was sampled (Appendix 15). Logs like these are probably an important source for charcoal in the archaeological deposit.

During winter fieldwork at Tunnel Cave, I observed storm-water run-off from the surface above the doline carry leaves, twigs, and charcoal fragments into the shelter (Appendix 15). I have no information on how far this water run-off carries large fragments of wood or charcoal across the forest floor, but the surface above the doline has a slope probably too shallow to allow long-distance water transport of charcoal fragments. Even if fragments were carried across the entire face of the slope, none would be derived beyond its peak, 200 m to the west, and the uneven surface would hinder long-distance water transport. Most fragments probably derive from the edge of the doline above Tunnel Cave, less than 20 m from the test-excavation.

These topographic influences are unlikely to have varied during the period in question, the last 22,000 years. In this period, Tunnel Cave was always accessible to people and the layers revealed by test-excavation show no change in slope or slope orientation, suggesting similar direction of deposition of all water-borne material, including charcoal.

Charcoal is common in all layers, suggesting that there have been fires around Tunnel Cave throughout most of the period of sediment deposition. Figure 9.1 suggests that more charcoal was produced in hearth-containing layers (5-lower, 7-upper, 7-lower, and 9-upper) but Holocene layers also contain large amounts of charcoal, showing that charcoal would have been deposited naturally at all times. The small amounts of charcoal in the non-hearth, Pleistocene layers, and the smaller and more fragile fragments in these layers, suggests that preservation of charcoal declines with age. Despite this gradual attrition of charcoal, adequate samples could be obtained for all periods (Table 9.2).

At Devil's Lair, charcoal probably also came from the immediate site locality, with charcoal from the uppermost layer deriving from outside the present entrance and charcoal in other layers from outside the former entrance (Burke 1997). Burke (1997) estimates that the maximum distance for water-borne transport of charcoal is 100 m.

Charcoal fragments are probably in reliable stratigraphic context, at least for the broad layer divisions used here (Chapter 6). A single burnt and subsequently fragmented log may dominate the apparent range of vegetation inferred from charcoal fragments, but analysing fragments picked randomly from several spits within a relatively thick layer may alleviate this problem.

In any case, inferences about natural factors that influence the ability of charcoal fragments to indicate local vegetation at sites are supported by comparison of the modern sample with observations of present-day vegetation.

Human factors introduce further considerations. Almost certainly, people brought wood to be burnt in the site, the bulk of which was firewood (wooden tools or other artefacts may have been burnt occasionally or accidentally). It was probably as easy in the past as it is now to collect enough firewood within 50-100 m of the site. So the resulting charcoal in hearths would probably not differ in species from charcoal deposited from non-human causes, which would have derived from fallen timber lying within a similar distance from the site. Admittedly, people may have chosen one species over another according to the different requirements of cooking or heating and depending on whether the wood burns well at all. But since people are as likely to collect what is most convenient, as much as what is most suitable to burn (Shackleton and Prins 1992), biases in the selection of wood, if they occurred, would not affect my overall results. In any case, since my purpose is to ascertain the nature of vegetation broadly contemporary with human occupation, rather than fire-building practices *per se*, the bulk of the charcoal sample derives from non-hearth layers and spits not adjacent to hearths.

Quantification

The simplest method of quantifying identifications is to compare counts of identified fragments and infer the importance of taxa from their abundance (*e.g.*, M.A. Smith *et al.* 1995). This procedure is potentially biased by contributions from a single burnt log, degradation of older samples, and variations in sample size (number of identified fragments or volume of deposit excavated), burning intensity, bio- perturbance, etc.

An alternative is to compare taxonomic abundance with the entire sample in the layer (*e.g.*, February 1992). Other researchers translate standardised counts to standardised weights, to avoid the biases of differential fragmentation (e.g. Cartwright and Parkington 1997). However, Thompson (1994) finds that variations in wood density bias relative weights of charcoal taxa. Moreover, Shackleton and Prins (1993) criticise standardised abundance, whether measured by weight or number, because it is biased by many of the same factors as raw counts. Differential degradation of samples and volume of deposit sampled must influence relative abundance, since both factors affect total sample size (Kadane 1988).

Some researchers therefore list simply the presence or absence of taxa (e.g. Dolby 1995, Hopkins *et al.* 1993). Unhappy with this loss of data, others have proposed measuring *ubiquity* throughout samples (Godwin 1956, Willcox 1974; and Braun-Blanquet 1932, Callen 1967, cited in Hubbard and Kvacek 1998). This method is to count the number of samples within a given period that contain the taxon of interest. It allows a comparison of each taxon that researchers feel is not biased by post-depositional variations in abundance. However, Kadane (1988) shows that statistically, the ubiquity method is just as affected by volume of deposit and sample degradation as the relative abundance method. Kadane's argument is considered carefully here.

Kadane argues that counts obtained from an assemblage of identified charcoal fragments, which constitutes a small sample of an enormous population (all the charcoal fragments in a deposit) are likely to follow a mathematical model, the Poisson distribution. This model has a parameter, λ, that defines the probability (P) that a taxon has a given count x, as follows:

$$P(x) = e^{-\lambda} \lambda^{x} / x!$$

where e is the base of the natural logarithm, and $x!$ is the product of the first x integers (Kadane 1988, Startup and Whittaker 1982). The parameter λ is the product of the factors n (the original population of charcoal fragments) and p_{sample} (the probability that a fragment from this population will be identified in an archaeological sample). Although difficult to quantify, n is probably enormous (at least at the sites studied here). While p_{sample} is probably fairly small (almost certainly less than 0.1), np_{sample} is still a large number. In a Poisson distribution, the parameter λ is also the mean of the counts and the variance of the counts around the mean. Thus if λ is small, the counts are likely to be small, but if λ is large (as is the case here) then the variance will be large. In short, the abundance of identified charcoal taxa in a deposit containing many thousands of charcoal fragments may not have much statistical significance.

Kadane (1988) goes on to express the count of a given taxon at a given provenance (C) as follows:

$$C = \lambda p v$$

where p is the probability that charcoal is preserved over time and v is the volume of deposit sampled.

Given that charcoal counts have a large variance, the ubiquity method makes intuitive sense and seems to avoid the pitfalls of counts. A taxon only has to have one specimen in each of five samples for a period, to have the same importance as a taxon with 100 specimens in those five samples. However, ubiquity is just as dependent on the variables λ, p, and v. Kadane (1988) shows that the probability of "presence in a sample" is expressed for a given sample as a probability of 1 (the sum of probabilities of all possible counts) minus the probability of a count of zero, that is, $1 - e^{-\lambda pv} \lambda^{0} / 0!$, or, $1 - e^{-\lambda pv}$.

Thus, in all methods the analyst must assess λ, p, and v. Here, I am easily able to determine v, which was recorded in the field. But I can only make a qualitative assessment of λ and p (respectively, the rate of identification of living taxa from contemporary charcoal, and the degradation rate of charcoal fragments).

Figure 9.1 shows that rate of charcoal recovery declines with age at Tunnel Cave. Because of the uncertainties of radiocarbon dating in the older layers, I cannot quantify the amount of charcoal degraded over time, but a significant change in the quantity of charcoal recovered is revealed by a rank-order comparison of the weight of charcoal recovered per kilogram of deposit, and the age of the layer (Spearman's $r_s = -0.92$, $p\ [H0] = 0.0001$; layer 1 excluded due to incomplete collection of charcoal). This negative correlation suggests either degradation of charcoal (strong p), or more charcoal deposition in younger layers because of a change in taxa (relating to λ). Both factors may be important here, but the former may be more so, because if hearth layers are excluded, layers ranked by decreasing quantity of charcoal are virtually in order of increasing age.

This finding implies that older samples will tend to have fewer taxa, since the sample is reduced. Again, the relationship cannot be yet be quantified, but identifying a wide range of taxa in older samples will be of more importance than identifying a similar range in younger ones. Uncertainty about λ and p means that there is little advantage in quantifying v in this analysis. Fortunately, the charcoal assemblages examined here derive from deposits of similar volume, since each sample within an assemblage (or period) comes from a spit of roughly similar size, and there are a similar number of samples in each period. The only exceptions are spits from the youngest periods, which were 5 cm thick (*c*.50 l) instead of the more usual 2 cm (20 l). This difference should act to increase sample size, again increasing the probability that more taxa will be represented in younger layers.

A final method is the Method of Normalised Lists (Boulter *et al.* 1993, Hubbard and Kvacek 1998). The method is to identify lists of associated taxa, as determined by previous ecological or palaeontological research, and count the number of taxa in each list in each sample. The counts are standardised by the length of the list of associate taxa, to give the relative importance of each association. But the method is inapplicable here, since all the taxa considered here appear in more than one plant association (Table 2.1).

In the analysis below, I compare the ubiquity and relative abundance methods. The latter at least gives that data that Kadane (1988) feels has potential for further statistical analyses. Following his cautions, I allow for biases from sampling and charcoal degradation.

Results

Devil's Lair

Burke (1997; pers. comm.) identified 129 charcoal fragments to a genus or species in Devil's Lair layers (Table 9.2). Large, dominant woody taxa dominate every sample, regardless of possible biases caused by different sampling units or rates of charcoal deposition of different taxa (Burke 1997: 45). This inference is confirmed by the modern sample, which represents the large woody dominant taxa growing at the site today. The only other woody plants at the site identified by Burke and myself during our reference wood collection are small shrubs, which are unlikely to survive as archaeological charcoal (Smart and Hoffman 1988).

Many taxa identified in older layers grow today 500-1,000 m from Devil's Lair, and could not have deposited their charcoal in the site unless they were much closer. One might argue that the vegetation boundaries discussed by Balme *et al.* (1978) and Merrilees (1979) shifted only by 500 m, but the presence of Tuart, which today grows no further south than Busselton (Beard 1981), suggests a more important climatic change. Busselton receives 10% less rainfall over fewer months of the year than the Margaret River and Cape Leeuwin districts (Table 2.6). Thus one change that would explain the presence of Tuart in the Devil's Lair locality in the Late Pleistocene is slightly lower or more seasonal rainfall. Such a small change in rainfall is within the parameters argued by Pickett (1997) in her assessment of south-western Australian climate at the LGM (see Chapter 3).

Table 9.2 Charcoal fragments identified at Devil's Lair (after Burke 1997: Figures 4.2-5.1, and Burke, pers. comm.).

Layer codes are abbreviations of layers given in full in Burke (1997). Scientific names for Karri, Jarrah, Marri, Tuart, and Peppermint are given in Table 2.1. *Banksia* refers to undifferentiated *Banksia* species, *Xanth.* is *Xanthorrhoea* sp. (grass-tree). "Unknown" specimens are those that had no match in reference material or published sources.

Period (yr BP)	Trench	Layer	Depth	No. coll.	No. id'd	Karri	Jarrah	Marri	Tuart	Pepper-mint	*Banksia*	She-oak	*Xanth.*	Un-known	Euc. sp.
modern	6	DE	89-95	32	30	20	2	6		2				1	
8,000-	1	OBCS	184-194	28	23		10	2	1	1	7		2	1	
12,000	1	OBCS	214-224	2	2		1		1						1
	7c	OB Mottl.	93	29	12		4	1	1	2	2		2	2	
	5	OBCSW	110-120	10	11		3	1	1	2	4			2	
	1	OBCS & R	233	1	1		1								
c. 18,000	5	OBCS 2	200-213	4	6				1	2		3		2	1
	8-2	6a	193-196	4	6		2		2	2					
	7b	7b	210-211	3	3		2		1						
	8-2	6	214-220	11	10		4		5	1				2	1
	2	LBCS	206-216	14	13		2		5	4	1	1		3	1
	5	LBCSW	189-200	9	12		4		1	4	1	2			
	2	LBCS	192-207	14	0										
Total				**161**	**129**	**20**	**35**	**10**	**19**	**20**	**15**	**6**	**4**	**13**	**4**

Burke (1997) found that most of the changes in taxa were statistically significant. In line with my own analysis of Tunnel Cave charcoal taxa, I have tested these changes in a slightly different way. My method is based on that used for significance testing of faunal distributions in Chapter 8, that is, I tested the significance of different taxonomic abundances using cross-classified chi-squared tests comparing the proportions of each taxon in each assemblage against all other taxa in all other assemblages. Chi-squared tests indicating independent samples permitted calculation of phi-squared estimates of strength of the difference (Table 9.3).

These tests confirm Burke's interpretations: the modern sample contains significantly more Karri and less Jarrah than other samples, the 8,000-12,000 year BP sample contains significantly more *Banksia* than other samples, and the 18,000 BP sample does not contain significantly different quantities of charcoal from the 8,000-12,000 BP sample, but it does contain significantly more Tuart than it. There is a perfect negative relationship between Karri and *Banksia*, that is, the two taxa are mutually exclusive. Documented associations of living plants already suggest that this negative relationship is strong, notwithstanding the intermingling of these species in "convergence zones" of tall open-forest and other formations (Beard 1981).

The results in Table 9.3 make clear that She-oak is better represented in the 18,000 BP sample. Given the distribution of some species of this genus (Beard 1981), its presence may confirm the indication given by Tuart, that is, this period was significantly (though slightly) drier or had more seasonal rainfall.

It is not possible to show whether the ubiquity of any charcoal taxon varies significantly over time. The modern period has only one sample, so cannot be included in the ubiquity analysis, and no species was sufficiently well represented in the two multi-sample periods to be significantly more "ubiquitous" than any other (Table 9.4; all Fisher probability estimates for H0 [that between the two multi-sample periods there is no difference in species ubiquity] > 0.05). However, the chi-squared significance testing is sufficient to indicate marked floristic changes at Devil's Lair in each of three periods.

Table 9.3 Phi-squared test results calculated for identified Devil's Lair charcoal fragments

Phi-squared test results indicate strength of the difference revealed by chi-squared tests on a scale of 0 to 1 (cf. Shennan 1988). Unbracketed values indicate that a taxon has a higher count than would be expected from random variation, bracketed values indicate that a taxon has a lower count than expected, ns = not significant.

Period (yr BP)	Karri	Jarrah	Marri	Tuart	Peppermint	*Banksia*	She-oak	Xanth.
modern	0.604	(0.068)	0.063	(0.054)	ns	(0.041)	ns	ns
8,000-12,000	(0.117)	0.040	ns	ns	ns	0.131	(0.031)	0.052
18,000	(0.114)	ns	(0.052)	0.127	0.038	(0.034)	0.082	ns

Table 9.4 Fisher tests of ubiquity of each taxon in Pleistocene samples from Devil's Lair

P = no. of samples in which taxon is present; A = no. of samples in which taxon is absent.
Because the modern sample contains only one sample, Fisher tests are conducted only between the "8000-12,000 BP" and "18,000 BP" samples.

Period	Karri		Jarrah		Marri		Tuart		Peppermint		*Banksia*		She-oak		*Xanth.*		Total
(years BP)	P	A	P	A	P	A	P	A	P	A	P	A	P	A	P	A	
8000-12,000		5	5		3	2	4	1	3	2	3	2		5	2	3	5
18,000		6	5	1		6	6		5	1	2	4	3	3		6	6
Total		10	10	1	3	8	10	1	8	3	5	6	3	8	2	9	11
Fisher estimate of H0	1.000		0.545		0.083		0.545		0.363		0.325		0.083		0.182		
Accept H0?	Yes		Can't reject		Can't reject		Can't reject		Can't reject		Can't reject		Can't reject		Can't reject		

Tunnel Cave

As this analysis is the first for Tunnel Cave plant remains, it is given in more detail than that of Devil's Lair. The Tunnel Cave samples comprise six taxa (Table 9.5), including the eucalypt species Karri, Jarrah, and Tuart. I identified 257 fragments, or 94% of the sample, to genus or species level. The reference charcoal from four *Banksia* species was so similar that I cannot make any species identification for the archaeological specimens. The three *Eucalyptus* specimens that are not identified to species are excluded from further analysis.

Table 9.5 Charcoal fragments identified at Tunnel Cave

For 14C ages, see notes, Table 9.1. "Early Holocene 1" includes 50 fragments examined under a light microscope and submitted for ^{14}C dating; the identified proportion includes 43 of the fragments examined under a light microscope. J=Jarrah; K=Karri; T=Tuart; BB=Blackbutt.

New layer*	Group ^{14}C age*	Sample name	Coll.	Id'd	Karri	Jarrah	Tuart	Pepper-mint	*Banksia*	*Xanth.*	*Eucalyptus* sp.****	Unid.
1	modern to 1,400	modern	27	25	20	5					2 J/T; K/T	
2	modern to 1,400	modern	3	3	3							
3	4,300-6,900	mid Holocene	10	10	10							
3	4,300-6,900	mid Holocene	10	10	7		3					
5-upper	6,900-8,300	early Holocene 2	10	10	8		2					
5-upper	6,900-8,300	early Holocene 2	15	15	9	6						
5-upper	9,800-9,900	early Holocene 1	65	58	27	13	5	1	12			7
5-lower	12,400-13,000	post-glacial	12	10		3	3	2	2			2
5-lower	12,400-13,000	post-glacial	11	11		7	1	2	1			
5-lower	12,400-13,000	post-glacial	10	10		4	2	1	3			
5-lower	12,400-13,000	post-glacial	32	31		8	5	6	10	2		1
7-lower	17,100-20,000	full glacial	5	5		4	1					
8	17,100-20,000	full glacial	30	25		14	2	1	7	1	1 J/BB	4
8	17,100-20,000	full glacial	4	1		1						3
9-upper	18,000-20,000	pre-glacial	10	10		7		1	2			
9-upper	18,000-20,000	pre-glacial	10	10		8			2			
9-upper	18,000-20,000	pre-glacial	10	10		6		1	3			
		Total	**274**	**257**	**84**	**86**	**24**	**15**	**42**	**3**	**3**	**17**

The modern sample suggests that in any sample the dominant tree growing above and around the entrance to Tunnel Cave is what would show up most heavily. Most other plants growing near the edge of the doline are smaller than Karri and they are all absent from the modern sample (notably, Peppermint and *Banksia* trees, and shrubs of the genera *Pimelea* and *Hakea*). Additionally, Marri, a large tree growing there, is also absent. All these taxa grow above the Tunnel Cave doline (W. Loneragan, Department of Botany, UWA, and K. Dixon, King's Park and Botanical Gardens, vouchered leaf and fruit specimens collected there).

The susceptibility of Marri to biological attack (see above) might have prevented burning of fallen Marri timber. Although Karri is only slightly less susceptible, its living population is relatively large, which probably explains its preponderance in the modern sample. Jarrah, not identified in survey of the doline edge, may occur in the modern sample simply because its wood is more durable. Small trees and shrubs identified in this survey are probably absent because their trunks are too thin to survive burning, or to produce fragments large enough to identify. Intermediate size taxa like *Banksia* are present on the doline edge today but do not dominate.

These observations suggest that at Tunnel Cave, only the largest, most prominent, and most durable trees, that are also in the doline or growing on the doline's edge, are preserved as charcoal in the cave deposit. The factor λ, rate of identification of living taxa from contemporary charcoal, probably has a positive relationship with tree size or trunk diameter, tree location, and wood durability. These factors are too complex to quantify in the present study. But in comparison with the modern sample, older samples include smaller trees (*Banksia* and *Agonis*) and more taxa, which would not be expected if there had been no taxonomic change. The taxonomic changes through the deposit are probably due to changes in the source taxa, not in the rates of deposition, preservation, or identification of charcoal.

Given the modern sample, older charcoal assemblages allow few inferences about prehistoric understorey vegetation, but they should reflect the dominant large trees in prehistoric plant formations. I have suggested some of these formations in Chapter 2 (Table 2.7).

Cross-classified chi-squared tests (Table 9.6, showing derived phi-squared estimates) suggest these inferences:

1. In Holocene samples, there are significantly more fragments of Karri, and significantly less of Jarrah.

2. The 10,000 BP sample also has significantly more Karri and less Jarrah. It also contains fragments of *Banksia* charcoal, the youngest sample to do so.

3. The phi-squared values for Karri and Jarrah indicate that their counts are most different in the very youngest and very oldest samples.

4. Tuart and Peppermint are significantly different in only one sample, dated 12,400-13,000 BP (layer 5 lower),

in which they are more numerous. *Banksia* is also significantly better represented in this sample.

5. *Xanthorrhoea* has no significantly different representation in any sample, probably because its counts are too small.

Comparisons of ubiquity support the chi-squared tests (Table 9.7). The "early Holocene 1" period cannot be analysed here, as it comprises only one sample, and *Xanthorrhoea* is not analysed owing to its small numbers. Fisher tests comparing the numbers of samples in which taxa are present or absent confirm that ubiquity of every other taxon changes significantly throughout the deposit, in much the same way as inferred from Table 9.6. An additional inference can be made by comparing Pleistocene and Holocene samples, as shown in the lower half of Table 9.7. In this part of the table, the only taxon whose ubiquity does not change significantly is Tuart. The frequency of Tuart is not significantly different between any pair of samples, unless one compares the post-glacial sample (12,400-13,000 BP) or the full glacial sample (17,100-20,000 BP) with the modern sample. Tuart only declined significantly in the modern period.
Pleistocene samples contain more taxa. But the Holocene layers contained more charcoal, charcoal samples, and a greater volume of sediment for sampling (than the selected Pleistocene layers), and so would have been expected to yield more variety of taxa. The fact that they do not confirms the difference in the vegetation of the two periods.

The taxa in the modern sample suggested that the rate of charcoal deposition (λ) favours large trees such as Karri, Jarrah, and Tuart, so either the Pleistocene vegetation included very large numbers of the small trees *Banksia* and Peppermint, or these small trees reached relatively large sizes in the Pleistocene. The former trend is seen in woodland where *Banksia* or Peppermint dominate; the trend towards large-sized individuals of these taxa is seen in open woodlands where the canopy is more open and shorter species can reach greater sizes (Beard 1981). Xanthorrhoea, present in Pleistocene layers, also favours an open habitat. Given faunal trends inferred in Chapter 8, Pleistocene vegetation was probably more open.

Table 9.6 Phi-squared test results calculated for identified Tunnel Cave charcoal fragments

Phi-squared test results indicate strength of the difference revealed by chi-squared tests on a scale of 0 to 1 (cf. Shennan 1988). Unbracketed values indicate that a taxon has a higher count than would be expected from random variation, bracketed values indicate that a taxon has a lower count than expected, ns = not significant.

Period (yr BP)	Karri	Jarrah	Tuart	Peppermint	*Banksia*	*Xanthorrhoea*	Total number in sample
modern	0.135	ns	ns	ns	(0.025)	ns	30
4,300-6,900	0.104	(0.044)	ns	ns	(0.017)	ns	20
6,900-8,300	0.060	ns	ns	ns	(0.022)	ns	25
10,000	0.024	(0.017)	ns	ns	ns	ns	58
12,400-13,000	(0.160)	ns	0.026	0.081	0.020	ns	62
17,100-20,000	(0.069)	0.047	ns	ns	ns	ns	31
18,000-20,000	(0.066)	0.078	ns	ns	ns	ns	30
Total number in taxon	84	86	24	15	42	3	254

Table 9.7 Fisher tests of ubiquity of each taxon in various periods at Tunnel Cave

P=no. of samples in which taxon is present; A=no. of samples in which taxon is absent. The period dated c.10,000 BP is excluded because it contains only one sample.

Period	Karri		Jarrah		Tuart		Peppermint		*Banksia*		Total
(years BP)	P	A	P	A	P	A	P	A	P	A	
modern	2		1	1		2		2		2	2
4,300-6,900	2			2	1	1		2		2	2
6,900-8,300	2		1	1	1	1		2		2	2
12,400-13,000		4	4		4		4		4		4
17,100-20,000		2	2		2		1	1	1	1	2
18,000-20,000		3	3			3	3		2	1	3
Total	6	9	11	4	8	7	8	7	7	8	15
Fisher estimate of H0	**2×10^{-4}**	**Reject H0**	**0.003**	**Reject H0**	**6×10^{-4}**	**Reject H0**	**3×10^{-4}**	**Reject H0**	**9×10^{-4}**	**Reject H0**	
Holocene	6		2	4	2	4		6		6	6
Pleistocene		9	9		6	3	8	1	7	2	9
Total	6	9	11	4	8	7	8	7	7	8	15
Fisher estimate of H0	**2×10^{-4}**	**Reject H0**	**0.011**	**Reject H0**	**0.196**	**Accept H0**	**0.001**	**Reject H0**	**0.006**	**Reject H0**	

Comparison of vessel size (February *et al.* 1995, Burke 1997) suggests the reason for the change. February *et al.* (1995) show that *Eucalyptus grandis* seedlings take up more water in wetter soils, and that they do this partly by increasing vessel size. A small percentage increase in vessel size has a big effect on the amount of water taken up. Burke's (1997) analysis showed that Jarrah charcoal from 18,000 BP to the period 8,000-12,000 BP increased from 96 μm to 110 μm, a small but significant increase (his two-tailed t-test shows $p < 0.0331$, df = 31). Burke (1997) attributed this increase in vessel size to increasing rainfall and soil moisture by *c.*8,000-12,000 BP, relative to the full-glacial samples. The difference in vessel size would be manifested in larger trees in the terminal Pleistocene.

Here I show the same relationship with Tunnel Cave Jarrah charcoal (Table 9.8). Like Burke (1997), I assume that Jarrah vessels grow bigger in response to wetter soil. The response has yet to be demonstrated in living specimens. I did not measure all vessel diameters in every specimen, so I make a comparison on the basis of the maximum vessel size in every specimen, a characteristic that I recorded for filtering out taxa in the CSIRO wood identification key (Appendix 15). Important thresholds for vessel diameter used in the key are 150 μm and 250 μm, so these diameters arbitrarily divide my samples. Chi-squared tests of observed and expected counts suggest significant changes in maximum vessel size in Jarrah (Table 9.8), but not in Karri ($p < 0.05$). Other taxa provided samples too small to test using this method.

Comparing individual periods or assemblages, there is a significant increase in the proportion of Jarrah specimens with large vessels over time. The biggest difference between pairs of adjacent assemblages is between the pre-glacial (18,000-20,000 BP) and the full-glacial (17,000-20,000 BP) assemblages ($\chi^2 = 10.181$, $p < 0.00142$, df = 1, $\phi^2 = 0.254$). The difference is almost as large between the post-glacial (12,400-13,000 BP) and the full-glacial assemblages ($\chi^2 = 10.201$, $p < 0.00140$, df = 1, $\phi^2 = 0.249$). These results suggest that the proportion of Jarrah specimens with large vessels is significantly smaller in the full-glacial assemblage. Rainfall would have increased after that time, favouring more closed vegetation.

Table 9.8 Chi-squared comparison of vessel diameters in Jarrah charcoal in different samples and combinations of samples at Tunnel Cave

Reference charcoal is treated as Holocene charcoal.

JARRAH CHARCOAL VESSEL DIAMETERS	Maximum vessel size				
	Observed			*Expected*	
Assemblages (yr BP)	<150 μm	150-250 μm	Total	<150 μm	150-250 μm
Reference collection	2	7	9	4.8	4.2
Modern	2	3	5	2.6	2.4
4300-6900	0	0	0	0	0
6900-8300	1	5	6	3.2	2.8
10,000	9	8	14	7.4	6.6
12,400-13,000	12	10	22	12	10
17,100-20,000	17	2	19	10	8.9
18,000-20,000	11	10	21	11	9.9
Total	54	42	96	54	42
	Chi-statistic 7.367 df = 1	*P* (H0) 0.00664 **Reject H0**	Phi-squared 0.0767		

	<150 μm	150-250 μm	Total	<150 μm	150-250 μm
Summed assemblages (1)					
Holocene assemblages	14	20	23	12.18	10.82
Pleistocene assemblages	40	22	62	32.82	29.18
Total	45	40	96	45	40
	Chi-statistic 4.861 df = 1	*P* (H0) 0.0275 **Reject H0**	Phi-squared 0.0506		

	<150 μm	150-250 μm	Total	<150 μm	150-250 μm
Summed assemblages (2)					
Down to early Holocene 2	5	15	20	12.18	10.82
Up to early Holocene 1	49	27	76	32.82	29.18
Total	54	42	96	45	40
	Chi-statistic 10.32428 df = 1	*P* (H0) 0.00131 **Reject H0**	Phi-squared 0.1075		

Continuing the vessel size comparison, there is an indication as to when rainfall increased most rapidly during the period when people stopped using Tunnel Cave (13,000 to 8,000 BP). After the full-glacial/post-glacial difference in vessel size, the next largest difference between adjacent layers is between the early Holocene 1 (10,000 BP) and the early Holocene 2 (6,900-8,300 BP) samples. The number of specimens in these samples is small, so to satisfy the requirements of the chi-squared test, I combine all the samples in the site into two groups and compare them, as follows:

In the middle part of Table 9.8, I show two groups divided around the conventional Pleistocene-Holocene boundary, 10,000 BP. In the lower part of the table, I show two groups divided at some time after 10,000 BP and before 8,300 BP, around either side of the two early Holocene samples (early Holocene 1 and 2). Comparison of the phi-squared results derived from each chi-squared test indicates that the conventional division gives a phi-squared result only half as strong as a division in the early Holocene, that is the greatest difference between Holocene samples occurs between 10,000 and 8,300 BP. It may have been at this time that rainfall increased the most during the whole period of transition.

Unlike the inferred, significant rainfall increases during the Pleistocene, the early Holocene rainfall increases (in both periods) correspond with significant changes in the trees growing at Tunnel Cave. Karri encroached on the site after 12,000 BP and before 10,000 BP; *Banksia* vanished from its immediate locality after 10,000 BP and before 8,300 BP, as hunter-gatherer occupation ceased. Did people cease to visit Tunnel Cave because, thanks to increasing rainfall, Karri forest encroached? Or was this encroachment caused partly by a cessation of Aboriginal firing? These questions are addressed in Chapter 10.

Discussion

Identified charcoal fragments dating back 19,000 years indicate vegetational changes at Tunnel Cave, consistent with palaeo-vegetational sequences for Devil's Lair and the Swan Coastal Plain, constructed by Burke (1997) and Pickett (1997). Significant change from Jarrah-dominated vegetation to Karri-dominated vegetation is identified in the period between 12,000 and 8,000 BP, and the changes are centred on 10,000 BP. The timing of this shift, too, is consistent with the above palaeo-vegetational sequences.

An important find in early Holocene and Pleistocene layers at Tunnel Cave, and Pleistocene layers at Devil's Lair, is Tuart. This tree now dominates tall woodland or associates with other trees in mixed woodland along the Swan Coastal Plain, which is drier than the Leeuwin-Naturaliste Region, "always clinging to the same substrate, sands over coastal limestone [calcarenite]" (Beard 1981: 135). The present range of Tuart today includes localities with annual rainfalls ranging from 10% to 40% less than now obtains at Tunnel Cave and Devil's Lair (see Chapter 2). Tuart woodland presently grows no further south than Busselton, at the southern end of the Swan Coastal Plain, 70 km north of Tunnel Cave.

Sands over calcarenite that now support Karri, as at Tunnel Cave and elsewhere on the Leeuwin Ridge, could probably have formed a suitable substrate for Tuart when rainfall was lower (cf. Beard 1981). The presence of Tuart at Tunnel Cave and Devil's Lair implies that it had invaded the southern part of the Leeuwin-Naturaliste Region at a period of lower rainfall. It is mostly too scarce to suggest that it was a dominant tree for long periods. At Tunnel Cave, Tuart is represented in the sample dated 4-8,000 BP, as if fragments of the woodland inferred for the terminal Pleistocene persisted at Tunnel Cave in the Holocene.

Peppermint or Jarrah-*Banksia* woodlands now found 200 m from Tunnel Cave did not contribute charcoal to the modern sample, so the presence of Peppermint and *Banksia* specimens in Pleistocene layers implies that these taxa were common near the doline, either as co-dominants with Jarrah trees or as a well-developed understorey layer in open woodland.

The most important taxonomic change at Tunnel Cave is the shift in the dominant trees at 10,000 BP, as Karri-Jarrah supplants Jarrah-*Banksia*. The sample dated to this age may represent a convergence zone, such as those seen today wherever Karri forest meets Jarrah-Banksia formations (Beard 1981, F.G. Smith 1973). Excluding this assemblage, there is a perfect negative relationship between *Banksia* and Karri, which finds a corollary in present-day ecology (Beard 1981) and in Burke's (1997) study for Devil's Lair.

In the Leeuwin-Naturaliste Region, as in many regions, one only has to walk a few hundred metres in any direction to find a new, markedly different vegetation association. At Tunnel Cave, the distance from the edge of the doline, in Karri forest, to the first Jarrah and *Banksia* trees, is 100 m, and at 200 m distance from the doline one is in Jarrah-*Banksia* woodland. The interval between 100 and 200 m from the doline is another convergence zone, with *Banksia* growing amongst scattered and relatively small Karri trees. As Burke (1997) found at Devil's Lair, were it not for the evidence of Tuart, Pleistocene-Holocene vegetational changes in the Leeuwin-Naturaliste Region could be summarised as shifts in vegetation boundaries of 200-500 m.

The Jarrah-*Banksia* change to Karri-Jarrah is nonetheless regionally important. It is contemporaneous with other regional environmental records (Balme *et al.* 1978, Burke 1997, Pickett 1997). Compared to other Karri regions in south-western Australia, present-day rainfall in the southern half of the Leeuwin-Naturaliste Region is low and confined to a shorter rainy season (Beard 1981). The Leeuwin-Naturaliste Region marks the northern extent of the range of Karri. Since it is absent in Tunnel Cave layers dated 19-12,000 BP, rainfall at these times was probably lower or more seasonal. Jarrah vessel size also suggests a lower rainfall regime, although more work is needed to demonstrate the response of Jarrah vessel size to wetter soils.

The appearance of Jarrah with other taxa in Pleistocene layers suggests mixed woodland, consistent with the structural changes inferred from the Devil's Lair faunal record (Balme *et al.* 1978: Figure 11). Chapter 8 showed that reptiles and xeric mammals (*Perameles*, *Bettongia lesueur*, and *Petrogale*) increase in number with dry conditions and open vegetation (Balme *et al.* 1978). Reptiles are numerous from 19-12,000 BP, reaching their peak at 15-14,000 BP. Non-forest mammal numbers are low 25-20,000 BP, high 18-12,000 BP, and low again from 12,000 BP. In contrast, two macropodids that are known today to inhabit thick vegetation around watercourses, are scarce 22-12,000 BP, and abundant afterwards. These macropodids are *Potorous* and *Setonix*. The remains of another macropodid that shelters in thickets, *Macropus eugenii*, are more common at Tunnel Cave after 12,000 BP.

Environmental changes may relate to continental and global climatic changes, or to glacio-eustatic marine transgression, which would have been rapid from 12,000 BP, reaching the present coastline about 6,000 BP (Balme *et al.* 1978, Chappell and Shackleton 1986, Harrison and Dodson 1993). From 12,000 BP, it is possible that the Leeuwin-Naturaliste Region experienced a more maritime climate, *i.e.*, less seasonal rainfall and moderate temperatures, with implications for vegetation (cf. Martin 1973). An increased maritime influence may have improved conditions for Karri forest, or at least more vegetation growth. Currently there is no vegetational record from inland parts of south-western Australia to compare with the Leeuwin-Naturaliste Region record. In the Quaternary vegetational record for south-western Australia, the relative importance of continental climatic shifts and marine transgression is questionable (Pickett 1997). However, both these potential causes of the vegetational shifts discussed here would have similar effects on local vegetation.

The appearance of Karri charcoal at 10,000 BP, the absence of *Banksia* charcoal in layers younger than 10,000 BP, and the apparent scarcity of Jarrah after 10,000 BP indicate marked changes in vegetation composition and possibly structure at Tunnel Cave at this critical time. Increases in vessel size in Jarrah specimens from Devil's Lair and Tunnel Cave suggest increasing size of trees, and hence increasing rainfall, from *c.*18,000 BP to after 10,000 BP.

The presence of Tuart charcoal in early and mid-Holocene assemblages suggests the presence of Tuart trees near Tunnel Cave, *c.*6,900-10,000 BP. Since Tuart prefers lower or more seasonal rainfall than received near Tunnel Cave today, the major change in vegetation *c.*10,000 BP was probably followed by additional climatic changes that eventually excluded Tuart. However, vegetational shifts may be caused by other factors besides rainfall, including changes in fire regime (Bell *et al.* 1989, McCaw 1986). The potential role of firing in these shifts is considered below.

Chapter 10 Conclusion

This chapter identifies the probable hunter-gatherer reactions to environmental change in south-western Australian forests. It summarises the main points of preceding chapters, and assesses the view that hunter-gatherer occupation persisted in forests because the mosaic form of forest vegetation allowed hunter-gatherers to follow geographic shifts in habitats, and because people can control unfavourable vegetation with fire. I then infer some of the wider implications of this view.

At the end of Chapter 2, I postulated that hunter-gatherer firing in south-western Australia influenced vegetation structure and composition and that the extent of this influence depends on local rainfall and soils. Holding the latter constant, frequent firing in marginal rainfall areas can convert Karri forest to mixed forest, by eliminating young Karri trees and fire-sensitive understorey shrubs before they produce viable seed. Conceivably, a combination of artificial and natural fire regimes creates a fine mosaic, consisting of intricately interlocked habitats and vegetation alliances that promote greater abundance and range of food species. In contrast, frequent firing in optimal rainfall areas promotes thick regrowth scrub that prohibits access to people and prey animals. Open habitat exists only immediately after fire.

Another issue noted in Chapter 2 and confirmed in Chapter 3 is that rainfall has probably never dropped so low as to totally exclude plants now growing in the Leeuwin-Naturaliste Region. Rather, there would have been shifts in the distribution of vegetation alliances. At times of low rainfall, Karri forest could have given ground to Jarrah-Marri forest, but never been totally eliminated from the region. Due to the resilience and drought tolerance of Jarrah forest and other vegetation alliances, no new vegetation alliances would have supplanted ones existing now.

Chapter 4, my review of archaeological research on Australian hunter-gatherers in forests, shows that hunter-gatherers are unlikely to have been able to resist or control regional scale forest expansions. However, they probably were never totally excluded from forest regions, even high latitude montane areas such as south-west Tasmania, because no single forest type covers the whole of any region. Adjustments in settlement patterns, if any, could have taken place at the local scale. The distribution of hunter-gatherers in relatively benign south-western Australian forests would reflect the distribution of interlocking Karri and Jarrah forest stands, with fewer people occupying the former.

Archaeological surveys and test-excavations reported in previous research and summarised in Chapter 5 identified the four archaeological sites analysed here, Devil's Lair, Tunnel Cave, Witchcliffe Rock Shelter, and Rainbow Cave. Analysis in Chapter 6 of radiocarbon determinations and stratigraphy at these sites indicates that people occupied Devil's Lair many times between 47,000 and 12,000 BP, but lack of concentrated occupational remains prevents identification of discrete episodes of occupation. At Tunnel Cave, concentrations of hearths and artefacts in what are identified as occupation layers suggest at least six separate occupations, all but one occurring between 20,000 and 12,000 BP. The last one, represented by hearth F1, occurred at 1,400 BP. Hearths and artefacts at Witchcliffe Rock Shelter and Rainbow Cave are all younger than this Tunnel Cave hearth layer. Late Holocene hearths at all three sites are associated with essentially the same type of occupational debris as late Pleistocene hearths at Tunnel Cave and Devil's Lair, that is, burnt and unburnt bone fragments, stone debitage, emu eggshell fragments, aquatic mollusc shell fragments, and ochre. The emu eggshell fragments are consistent with wintertime occupation, which in historical times involved a dispersed population foraging away from the coast. Consistent with this seasonal settlement pattern, Rainbow Cave is the only cave site to lack emu eggshell, and the closest to the open shoreline, which was least visited by historic hunter-gatherers in winter.

To characterise site occupation patterns further, Chapter 7 attempted to identify changing scale and duration of human occupations from characteristics of stone debitage. Allowing for changes in raw material accessibility (caused by glacio-eustatic marine transgression that submerged stone outcrops), no changes in the scale of occupations could be discerned at Devil's Lair. The evidence from Tunnel Cave artefacts suggested much the same lack of change in occupation intensity throughout the last 10,000 years of the Pleistocene. An exceptional period was at the LGM when chert artefact re-use intensified and several occupations probably followed in rapid sequence - this was already suggested by numerous hearths present in a single layer dated 17-16,000 BP (Chapter 6).

The small sample of stone artefacts from hearth F1 prevents analysis of occupation patterns at Tunnel Cave in the late Holocene. However, the apparent use of nearby stone sources for artefact manufacture at Witchcliffe Rock Shelter and Rainbow Cave suggests small scale or short occupations at this time also. The four human deciduous teeth found at Devil's Lair and Tunnel Cave suggest that cave and rock shelter occupations were by family groups. These people probably foraged in a localised area before moving onto other occupation sites elsewhere. The only prolonged or recurrent visits to Tunnel Cave occurred at the LGM, when winter windspeeds and open-air temperatures would have been least salubrious.

Chapters 8 and 9 show that in the terminal Pleistocene, there were consistent changes in vegetation at Devil's Lair and Tunnel Cave. These changes were in vegetation structure and floristic composition. Faunal remains show that vegetation structure probably started to become more closed from 13,000 BP. Excluding species that human predation contributed in unduly large proportions, both sites show an increase in thicket-dwelling mammals *Potorous* and *Setonix*, and a decrease in xeric fauna (lizards and at Devil's Lair, *Perameles*). The regional predominance of *Setonix* in mid to late Holocene time is confirmed at Witchcliffe Rock Shelter and Rainbow Cave. Fauna preferring open habitat, including the major human prey animals, are generally absent from late Holocene layers at all sites, except for *Petrogale*, whose presence in the Leeuwin-Naturaliste Region at this time was not previously suspected (Merrilees 1979a).

Charcoal deriving from trees that probably grew within a few tens of metres from Devil's Lair and Tunnel Cave indicate that climate was getting wetter at *c*.13,000 BP. At this time, and at both sites, vessel size in Jarrah charcoal increased, suggesting a growth in tree size in response to wetter conditions. There is a total absence of Karri charcoal in Pleistocene layers at both sites, which instead contain a mixture of charcoal from other eucalypts (mainly Jarrah and Tuart), Peppermint, Banksia, She-oak, and grass-tree (*Xanthorrhoea*). Today, these species are found in woodland or open-forest, and with the exception of Tuart, within a few hundred metres of either site. After 12,000 BP, flowstone sealed the Devil's Lair deposit, whereas at Tunnel Cave, biotic remains continued to accumulate. Karri charcoal appeared in the latter deposit at 10,000 BP. With the loss of *Banksia* before 8,000 BP, Karri would have dominated the canopy surrounding Tunnel Cave, and when Tuart disappeared, 7-4,000 BP, this canopy probably resembled the present-day one.

Hunter-gatherer forest occupation and vegetation change

Archaeological evidence for forest occupation presented here is consistent with Aboriginal use of south-western Australian forests summarised in Chapter 4 and above. Hunter-gatherers using these sites probably rarely occupied a site for more than a few days, exploited very localised resources, and usually avoided occupying locations in difficult forest types. Their avoidance of such locations need not imply that large areas were abandoned, since palaeo-environmental reconstructions (notably, Pickett 1997) give no reason to believe that the mosaic of vegetation communities around these sites was any less intricate or varied at any time during the period of human occupation of the region. Hunter-gatherers would have considered vegetation community as a prime factor in choice of site. Since glacial and post-glacial climatic changes probably caused minor geographic movements of vegetation communities, one cannot envisage much dislocation in occupation patterns.

As far as the archaeological evidence for human reactions to Pleistocene-Holocene vegetation shifts, Devil's Lair offers no evidence dating to the period of greatest interest, 13,000-8,000 BP, since it was at this approximate time that the cave was sealed. At Tunnel Cave, however, fewer archaeological remains were deposited after 12,000 BP, suggesting fewer or shorter occupations after that time. By the time of Karri dominance, 8,000 BP, human occupation at Tunnel Cave appears to have ceased. Sites listed in Table 4.1 show that people were still in the region at this time and afterwards (*e.g.*, Arumvale, Calgardup Brook). Archaeological evidence from Witchcliffe Rock Shelter and Rainbow Cave show human occupations in the last few centuries, in woodland or coastal scrub less than one kilometre from present-day Karri forest (cf. F.G. Smith 1973). These occupations were based around exploitation of coastal, riverine, and forest resources. Notwithstanding any expansions of Karri forest, people clearly did occupy other types of vegetation in the region.

There is another complication with the view that vegetation changes forced human reactions. To hypothetically reverse the direction of cause, vegetation change could have itself been a reaction to human activities. Absence or alteration of a human firing regime can itself lead to vegetation change (cf. McNicol 1999). Since the ending of traditional Aboriginal foraging and firing patterns in many parts of Australia over the last 150 years, ecologists have documented large-scale changes in both forest structure and composition.

A decrease in the frequency of hunter-gatherer site occupations in the terminal Pleistocene might have had as much effect on local habitats and flora as the vegetation changes putatively caused by the cessation of hunter-gatherer firing in south-western Australia *c*.AD 1850 (Chapter 2). However, a terminal Pleistocene change caused predominantly by hunter-gatherers is difficult to reconcile with the increase in vessel size in trees and Australia-wide palaeo-climatic evidence for increases in rainfall from *c*.13,000 BP. Rainfall increase, not changes in human fire regime, was probably the prime cause of Karri forest expansion, for it began five millennia before people ceased visiting Tunnel Cave.

Paradoxically, the *continuation* of hunter-gatherer frequent firing at the time of increasing rainfall may have been an additional, critical factor in the decision to cease using Tunnel Cave. Frequently firing vegetation in *high* rainfall conditions can promote dense understoreys. Around Tunnel Cave at the time of increases in rainfall, 13,000-10,000 BP, frequent firing well-suited to the existing mixed Karri forest in dry conditions might have inadvertently promoted rapid growth of understorey plants. Hypothetically, human firing at a time of relatively fast climatic change had the result that hunter-gatherers wanted least.

Because of the compression of three to five millennia of change into a single archaeological layer at Tunnel Cave (5 upper), evidence to confirm this hypothesis is lacking. In any case, other research suggests that these events were not inexorably repeated elsewhere (J. Dortch *et al.* 1998). At Dombakup Compartment 24, the density of the Karri forest is impressive. Under a Karri canopy more than 50 m high, a five metre high understorey of Karri wattles (*Acacia pentadenia*) is packed so tightly that a person can only move through them by slashing them with a machete or hurling themselves into the mass of vegetation. On transect lines crossing an area of *c.*150 ha, researchers excavated 32 test-pits and in 17, found stone artefacts near the surface and down to 50 cm depth. The broad scatter of artefacts, their often-shallow depths below surface, and their likely derivation from Holocene use of the adjacent Northcliffe Silcrete Quarry (Dortch and Gardner 1977), suggests that people occupied small sites throughout this forest when the overstorey included Karri trees. If only for access, they must have controlled the understorey by firing it.

A Karri forest understorey not nearly as dense as that at Dombakup 24 now extends from the entrance to Tunnel Cave. Given the findings from Dombakup 24, people would not have been kept away from Tunnel Cave because of the understorey: it could have been controlled at some point in time. However, as argued in eastern Australia (Hopkins *et al.* 1993; MacPhail 1980, 1984), increasing rainfall and vegetation density in the early Holocene could have impressed people with the difficulty of altering the forest around Tunnel Cave. For several millennia afterwards, people relocated campsites to areas where soils or exposure favoured open vegetation.

Hearth F1 at Tunnel Cave shows that people returned to the site at least once, at 1,400 BP. By this time the climate might have become drier, permitting the resumption of frequent firing, although there is yet no evidence that late Holocene vegetation was more open or more manageable. The prolonged absence of people from Tunnel Cave, and perhaps other places in Karri forest, probably reflects their pragmatic decision to follow small geographic shifts in vegetation boundaries, but not their regional retreat (cf. Ferguson 1985).

Wider implications

This study indicates that environmental changes probably did not force late Pleistocene and Holocene hunter-gatherers to abandon entire forested regions. Where environmental changes involved no extremes of cold or aridity, hunter-gatherers probably relocated occupation sites according to geographic shifts in favourable habitats, or where access or productivity had to be improved, by altering unfavourable habitats with fire. Hunter-gatherers especially need not have abandoned regions characterised by mosaic vegetation. As this study has done for south-western Australian sites, localised palaeo-ecological records in Queensland and Tasmanian forests show that fine-scale vegetation mosaics shifted in relation to changes in regional climate and local fire regime (Hopkins *et al.* 1993, MacPhail 1984). The latter could have been partially controlled by Aboriginal hunter-gatherers. Factors of latitude and altitude can hinder broad generalisation about human-vegetation interactions, but certainly, where mosaic vegetation exists, archaeologists should consider these interactions at the small scale.

In low latitude forests, the limits on hunter-gatherer populations, as indicated by poor soils and low food productivity (Attiwill and Leeper 1990), suggest the need for flexible occupation of forests (Endicott and Bellwood 1991). Departures from the settlement pattern are to be expected. The Gleasonian view in ecology (Colinvaux 1986) suggests palaeo-environments as actually experienced by human groups living in them cannot be reconstructed in detail, because vegetation alliances are too short-lived to be identified in archaeological records. Because archaeologists must use reliable chronologies, such as those inferred in Chapter 6, they often study palimpsests of biotic remains corresponding to millennia of deposition. Moreover, human responses to vegetation alliances must themselves have a transitory nature. One can only assume that broad regional interpretations allow for such ephemera by outlining the superimposed patterns of decades or centuries over millennia.

Nevertheless, the small aberrations from such broad outlines, such as the scant Karri forest occupational record, are as interesting as the norm. They are not only expressions of individuals. They also show that hunter-gatherers prioritised their use of forest types, rather than finding themselves compelled to avoid one or another. The approach here has been to try to assess general trends from localised evidence. The implication of this case study is that in temperate regions with mosaic vegetation – a large part of the world – Late Quaternary hunter-gatherers maintained settlement patterns in which the location of various plant communities was highly significant. Global climate changes may have caused new plant alliances to form and existing ones to expand or retreat, but the diversity inherent in such regions implies that at times natural regional variation could have rivalled the variations that were experienced over the long term. The records of site occupation and abandonment presented here suggest that at times of major climate change, human populations coped easily with shifts in vegetation communities, making only slight adjustments to existing settlement strategies.

References

Adie, L., H. Cocks, P. Fleay, T. Lopes, N. Mutch, P. Valentine, and A. Wilson 1990 Field report of excavations at Rainbow Cave, Margaret River, Western Australia. Unpublished student report, Department of Archaeology, University of Western Australia, Perth.

Aharan, P., and J. Chappell 1986 Oxygen isotopes, sea level changes and the temperature history of a coral reef environment in New Guinea over the past 10^5 years. *Palaeogeography, Palaeoclimatology, Palaeoecology* 56: 337-379.

Allbrook, D. 1976 A human hip-bone from Devil's Lair, Western Australia *Archaeology and Physical Anthroplogy in Oceania* 11: 48-50.

Allen, J. (ed) 1996 Report of the Southern Forests Archaeological Project. Vol. 1. Site descriptions, stratigraphies and chronologies. School of Archaeology, La Trobe University, Melbourne.

Allen, J., and R. Cosgrove 1996 Background history of the Southern Forests Archaeological Project. In Allen, J. (ed) 1996 *Report of the Southern Forests Archaeological Project. Vol. 1. Site descriptions, stratigraphies and chronologies*, pp 3-19. School of Archaeology, La Trobe University, Melbourne.

Allen, J., and J.F. O'Connell (eds) 1995 Transitions: Pleistocene to Holocene in Australia and Papua New Guinea. *Antiquity* 69 Special Number 265.

Amick, D.S., and R.P Mauldin 1989 Comments on Sullivan and Rozen's "Debitage analysis and archaeological interpretation". *American Antiquity* 54: 166-168.

Anderson, J. 1984 *Between plateau and plain.* Ocasional papers in Prehistory 4. Research School of Pacific Studies, Australian National University, Canberra.

Anon. [Collie] 1834 Anecdotes remarks relative to the Aborigines at King George's Sound. *The Perth Gazette and Western Journal*, pp 315-339. [original not seen, cited in Meagher 1974]

Archer, M. 1977 Faunal remains from the excavation at Puntutjarpa Rock-shelter, *Anthropological Papers of the American Museum of Natural History* 54: 158-65.

Archer, M., and A. Baynes 1974 Prehistoric mammal faunas from two small caves in the extreme south-west of Western Australia. *Journal of the Royal Society of Western Australia* 55: 80-89.

Archer, M., I. M. Crawford, and D. Merrilees 1980 Incisions, breakages and charring, some probably man-made, in fossil bones from Mammoth Cave, Western Australia. *Alcheringa* 4: 115-131.

Ashton, D.H., and P.M. Attiwill 1994 Tall open-forests. In Groves, R.H., (ed) *Australian vegetation*, pp 157-196. Second Edition. Cambridge University Press, Cambridge.

Attenbrow, V. 1982 The archaeology of the Upper Mangrove Creek catchment: research in progress. In Bowdler, S. (ed) Coastal archaeology in eastern Australia. *Proceedings of the 1980 Valla Conference on Australian prehistory*, pp 67-78. Department of Prehistory, Research School of Pacific Studies, ANU, Canberra.

Attiwill, P.M. 1994 Ecological disturbance and the conservative management of eucalypt forests in Australia. *Forest Ecology and Management* 63: 301-346.

Attiwill, P.M., and G.W. Leeper 1990 *Forest soils and nutrient cycles*. Paperback edition. Melbourne University Press, Melbourne.

Bailey, R.C., G. Head, M. Jenike, B. Owen, R. Rechtman, and E. Zechenter 1989 Hunting and gathering in tropical rainforest: is it possible? *American Anthropologist* 91: 59-82.

Baird, R.F. 1991 Holocene avian assemblage from Skull Cave, south-western Western Australia. *Records of the Western Australian Museum* 15: 267-286.

Baird, R.F. no date [1992] Late Quaternary avian assemblage from Devil's Lair (W-61E), south-western Western Australia. Unpublished manuscript.

Balme, J. 1978 An apparent association of artefacts and extinct fauna at Devil's Lair, Western Australia. *The Artefact* 3(3): 111-116.

Balme, J. 1979 Artificial bias in a sample of kangaroo incisors from Devil's Lair, Western Australia. *Records of the Western Australian Museum* 7(2): 229-244.

Balme, J. 1980a An analysis of charred bone from Devil's Lair, Western Australia. *Archaeology and Physical Anthropology in Oceania* 15(2): 81-85.

Balme, J. 1980b Some archaeological studies on a bone accumulation from Devil's Lair, Western Australia. *Australian Institute of Aboriginal Studies Newsletter* 13: 49-51.

Balme, J., D. Merrilees and J.K. Porter 1978 Late Quaternary mammal remains spanning about 30,000 years from excavations in Devil's Lair, Western Australia. *Journal of the Royal Society of Western Australia* 6(2): 33-65.

Bamforth, D. 1986 Technological efficiency and tool curation. *American Antiquity* 51: 38-50.

Bannister, T. 1833 A report of Captain Bannister's journey to King George's Sound, over land. In Cross, J. (ed), Journals of several expeditions made in Western Australia during the years 1829, 1830, 1831, and 1832; under the sanction of the Governor, Sir James Stirling, containing the latest authentic information relative to that country, pp. 98-109. J. Cross, London.

Bar-Yosef, O., B. Vandermeersch, B. Arensburg, A. Belfer-Cohen, P. Goldberg, H. Laville, L. Meignen, Y. Rak, J.D. Speth, E. Tchernov, A.-M. Tillier, and S. Weiner 1992 The excavations in Kebara Cave, Mt Carmel. *Current Anthropology* 33: 497-550.

Bastian, L. 1964 Morphology and development of caves in the southwest of Western Australia. *Helictite* (July 1964): 105-119.

Bastian, L.V. 1985 In the beginning: the prehistpry and origin of W.A.S.G. *The Western Caver* 25: 6-9.

Baynes, A., D, Merrilees, and J.K. Porter 1975 Mammal remains from the upper levels of a late Pleistocene deposit in Devil's Lair, Western Australia. *Journal of the Royal Society of Western Australia* 58: 97-126.

Bé, A.W.H., and J.-C. Duplessy 1976 Subtropical convergence fluctuations and Quaternary climates in the middle latitudes of the Indian Ocean. *Science* 194: 419-422.

Beard, J. 1981 The Vegetation of the Swan Area. Swan 1: 1 000 000 Vegetation Series. Explanatory Notes to Sheet 7. University of Western Australia Press, Perth.

Beard, J. 1983 Ecological control of the vegetation of southwestern Australia: moisture versus nutrients. In Kruger, F.J., D.T. Mitchell, and J.V.M. Jarvis (eds) *Mediterranean-type ecosystems: the role of nutrients*, pp 66-73. Springer-Verlag, Berlin.

Beaton, J. 1977 Dangerous harvest: investigations in late prehistoric occupation of upland south-eastern Queensland. Unpublished PhD thesis, ANU.

Beck, W., A. Clarke, and L. Head 1989 *Plants in Australian archaeology*. Tempus 1, Archaeology and material culture studies in anthropology. Anthropology Museum, University of Queensland, St Lucia.

Bednarik, R. 1997 The role of Pleistocene beads in documenting hominid cognition. *Rock Art Research* 14: 27-41.

Bell, D.T., and E.M. Heddle 1989 Floristic, morphologic, and vegetational diversity. In Dell, B., J.J. Havel, and N. Malajczuk (eds) *The Jarrah forest: a complex Mediterranean ecosystem*, pp 53-66. Kluwer, Dordrecht.

Bell, D.T., W.L. McCaw, and N.D. Burrows 1989 Influence of fire on Jarrah forest vegetation. In Dell, B., J.J. Havel, and N. Malajczuk (eds) *The Jarrah forest: a complex Mediterranean ecosystem*, pp 203-215. Kluwer, Dordrecht.

Bellomo, R.V. 1993 A methodological approach for identifying archaeological evidence of fire resulting from human activities. *Journal of Archaeological Science* 20: 525-533.

Belshaw, J. 1978 Population distribution and the pattern of seasonal movement in northern New South Wales. In MacBryde, I. (ed) *Records of times past*, pp 65-81. Australian Institute of Aboriginal Studies, Canberra.

Bender, B. 1978 Gatherer-hunter to farmer: a social perspective. *World Archaeology* 10: 204-222.

Bie, M. de 1997 Book review: L.G. Straus, B.V. Eriksen, J.M. Erlandson, and D.R. Yesner (eds) 1996 Humans at the end of the Ice Age - the archaeology of the Pleistocene-Holocene transition. Plenum Press, New York. *World Archaeology* 24: 495-500.

Bindon, P.R. 1986 Old campsites, new tenants. A study of recent and prehistoric fireplaces near Wiluna, Western Australia. Unpublished MA thesis, Centre for Prehistory, University of Western Australia, Perth.

Bindon, P. 1996 *Useful bush plants*. Western Australian Museum, Perth.

Bindon, P., and C. Dortch 1982 Dating problems at the Ellen Brook site. *Australian Archaeology* 14: 13-17.

Binford, L.R. 1968 Post-Pleistocene adaptations. In Binford, S.R., and L.R. Binford *New perspectives in archaeology*, pp313-341. Seminar Press, New York.

Binford, L.R. 1980 Willow smoke and dog's tails: hunter-gatherer settlement systems and archaeological site formation. *American Antiquity* 45: 4-20.

Binford, L.R. 1983 *In pursuit of the past: decoding the archaeological record*. Paperback edition. Thames and Hudson, London.

Bird C. 1985 Prehistoric lithic resource utilization: a case study from the southwest of Western Astralia. Unpublished PhD thesis, Centre for Prehistory, University of Western Australia.

Bird, C., and C. Beeck 1988 Traditional plant foods in the southwest of Western Australia: the evidence from salvage ethnography. In Meehan, B., and R. Jones (eds) *Archaeology with ethnography: an Australian perspective*, pp 113-122. Research School of Pacific Studies, ANU, Canberra.

Boulter, M.C., R.N.L.B. Hubbard, and Z. Kvacek 1993 A comparison of intuitive and objective interpretations of Miocene plant assemblages from north Bohemia. *Palaeogeography, Palaeoclimatology, Palaeoecology* 101: 81-96.

Bowdler, S. 1974 Pleistocene date for man in Tasmania. *Nature* 252:697-698.

Bowdler, S. 1979 Hunter Hill, Hunter Island. Unpublished PhD thesis, ANU, Canberra.

Bowdler, S. 1983a Rainforest: colonised or coloniser? *Australian Archaeology* 17: 59-66.

Bowdler, S. 1983b *Aboriginal sites on the Crown-timber lands of New South Wales*. Forestry Commission of New South Wales, Sydney.

Bowdler, S. 1984 Hunter Hill, Hunter Island. *Terra Australis* 8. ANU Press, Canberra.

Bowler, J.M. 1976 Aridity in Australia: age, origins and expression in aeolian landforms and sediments. *Earth Science Reviews* 12:279-310.

Bowler, J.M., and R.J. Wasson 1984 Glacial age environments of inland Australia. In Vogel, J.C. (ed) *Late Cainozoic palaeoclimates of the Southern Hemisphere*, pp 183-208. Balkema, Rotterdam.

Bowman, D.M.J.S. 1998 The impact of Aboriginal landscape burning on the Australian biota. Tansley Review No. 101. *New Phytologist* 140: 385-410.

Bowman, D.M.J.S., and M.J. Brown 1986 Bushfires in Tasmania: a botanical approach to anthropological questions. *Archaeology in Oceania* 21: 166-171.

Bowman, D.M.J.S., and W.J. Panton 1993 Decline of *Callitris intratropica* R.T. Baker & H.G. Smith in the Northern Territory: implications for pre- and post-European colonization fire regimes. *Journal of Biogeography* 20: 373-381.

Bradshaw, F.J. 1985 Silvicultural guidelines for the treatment of even-aged regrowth and two-tiered karri forests. *Department of Conservation and Land Management Technical Report* 1.

Brobowsky, P.T., and B.F. Ball 1989 The theory and mechanics of ecological diversity in archaeology. In Leonard, R.D., and G.T. Jones (eds) *Quantifying diversity in archaeology*, pp 4-12. Cambridge University Press, Cambridge.

Brock, R. 1998 An investigation into the integrity of prehistoric sites located in Lake Jasper, Western Australia. Unpublished Honours thesis, Centre for Archaeology, University of Western Australia.

Brooker, M.I.H., and D.A. Kleinig 1990 Field guide to the eucalypts: 2. Southwestern and southern Australia. Inkata, Melbourne.

Brown, R. G. 1983 Sea level history over the past 15,000 years along the Western Australian coastline. In Hopley, D., (ed) *Australian sea levels in the last 15,000 years: a review*, pp 29-36. Monograph Series Occasional Paper 3, Department of Geography, James Cook University.

Brown, S. 1993 Mannalargenna Cave: a Pleistocene site in Bass Strait. In M.A. Smith, M. Spriggs, and B. Fankhauser (eds) *Sahul in review*, pp 258-271. Occasional Papers in Prehistory 24, Department of Prehistory, Research School of Pacific and Asian Studies, The ANU, Canberra.

Bryant, V.M. 1996 Paleoenvironmental reconstruction. In Fagan, B.M. (ed) *The Oxford Companion to Archaeology*, pp 551-552. Oxford University Press, New York.

Burbridge, A.A., K. A. Johnson, P.J. Fuller, and R.I. Southgate 1988 Aboriginal knowledge of the mammals of the central deserts of Australia. *Australian Wildlife Research* 15: 9-39.

Bureau of Meteorology 1975 *Climatic averages Australia*. Department of Science and Consumer Affairs, Bureau of Meteorology. Australian Government Publishing Service, Canberra.

Burke, S. 1997 Human responses to vegetation change as identified from carbonised plant remains from Devil's Lair, south-western Western Australia. Unpublished Honours thesis, Centre for Archaeology, University of Western Australia.

Burrows, N.D. 1987 The soil dryness index for use in fire control in the south-west of Western Australia. Technical Report No. 17. Department of Conservation and Land Management, Como, Western Australia.

Burrows, N.D., B. Ward, and A.D. Robinson 1995 Jarrah forest fire history from stem analysis and anthropological evidence. *Australian Forestry* 58: 7-16.

Burton, J. 1980 Making sense of waste flakes: new methods for investigating the technology and economics behind chipped stone assemblages. *Journal of Archaeological Science* 7: 131-148.

Bussell, A.P. no date [1834] "Vasse Journal" (letters to relatives), typescript. State Archives, MN 586 337A/784, Battye Library, Perth, Western Australia.

Bussell, A.J. no date [1941] South-west Aboriginal language or dialect- what the Aboriginals term 'Dornderup Wongie' and other things concerning Australia generally, typescript. Battye Library, Perth.

Bussell, J.C. 1833 Report of an excursion to 7the Northward from Augusta. In Cross, J., (ed) Journals of several expeditions made in Western Australia during the years 1829, 1830, 1831, and 1832; under the sanction of the Governor, Sir James Stirling, containing the latest authentic information relative to that country, pp 178-185. J. Cross, London.

Byrne, D. 1980 Dynamics of dispersion: the place of silcrete in archaeological assemblages from the-lower Murchison, Western Australia. *Archaeology and physical anthropology in Oceania* 15: 110-119.

Byrne, D. 1983 *The Five Forests: an archaeological and anthropological investigation*, v. 1-2. National Parks and Wildlife Service, Sydney.

Byrne, D. 1987 *The Aboriginal and archaeological significance of the New South Wales rain-forests.* Report to the Forestry Commission of New South Wales and the Australian Heritage Commission. Forestry Commission of New South Wales, Sydney.

Callow, P. 1986 Raw materials and sources. In Callow, P., and J.M. Cornford (eds) *La Cotte de St Brelade 1961-1978. Excavations by C.B.M. McBurney*, pp 203-211. GeoBooks, Norwich.

Capaldo, S.D. 1998 Simulating the formation of dual-patterned archaeofaunal assemblages with experimental control samples. *Journal of Archaeological Science* 25: 311-330.

Campbell, V. 1978 Ethnohistorical evidence on the diet and economy of the Aborigines of the Macleay River Valley. In MacBryde, I. (ed) *Records of times past*, pp 82-100. Australian Institute of Aboriginal Studies, Canberra.

Cartwright, C., and J. Parkington 1997 The wood charcoal assemblage from Elands Bay Cave, southwestern Cape: principles, procedures, and preliminary interpretation. *South African Archaeological Bulletin* 52: 59-72.

Chapman, J., P.A. Delcourt, P.A. Cridlebaugh, A.B. Shea, and H.R. Delcourt 1982 Man-land interaction: 10,000 years of Indian impact on native ecosystems in the lower Little Tennessee River Valley, eastern Tennessee. *Southeastern archaeology* 1: 115-121.

Chappell, J. 1993 Late Pleistocene coasts and human migrations in the Austral region. In Spriggs, M., D.E. Yen, W. Ambrose, R. Jones, A. Thorne, and A. Andrews (eds) *A community of culture: the peopling and prehistory of the Pacific*, pp 43-48. Occasional Papers in Prehistory 21, Department of Prehistory, Research School of Pacific and Asian Studies, ANU, Canberra.

Chappell, J., and N.J. Shackleton 1986 Oxygen isotopes and sea level. *Nature* 324: 137-139.

Christensen, P.E. 1980 The biology of *Bettongia penicillata* Gray, 1837, and *Macropus eugenii*

Desmarest, 1817, in relation to fire. *Forests Department of Western Australia Bulletin* 91.

Christensen, P.E. 1992 *The Karri forest*. Department of Conservation and Land Management, Como, Western Australia.

Christensen, P.E., and I. Abbott 1989 Impact of fire in the eucalypt forest ecosystem of southern Western Australia: a critical review. *Australian Forestry* 52(2): 103-121.

Christensen, P.E., and A. Annels 1985 Fire in southern tall forests.In Ford, J.R. (ed) *Fire ecology and management in Western Australian ecosystems.* Symposium, Perth. Western Australian Institute of Technology: 67-82.

Christensen, P., A. Annels, G. Liddelow, and P. Skinner 1985 *Vertebrate fauna in the southern forests of Western Australia.* Forests Department of Western Australia, Perth.

Christensen, P.E., and P.C. Kimber 1975 Effect of prescribed burning on the flora and fauna of south-western Australian forests. In Kikkawa, J., and H.A. Nix (eds) Managing terrestrial ecosystems. *Proceedings of the Ecological Society* Vol. 9, pp 85-106. Symposium, Brisbane, 15-16 July 1975.

Churchill, D.M. 1968 The distribution and prehistory of *Eucalyptus diversicolor* F. Muell, *E. marginata* Donn ex SM and *E. calophylla* R. Br. in relation to rainfall. *Australian Journal of Botany* 16: 125-151.

Clark, G. 1975 *The earlier stone age settlement of Scandinavia.* Cambridge University Press, Cambridge.

Clarke, J. 1983 An Aboriginal engraving site in the south-west of Western Australia. *Records of the Western Australian Museum* 11:63-67.

Clarke, J., and C.E. Dortch 1977 A 10,000 year BP radiocarbon date for archaeological finds within a soil of the Spearwood Dune System, Mosman Park, WA. *Search* 8: 36-38.

Cocks, H. 1993 An analysis of the stone artefact assemblage from Rainbow Cave, south-west Western Australia. Unpublished Honours thesis, Centre for Archaeology, UWA.

Colhoun, E.A. 1991 Climate during the last glacial maximum in Australia and New Guinea. Evidence inferred from biogeographical and geomorphological data. *Australia and New Zealand Geomorphology Group Special Publication No. 2.*

Colhoun, E.A., and J.A. Peterson 1986 Quaternary landscape evolution and the evolution of the cryosphere: research progress from Sahul to Australian Antarctica. *Australian Geographical Studies* 24:145-167.

Colinvaux, P.A. 1987 The changing forests: ephemeral communities, climate and refugia. *Quarterly Review of Archaeology* 8:1-7.

Connolly, J.R. 1967 Postglacial-glacial change in climate in the Indian Ocean. *Nature* 214: 873-875.

Cork, S.J., and P.C. Catling 1996 Modelling distributions of arboreal and ground-dwelling mammals in relation to climate, nutrients, plant chemical defences and vegetation structure in the eucalypt forests of southeastern Australia. *Forest Ecology and Management* 85: 163-175.

Cosgrove, R. 1990 The archaeological resources of Tasmanian forests: past Aboriginal use of forested environments. *Occasional Paper No. 27.* Department of Parks, Wildlife, and Heritage, Hobart.

Cosgrove, R. 1995 Late Pleistocene behavioural variation and time trends: the case from Tasmania. *Archaeology in Oceania* 30: 83-104.

Cosgrove, R. 1996 Origin and development of Australian Aboriginal tropical rainforest culture: a reconsideration. *Antiquity* 70: 900-912.

Cosgrove, R., J. Allen, and B. Marshall 1990 Palaeo-ecology and Pleistocene human occupation in south central Tasmania. *Antiquity* 64: 59-78.

Cotterell, B., and J. Kamminga 1990 *Mechanics of pre-industrial technology.* Oxford University Press, Oxford.

Courty, M.A., P. Goldberg, and R. Macphail 1989 *Soils and micromorphology in archaeology. Cambridge Manuals in Archaeology.* Cambridge University Press, Cambridge.

Coutts, P.J.F., and D.C. Witter 1977 New radiocarbon dates for Victorian archaeological sites. *Records of the Victoria Archaeological Survey* 4: 59-73.

Coventry, G.1998 Site function and mobility in the Leeuwin-Naturaliste Region of Western Australia during the Late Holocene. Unpublished Honours thesis, Centre for Archaeology, University of Western Australia, Perth.

Cruz-Uribe, K. 1988 The use and meaning of species diversity and richness in archaeological faunas. *Journal of Archaeological Science* 15: 179-196.

Dadswell, H.E. 1972 *The anatomy of eucalypt woods.* Commonwealth Scientific and Industrial Research Organisation (CSIRO), Melbourne.

Dadswell, H.E., and A.M. Eckersley 1935 *The identification of the principal commercial Australian timbers other than eucalypts.* Bulletin 90. Council for Scientific and Industrial Research, Melbourne.

David, B. 1990 How was this bone burnt? In Solomon, S., I. Davidson, and D. Watson (eds) *Problem solving in taphonomy. Archaeological and palaeontological studies from Europe, Africa, and Oceania*, pp 65-79. Tempus 2, Archaeology and material culture studies in anthropology. Anthropology Museum, University of Queensland, St Lucia.

David, M.A. 1993 Postcards from the edge: an analysis of tasks undertaken during the Late Pleistocene at Devil's Lair, south-western Australia. Unpublished M.L. dissertation, University of New England.

Davies, P.L. 1968 An 8,000 to 12,000 year old human tooth from Western Australia. *Archaeology and Physical Anthropology in Oceania* 3: 33-40.

Daw, B., T. Walley, and G. Keighery 1997 *Bush tucker plants of the south-west.* Department of Conservation and Land Management, Perth.

Deacon, H.J. 1995 Two Late Pleistocene-Holocene archaeological depositories from the Southern Cape, South Africa. *South African Archaeological Bulletin* 50: 121-131.

Dell, B., A.J.M. Hopkins, and B.B. Lamont 1986 Introduction. In Dell, B., A.J.M. Hopkins, and B.B. Lamont (eds) *Resilience in Mediterranean-type ecosystems*, pp 1-4. Tasks for vegetation science, Series 16. Kluwer, Dordrecht.

Dell, B., J.J. Havel, and N. Malajczuk (eds) 1989 *The Jarrah forest: a complex Mediterranean ecosystem.* Kluwer, Dordrecht.

Dibble, H. 1987 The interpretation of Middle Paleolithic scraper morphology. *American Antiquity* 52: 109-117.

Dimbleby, G.W. 1961 The ancient forest of Blackamore. *Antiquity* 35: 123-128.

Dixon, K.W., S. Roche, and J.S. Pate 1995 The promotive effect of smoke derived from burn native vegetation on seed germination of Western Australian plants. *Oecologia* 101: 185-192.

Dodson, J.R. 1987 Mire development and environmental change, Barrington Tops, New South Wales. *Quaternary Research* 27: 561-585.

Dodson, J.R., R. Fullagar, and L. Head 1992 Dynamics of environment and people in the forested crescents of temperate Australia. In Dodson, J. (ed) *The Naïve Lands: prehistory and environmental change in Australia and the southwest Pacific*, pp 115-159. Longman Cheshire, Melbourne.

Dodson, P., and D. Wexlar 1979 Taphonomic investigations of owl pellets. *Paleobiology* 5: 275-284.

Dolby, N. 1995 A unique perspective: charcoal, environment and fuel use at the Pleistocene archaeological site of Nunamira, southwest Tasmania. Unpublished Honours thesis, School of Archaeology, La Trobe University, Melbourne.

Donoghue, D. 1989 Carbonised plant macrofossils. In Beck, W., A. Clarke, and L. Head (eds) *Plants in Australian archaeology*, pp. 90-110. Tempus 1. Archaeology and material culture studies in anthropology. Anthropology Museum, University of Queensland, St. Lucia, Queensland.

Dortch, C.E. 1974 A twelve thousand year old occupation floor in Devil's Lair, Western Australia. *Mankind* 9: 195-205.

Dortch, C.E. 1976a Two engraved stone plaques of Late Pleistocene age from Devil's Lair, Western Australia. *Archaeology and Physical Anthropology in Oceania* 11: 32-44.

Dortch, C.E. 1976b *Devil's Lair: a search for ancient man in Western Australia.* Western Australian Museum Press, Perth.

Dortch, C.E. 1977 Early and late stone industrial phases in Western Australia. In Wright, R.V.S., (ed) *Stone tools as cultural markers*, pp 104-132. Australian Institute of Aboriginal Studies Press, Canberra.

Dortch, C.E. 1979a Devil's Lair, an example of prolonged cave use in Western Australia. *World Archaeology* 10 (3): 258-281.

Dortch, C.E. 1979b Australia's oldest ornaments. *Antiquity* 53: 39-43.

Dortch, C.E. 1979c 33,000 year old stone and bone artifacts from Devil's Lair. *Records of the Western Australian Museum* 7: 329-367.

Dortch, C.E. 1980 A possible pendant of marl from Devil's Lair, Western Australia. *Records of the Western Australian Museum* 7: 329-367.

Dortch, C.E. 1984 *Devil's Lair- a study in prehistory.* Western Australian Museum, Perth.

Dortch, C.E. 1985 The Malimup middens: evidence for mollusc eating in prehistoric south-western Australia. In Misra, V.N., and P. Bellwood (eds) *Recent advances in Indo-Pacific prehistory*, pp 251-256. Oxford and IBH, New Delhi.

Dortch, C.E. 1986a Excavations outside the former (north) entrance to Devil's Lair, southwestern Western Australia. *Australian Archaeology* 23:62-69.

Dortch, C.E. 1986b An 11,000 BP radiocarbon date from the Arumvale site, southwestern Australia. *Australian Archaeology* 23: 70-72.

Dortch, C.E. 1997 New perceptions of the chronology and development of Aboriginal fishing in south-western Australia. *World Archaeology* 29: 15-35.

Dortch, C.E. 1999 Archaeological assessment of Aboriginal estuarine fishing on the Southern Ocean coast of Western Australia. In Hall, J., and I. McNiven (eds) *Australian coastal archaeology*, pp 25-35. Department of Archaeology and Natural History, Research School of Pacific and Asian Studies, The ANU, Canberra.

Dortch, C.E., and J. Dortch 1996 Review of Devil's Lair artifact classification and radiocarbon chronology. *Australian Archaeology* 43: 28-31.

Dortch, C.E., and J. Dortch 1997 Aboriginal occupation in the limestone caves and rock shelters of the Leeuwin-Naturaliste Region, Western Australia: research background and archaeological perspective. *The Western Australian Naturalist* 21: 191-206.

Dortch, C.E., and G. Gardner 1976 Archaeological investigations in the Northcliffe District, Western Australia. *Records of the Western Australian Museum* 4(3): 257-293.

Dortch, C.E., G. Kendrick, and K. Morse 1984 Aboriginal mollusc exploitation in south-western Australia. *Archaeology in Oceania* 19:81-104.

Dortch, C. and W.M. McArthur 1985 Apparent association of Bryozoan chert artefacts and quartz geometric microliths at an open-air site, Arumvale, southwestern Western Australia. *Australian Archaeology* 21: 74-90.

Dortch, C.E., and D. Merrilees 1971 A salvage excavation in Devil's Lair, Western Australia. *Journal of the Royal Society of Western Australia* 54: 103-113.

Dortch, C.E., and D. Merrilees 1973 Human occupation of Devil's Lair, Western Australia, during the Pleistocene. *Archaeology and Physical Anthroplogy in Oceania* 8: 89-115.

Dortch, C.E., and M.V. Smith 2001 Appraisal of palaeodemographic modelling in South-western prehistory, Western Australia. *Archaeology in Oceania.*

Dortch, J. 1994 Pleistocene radiocarbon dates for hearths at Tunnel Cave, south-western Australia. *Australian Archaeology* 38: 45-46.

Dortch, J. 1995 Report of an archaeological excavation at the Dunsborough 2 site, Busselton Shire, Western Australia. Unpublished report submitted to Water Authority of Western Australia, Bunbury.

Dortch, J. 1996 Late Pleistocene and recent Aboriginal occupation of Tunnel Cave and Witchcliffe Rock Shelter, south-western Australia. *Australian Aboriginal Studies* 1996 (2): 51-60.

Dortch, J., and G. Kelly 1996 Test-excavation at Nookanellup Rock Shelter, Point D'Entrecasteaux, south-western Australia. Unpublished report to the Manjimup Aboriginal Corporation and the Australian Institute of Aboriginal and Torres Strait Islander Studies.

Dortch, J., and G. Kelly 1997 Further test-excavations at Nookanellup Rock Shelter, Point D'Entrecasteaux, south-western Australia. Unpublished report to the Manjimup Aboriginal Corporation and the Australian Institute of Aboriginal and Torres Strait Islander Studies.

Dortch, J., A. Gardos, and T. O'Reilly 1998 Archaeological survey at the Dombakup 24 logging coupe, Manjimup Shire, south-western Australia. Report to the Department of Conservation and Land Management (CALM). Centre for Archaeology, University of Western Australia.

Douglas, A.M., G.W. Kendrick, and D. Merrilees 1966 A fossil bone deposit near Perth, Western Australia, interpreted as a carnivore's den after feeding tests on living *Sarcophilus* (Marsupialia, Dasyruridiae). *Journal of the Royal Society of Western Australia* 49: 88-90.

Driver, J.C., M. Handly, K.R. Fladmark, D.E. Nelson, G.M. Sullivan, R. Preston 1996 Stratigraphy, radiocarbon dating, and culture history of Charlie Lake Cave, British Columbia. *Arctic* 49: 265-277.

Drummond, J. 1843 On the botany of Western Australia. Letter to the Editor of the Inquirer. *The Inquirer, a Western Australian Journal of Politics and Literature*, 137, 22 March 1843 [original not seen, cited in Meagher 1974].

Endicott, K., and P. Bellwood 1991 The possibility of independent foraging in the rain forest of peninsular Malaysia. *Human ecology* 19: 151-185.

Ensor, H.B., and E. Roemer 1989 Comments on Sullivan and Rozen's debitage analysis and archaeological interpretation. *American Antiquity* 54: 175-178.

Fahrner, C. 1993 Links between the Leeuwin Current and rainfall in Western Australia. Unpublished Honours thesis, Centre for Water Research, University of Western Australia.

February, E. 1992 Archaeological charcoals as indicators of vegetation change and human fuel choice in the Late Holocene at Elands Bay, Western Cape Province, South Africa. *Journal of Archaeological Science* 19: 347-354.

February, E. 1994 Rainfall reconstruction using wood charcoal from two archaeological sites in South Africa. *Quaternary Research* 42: 100-107.

February, E.C., W.D. Stock, W.J. Bond, and D.J. Le Roux 1995 Relationships between water availability and selected vessel characteristics in *Eucalyptus grandis* and two hybrids. *International Association of Wood Anatomists Journal* 16: 269-276.

Ferguson, W.C. 1980a Fossiliferous chert in southwestern Australia after the Holocene transgression: a behavioural hypothesis.*The Artefact* 5: 155-169.

Ferguson, W.C. 1980b Edge-angle classification of the Quininup Brook implements: testing the ethnographic analogy. *Archaeology and physical anthropology in Oceania* 15: 56-72.

Ferguson, W.C. 1981 Archaeological investigations at the Quininup Brook site complex, Western Australia. *Records of the Western Australian Museum* 8: 609-637.

Ferguson, W.C. 1985 A mid-Holocene de-population of the Australian south-west. 2 vols. Unpublished PhD thesis, The ANU, Canberra.

Figueiral, I. 1995 Evidence from chracoal analysis for environmental change during the interval late Bronze Age to Roman, at the archaeological site of Castro de Penices, N.W. Portugal. *Vegetation History and Archaeobotany* 4: 93-100.

Flannery, T.F. 1995 *The future eaters*. Paperback edition. Reed Books, Port Melbourne.

Flenniken, J.J., and J. P. White 1985 Australian flaked stone tools: a technological perspective. *Records of the Australian Museum* 36: 131-151.

Fox, B.J., and M.D. Fox 1986 Resilience of animal and plant communities to human disturbance. In Dell, B., A.J.M. Hopkins, and B.B. Lamont (eds) *Resilience in Mediterranean-type ecosystems*, pp 39-65. Tasks for vegetation science, Series 16. Kluwer, Dordrecht.

Frankel, D. 1995 The Australian transition: real and perceived boundaries. *Antiquity* 69: 649-655.

Frankel, D., and S. Holdaway 1993 Diverse: a program to calculate various measures of diversity. Computer program, version 3.0. Department of Archaeology, La Trobe University.

Freedman, L. 1976 A deciduous human incisor tooth from Devil's Lair, Western Australia. *Archaeology and Physical Anthropology in Oceania* 11: 45-47.

Freslov, J. 1993 The role of open sites in the investigation of Pleistocene phenomena in the inland southwest of Tasmania. In M.A. Smith, M. Spriggs, and B.

Fankhauser (eds) *Sahul in review*, pp 233-239. Occasional Papers in Prehistory 24, Department of Prehistory, Research School of Pacific and Asian Studies, The ANU, Canberra.

Gamble, C. 1995 *Timewalkers - the prehistory of global colonization*. Paperback edition. Penguin Books, London.

Gardos, A. 1997 Stone technological analysis at Kalgan Hall, south-western Australia: a test of a mid-Holocene de-population model. Unpublished BA Honours thesis, Centre for Archaeology, University of Western Australia.

Gell, P.A., I.-M. Stuart, and J.D. Smith 1993 The response of vegetation to changing fire regimes and human activity in East Gippsland, Victoria, Australia. *The Holocene* 3: 150-160.

Gentilli, J. 1989 Climate of the Jarrah forest. In Dell, B., J.J. Havel, and N. Malajczuk (eds) *The Jarrah forest: a complex Mediterranean ecosystem*, pp 23-40. Kluwer, Dordrecht.

Gentilli, J. 1991 Homologous peri-oceanic west coast climates in the southern hemisphere. *Journal of the Royal Society of Western Australia* 74:15-33.

Geological Survey of Western Australia 1990 Geology and Mineral Resources of Western Australia: Geological Survey of Western Australia, Memoir 3. Department of Mines, Perth.

Gibbs, M. 1987 Aboriginal gatherings in the west coastal region of southwest Western Australia: an ethnohistorical study. Unpublished Honours thesis, University of Western Australia.

Gifford, D.P. 1981 Taphonomy and paleoecology: a critical review of archaeology's sister disciplines. *Advances in archaeological method and theory* 4: 365-437.

Gifford-Gonzalez, D.P. 1989 Ethnographic analogues for interpreting modified bones: some cases from East Africa. In Bonnichsen, R., and M.H. Sorg (eds) *Bone modification*, pp 179-246. University of Maine Center for Study of the First Americans, Orono.

Gill, A.M. 1994 Patterns and processes on open-forest of *Eucalyptus* in southern Australia. In Groves, R.H., (ed) *Australian vegetation*, pp 197-226. Second Edition. Cambridge University Press, Cambridge.

Gill, A.M., and P.H.R. Moore 1997 *Contemporary fire regimes in the forests of southwestern Australia*. Centre for Plant Biodiversity Research, CSIRO, Canberra.

Gill, A.M., R.H. Groves, and I.R. Noble (eds) 1981 *Fire and the Australian biota*. Australian Academy of Science, Canberra.

Glauert, L. 1910 The Mammoth Cave. Records of the Western Australian Museum and Art Gallery 1: 11-36.

Glauert, L. 1948 The cave fossils of the south-west. *The Western Australian Naturalist* 1: 100-104.

Glover, J.E. 1974 Petrology of chert artefacts from Devil's Lair, Western Australia. *Journal of the Royal Society of Western Australia* 57: 51-53.

Glover, J.E. 1979 The mineral properties and probable provenance of a 33,000 year old opaline artefact from Devil's Lair, southwestern Australia. *Records of the Western Australian Museum* 7: 369-374

Glover, J.E. 1984 The geological sources of stone for artefacts in the Perth Basin and nearby areas. *Australian Aboriginal Studies* 1984 (1): 17-25.

Glover, J.E., A.N. Bint, and C.E. Dortch 1993 Typology, petrology, and palynology of the Broke Inlet Biface, a large flaked chert artifact from south-western Australia. *Journal of the Royal Society of Western Australia* 76: 41-47.

Glover, J.E., and R. Lee 1984 Geochemistry and provenance of Eocene chert artifacts, southwestern Australia. *Archaeology and physical anthropology in Oceania* 19: 16-20.

Goede, A., and P. Murray 1977 Pleistocene man in south central Tasmania: evidence from a cave site in the Florentine Valley. *Mankind* 11: 2-10.

Goede, A., P. Murray, and R. Harmon 1978 Pleistocene man and megafauna in Tasmania: dated evidence from cave sites. *The Artefact* 3: 139-149.

Gollan, K. 1984 The Australian dingo: in the shadow of Man. In Archer, M., and G. Clayton (eds) *Vertebrate zoogeography and evolution in Australasia (animals in space and time)*, pp 921-927. Hesperian Press, Carlisle, Western Australia.

Gosden, C., and L. Head 1994 Landscape - a usefully ambiguous concept. *Archaeology in Oceania* 29: 113-116.

Gould, J. 1863 *The mammals of Australia*. Vol 1. J. Gould, London.

Gould, R.A. 1971 Uses and effects of fire among the Western Desert Aborigines of Australia. *Mankind* 8: 14-24.

Gould, R.A. 1977a Puntutjarpa rockshelter and the Australian desert culture. American Museum of Natural History, New York, *Anthropological Papers* 54: 1-187.

Gould, R.A. 1977b Ethno-archaeology; or where do models come from? In Wright, R.V.S., (ed) *Stone tools as cultural markers*, pp 162-168. Australian Institute of Aboriginal Studies, Canberra.

Gould, R.A. 1980 *Living Archaeology*. Cambridge University Press, Cambridge.

Gould, R.A. 1996 Faunal reduction at Pututjarpa rockshelter, Warburton Ranges, Western Australia. *Archaeology in Oceania* 31: 72-86.

Grace, R. 1996 Use-wear analysis: the state of the art. *Archaeometry* 38: 209-229.

Grace, R. 1997 The *chaîne opératoire* approach to lithic analysis. Published in Issue 2 of *Internet Archaeology*. Retrieved 3 February, 1999 from the World Wide Web: http://www.hf.uio.no/iakn/roger/lithic/opchainpaper.html

Grayson, D.K. 1984 Quantitative zooarchaeology. Topics in the analysis of archaeological faunas. Academic Press, Orlando, Florida.

Green, R.H. 1967 Notes on the devil (*Sarcophilus harrisii*) and the quoll (*Dasyurus viverrinus*) in north-eastern Tasmania. *Records of the Queen Victoria Museum* 27: 1-12.

Grey, G. 1841 Journals of two expeditions of discovery in north-west and western Australia. Boone, London.

Griffiths, D.R., C.A. Bergman, C.J. Clayton, K. Ohnuma, G.V. Robins, and N.J. Seeley 1987 Experimental investigation of the heat treatment of flint. *In* Sieveking, G. de G., and M.H. Newcomer (eds) *The human uses of flint and chert. Proceedings of the Fourth International Flint Symposium, British Polytechnic 10-15 April 1983*, pp 43-52. Cambridge University Press, Cambridge.

Grove, T.S., and N. Malajczuk 1985 Biomass production by trees and understorey shrubs in an age-series of *Eucalyptus diversicolor* F.Muell. stands. *Forest Ecology and Management* 32: 117-134.

Groves, R.H. 1994 *Australian vegetation.* Second Edition. Cambridge University Press, Cambridge.

Guiler, E.R. 1970 Observations on the Tasmanian devil, *Sarcophilus harisii* (Marsupialia: Dasyuridae). 1. Numbers, home range, movements, and food in two populations. *Australian Journal of Zoology* 18: 49-62.

Guiler, E.R. 1982 Temporal and spatial distribution of the Tasmanian devil, *Sarcophilus harisii* (Dasyuridae: Marsupialia). *Papers and Proceedings of the Royal Society of Tasmania* 116: 153-163.

Hallam, S.J. 1974 Excavations in the Orchestra Shell Cave, Wanneroo, Western Australia, part I: ethnographic and environmental background; part II: archaeology. *Archaeology and physical anthropology in Oceania* 9:66-84, 134-155.

Hallam, S.J. 1975 *Fire and hearth.* Australian Institute of Aboriginal Studies, Canberra.

Hallam, S.J. 1987 Coastal does not equal littoral. *Australian Archaeology* 25: 10-29.

Halse, S.A., and J.D. Blyth 1992 Aquatic fauna of the karri forest. In Department of Conservation and Land Management (ed) *Research on the impact of forest management in south-west Western Australia. Occasional Paper 2/92*, pp 115-138. Department of Conservation and Land Management, Como, Western Australia.

Hammond, J.E. 1933 [1980] *Winjan's People.* Facsimile edition, Hesperian Press, Victoria Park, Western Australia (original published by Imperial Printing Company, Perth).

Hardy, B.L., and G.T. Garufi 1998 Identification of woodworking on stone tools through residue and use-wear analyses: experimental results. *Journal of Archaeological Science* 25: 177-184.

Harris, D. and G.C. Hillman (eds) 1989 *Foraging and farming: the evolution of plant exploitation.* Unwin Hyman, London.

Harrison, S.P., and J. Dodson 1993 Climates of Australia and New Guinea since 18,000 years BP. In Wright, H.E., Jr, J.E. Kutzbach, T. Webb III, W.F. Ruddiman, F.A. Street-Perrott, and P.J. Bartlein (eds) *Global climates since the Last Glacial Maximum*, pp 265-93. University of Minnesota Press, Minneapolis.

Havel, J.J. 1975 Site-vegetation mapping in the northern jarrah forest. II. Location and mapping of site-vegetation types. *Forests Department of WA Bulletin* 87.

Hayden, B. 1975 Dingoes: pets or producers? *Mankind* 10: 11-15.

Hayden, B. 1977 Stone tool functions in the Western Desert. *In* R.V.S. Wright (ed) *Stone tools as cultural markers*, pp 178-188. Australian Institute of Aboriginal Studies, Canberra.

Hayden, B., and J. Kamminga 1979 An introduction to use-wear: the first CLUW. In Hayden, B. (ed) *Lithic use-wear analysis*, pp 1-13. Academic Press, New York.

Head, L. 1983 Environment as artefact: a geomorphic perspective on the Holocene occupation of south-western Victoria. *Archaeology in Oceania* 18: 73-80.

Head, L. 1988 Holocene vegetation, fire, and environmental history of the Discovery Bay region, south-western Victoria. *Australian Journal of Ecology* 13:21-49.

Head, L. 1989 Prehistoric Aboriginal impacts on Australian vegetation: an assessment of the evidence. *Australian Geographer* 20: 37-45.

Head, L. 1994 Landscapes socialised by fire: post-contact changes in Aboriginal fire use in northern Australia, and implications for prehistory. *Archaeology in Oceania* 29: 172-181.

Headland, T.N. 1987 The wild yam question: how well could independent hunter-gatherers live in a tropical rainforest environment? *Human Ecology* 15: 463-491.

Hingston, F.J., A.M. O'Connell, and T.S. Grove 1989 Nutrient cycling in jarrah forest. In Dell, B., J.J. Havel, and N. Malajczuk (eds) *The Jarrah forest: a complex Mediterranean ecosystem*, pp 155-177. Kluwer, Dordrecht.

Hiscock, P. 1981 Comments on the use of chipped stone artefacts as a measure of 'intensity of site usage'. *Australian Archaeology* 13: 30-34.

Hiscock, P. 1982 A technological analysis of quartz assemblages from the south coast. In Bowdler, S. (ed) *Coastal archaeology in eastern Australia: Proceedings of the 1980 Valla Conference on Australian prehistory*, pp 32-45. Department of Prehistory, Research School of Pacific Studies, The ANU, Canberra. *Occasional Papers in Prehistory* 11.

Hiscock, P. 1985 The need for a taphonomic perspective in stone artefact analysis. *Queensland Archaeological Research* 2: 82-95.

Hiscock, P. 1986 Technological change in the Hunter River valley and the interpretation of late Holocene change in Australia. *Archaeology in Oceania* 21: 40-50.

Hiscock, P. 1990 A study in scarlet: taphonomy of inorganic artefacts. *In* Solomon, S., I. Davidson, and

D. Watson (eds) Problem solving in taphonomy. *Tempus* 2: 34-49.

Hiscock, P. 1993 Bondaian technology in the Hunter Valley, New South Wales. *Archaeology in Oceania* 28: 65-76.

Hiscock, P. 1994 Technological responses to risk in Holocene Australia. *Journal of World Prehistory* 8: 267-292.

Hiscock, P. 1996a Mobility and technology in the Kakadu coastal wetlands. *Indo-Pacific Prehistory Association Bulletin* 15 (Chiang Mai Papers Volume 2): 151-156.

Hiscock, P. 1996b Transformations of Upper Palaeolithic implements in the Dabba industry from Haua Fteah (Libya). *Antiquity* 70: 657-664.

Hiscock, P., and V. Attenbrow 1998 Early Holocene backed artefacts from Australia. *Archaeology in Oceania* 33:49-62.

Hiscock, P., and S. Mitchell 1993 Stone artefact quarries and reduction sites in Australia: towards a type profile. *Australian Heritage Commission Technical Publications Series No. 4*. Australian Government Publishing Service, Canberra.

Ho Ho Committee 1979 The Ho Ho Classification and Nomenclature Committee Report. In Hayden, B. (ed) *Lithic use-wear analysis*, pp 133-135. Academic Press, New York.

Hoffman, N., and A. Brown 1992 *Orchids of south-west Australia*. Second edition. University of Western Australia Press, Perth.

Holdaway, S. 1995 Stone artefacts and the transition. In Allen, J., and J.F. O'Connell (eds) Transitions: Pleistocene to Holocene in Australia and Papua New Guinea. *Antiquity* 69 Special Number 265: 784-797.

Holdaway, S., and N. Porch 1996 Dates as data: an alternative approach to the construction of chronologies for Pleistocene sites in southwest Tasmania. In Allen, J. (ed) 1996 *Report of the Southern Forests Archaeological Project. Vol. 1. Site descriptions, stratigraphies and chronologies*, pp 251-277. School of Archaeology, La Trobe University, Melbourne.

Hope, G.S. 1978 The Late Pleistocene and Holocene vegetational history of Hunter Island, north-western Tasmania. *Australian Journal of Botany* 26:493-514.

Hope, G. 1989 Climatic interpretations of timberline changes in Australasia from 30 000 yr BP to present. In Donnelly, T.H., and R.J. Wasson (eds) *CLIMANZ 3: proceedings of symposium*, pp 91-99. CSIRO Division of Water Resources, Canberra.

Hope, G. (ed) 1998 *Identifying wood charcoal remains as palaeo evidence for regions of central and northeast Australia*. Archaeology and Natural History Publications, Research School of Pacific and Asian Studies, The ANU, Canberra.

Hope, J.H., R.J. Lampert, E. Edmondson, M. J. Smith, and G. F. Van Tets 1977 Late Pleistocene faunal remains from Seton Rock Shelter, Kangaroo Island, South Australia. *Journal of Biogeography* 4: 363-385.

Hopkins, M.S., J. Ash, A.W. Graham, J. Head, and R. Hewett 1993 Charcoal evidence of the spatial extent of the Eucalyptus woodland expansions and rainforest contractions in North Queensland during the Late Pleistocene. *Journal of Biogeography* 20: 357-372.

Horsfall, N. and J. Hall 1990 People and the rainforest: an archaeological perspective. In Webb, L.J., and J. Kikkawa (eds) *Australian tropical rainforests: science - values - meaning*, pp 33-39. CSIRO, Melbourne.

Horton, D. 1982 The burning question: Aborigines, fire, and Australian ecosystems. *Mankind* 13(3): 237-251.

Horton, D.R. 1984 Minimum numbers: a consideration. *Journal of Archaeological Science* 11: 255-271.

How, R.A., J. Dell, and W.F. Humphreys 1987 The ground vertebrate fauna of coastal areas between Busselton and Albany, Western Australia. *Records of the Western Australian Museum* 13: 553-574.

Howitt, A.W. 1890 The eucalypts of East Gippsland. *Transactions of the Royal Society of Victoria* 2: 83-120.

Howitt, A.W. 1904 The native tribes of south-east Australia. MacMillan, London.

Hubbard, N. 1995a An integrated method for reconstructing regional palaeoclimates: Australia (18,000 yr BP). *Palaeogeography, Palaeoclimatology, Palaeoecology* 116:141-166.

Hubbard, N. 1995b In search of regional palaeoclimates: Australia, 18,000 yr BP. *Palaeogeography, Palaeoclimatology, Palaeoecology* 116:167-188.

Hubbard, R.N.L.B. 1993 The palaeoecology of the large herbivores of Europe and the Near East. Privately circulated manuscript.

Hubbard, R.N.L.B., and Z. Kvacek 1998 Reconstruction of a Czech Early Miocene vegetation. *Proceedings of the Geologist's Association* 109: 33-50.

Hutchins, B., and M. Thompson 1983 *The marine and estuarine fishes of south-western Australia*. Western Australian Museum, Perth.

Ilic, J. 1991 *CSIRO Atlas of Hardwoods*. Crawford House Press and CSIRO Publications, Melbourne.

Ilic, J. 1993 Computer aided wood identification using CSIROID. *International Association of Wood Anatomists Journal* 14: 333-340.

Ilic, J. 1996 CSIRO ID. Wood Identification [Computer] Program Version 3.2.

Ingold, T. 1984 Time, social relationships, and the exploitation of animals: anthroplogical reflections on prehistory. In Clutton-Brock, J., and C. Grigson (eds) *Animals and Archaeology 3. Early herders and their flocks*, pp 3-12. BAR International Series 202, Oxford.

Inizan, M.-L., H. Roche, and J. Tixier 1992 *Technology of knapped stone*. (*Préhistoire de la pierre taillée*, v. 3). Cercle de Recherches et d'Etudes Préhistorique, Meudon.

Iredale, T. 1939 A review of the land mollusca of Western Australia. *Records of the Western Australian Museum and Art Gallery* 2: 1-88.

Jackson, G.A. 1992 Ratiocination: a taphonomic analysis of rodent and oher bones from Rainbow Cave, southwest Western Australia. Unpublished Honours thesis, Centre for Archaeology, University of Western Australia.

Jackson, W.D. 1965 Vegetation. In Davies, J.L., (ed) *Atlas of Tasmania*, pp 30-35. Mercury Press, Hobart.

Jackson, W.D. 1968 Fire, air, water, and earth - an elemental ecology of Tasmania. *Proceedings of the Ecological Society of Australia* 3: 9-16.

Jane, F.W. 1970 *The structure of wood.* Second edition. Adam and Charles Black, London.

Jochim, M.A. 1996 Surprises, recurring themes, and new questions in the study of the late glacial and early postglacial. In L.G. Straus, B.V. Eriksen, J.M. Erlandson, and D.R. Yesner (eds) *Humans at the end of the Ice Age - the archaeology of the Pleistocene-Holocene transition*, pp 357-363. Plenum Press, New York.

Johnson, E. 1985 Current developments in bone technology. In Schiffer, M.B., (ed) *Advances in archaeological method and theory* 8: 157-235. Academic Press, New York.

Jones, B.D. 1994 Hunter-gatherer adaptations to the terminal Pleistocene and early Holocene of northeastern North America: effects on site visibility. Retrieved 10 October, 1997 from: http://spirit.lib.uconn.edu/ArchNet/Regions/Northeast/72163/neaa.html

Jones, R. 1969 Fire-stick farming. *Australian Natural History* 16: 224-228.

Jones, R. 1971 Rocky Cape and the problem of the Tasmanians. Unpublished PhD thesis, Unversity of Sydney.

Jones, R. 1977 The Tasmanian paradox. In Wright, R.V.S. (ed) *Stone tools as cultural markers: change, evolution, complexity*, pp 189-204. Australian Institute of Aboriginal Studies, Canberra.

Jones, R., and N. White 1988 Point blank: stone tool manufacture at the Ngilipitji Quarry, Arnhem Land, 1981. *In* Meehan, B., and R. Jones (eds) *Archaeology with ethnography: an Australian perspective*, pp 51-88. Department of Prehistory, Research School of Pacific Studies, The ANU, Canberra.

Jones, R. 1990 From Kakadu to Kutikina: the southern continent at 18 000 years ago. In Gamble, C., and O. Soffer (eds) *The world at 18 000 BP. Volume 2 Low latitudes*, pp 264-295. Unwin Hyman, London.

Jouzel, J., C. Lorius, J.R. Petit, N.I. Barkov, and V.M. Kotlyakov 1994 Vostok isotopic temperature record. In Boden, T.A., D.P. Kaiser, R.J. Sepanski, F.W. Stoss, and F.W. Logsdon (eds) *Trends '93; a compendium of data on global change*, pp 590-602. Carbon Dioxide Information Analysis Centre, Oak Ridge, Tennessee.

Kadane, J.B. 1988 Possible statistical contributions to paleoethnobotany. In Hastorf, C.A., and V.S. Popper (eds) *Current paleoethnobotany*, pp. 206-214. University of Chicago Press, Chicago.

Kamminga, J. 1982 Over the edge: functional analysis of Australian stone tools. *Occasional papers in anthropology* 12. Anthropology Museum, University of Queensland, St Lucia.

Kass, G. no date Rimfire Headstamp Guide, parts 1-9. Manuscript, produced by George Kass, Forensic Ammo, Okemos, Michigan, USA.

Kemp, E. 1981 Evolution of the Australian biota. In Gill, A.M., R.H. Groves, and I.R. Noble (eds) *Fire and the Australian biota*, pp 3-21. Australian Academy of Science, Canberra.

Kendrick, G.W. 1977 Middle Holocene marine molluscs from near Guildford, Western Australia, and evidence for climatic change. *Journal of the Royal Society of Western Australia* 59:97-104.

Kershaw, A.P. 1976 A Late Pleistocene and Holocene pollen diagram from Lynch's Crater, north-east Queensland, Australia. *New Phytologist* 77: 469-498.

Kershaw, A.P. 1986 Climatic change and Aboriginal burning in north-east Australia during the last two glacial/interglacial cycles. *Nature* 322: 47-49.

Kershaw, A.P. 1995 Environmental change in Greater Australia. *Antiquity* 69: 656-75.

Kiernan, K., R. Jones, and D. Ranson 1983 New age for glacial age man in south-west Tasmania. *Nature* 301: 28-32.

King, P.P. 1827 Narrative of a survey of the intertropical and western coasts of Australia, performed between the years 1818 and 1822. Vol. 2. John Murray, London.

Kintigh, K.W. 1989 Sample size, significance, and measures of diversity. In Leonard, R.D., and G.T. Jones (eds) *Quantifying diversity in archaeology*, pp 25-36. Cambridge University Press, Cambridge.

Kitchener, D.J. 1995 Quokka, *Setonix brachyurus*. In Strahan, R. (ed) *The mammals of Australia: the national photographic index of Australian wildlife*, pp 401-403. The Australian Museum with Angus and Robertson, Sydney.

Krakker, J.J., M.J. Shott, and P.D. Welch Design and evaluation of shovel-test sampling in regional archaeological survey. *Journal of Field Archaeology* 10: 469-480.

Krebs, C.J. 1985 Ecology: the experimental analysis of distribution and abundance. Third edition. Harper and Row, New York.

Kuhn, S.L. 1992 On planning and curated technologies in the Middle Paleolithic. *Journal of Anthropological Research* 48: 185-214.

Kutzbach, J.E., and T. Webb III 1993 Conceptual basis for understanding late-Quaternary climates. In Wright, H.E., Jr, J.E. Kutzbach, T. Webb III, W.F. Ruddiman, F.A. Street-Perrott, and P.J. Bartlein (eds) *Global climates since the Last Glacial Maximum*, pp12-23. University of Minnesota Press, Minneapolis.

Lamont, B.B. 1995 Mineral nutrient relations in Mediterranean regions of California, Chile, and Australia. In Arroyo, M.T., P.H. Zedler, and M.D. Fox (eds) *Ecology and biogeography of Mediterranean ecosystems in Chile, California, and Australia*, pp 211-238. Ecological studies v. 108. Springer-Verlag, Berlin.

Lang, S. 1997 Burning the bush: a spatio-temporal analysis of Jarrah forest fire regimes. Unpublished Honours thesis, Department of Geography, The ANU, Canberra.

Lawrence, R.A. 1979 Experimental evidence for the significance of attributes used in edge-damage analysis. In Hayden, B. (ed) *Lithic use-wear analysis*, pp 113-121. Academic Press, New York.

Leney, L., and R.W. Casteel 1975 Simplified procedure for examining charcoal specimens for identification. *Journal of Archaeological Science* 2: 153-159.

Leonard, R.D., and G.T. Jones (eds) 1989 *Quantifying diversity in archaeology*. Cambridge University Press, Cambridge.

Lewis, H.T. 1973 Patterns of Indian burning in California: ecology and ethnohistory. *Anthropological Papers* 1: 101, Ballena Press, Ramona, California.

Lightfoot, K.G. 1986 A defense of shovel-test sampling: a reply to Shott. *American Antiquity* 54: 413-416.

Lilley, I. 1993 Recent research in southwestern Western Australia: a summary of recent findings. *Australian Archaeology* 36: 34-41.

Loneragan, O.W., and J.F. Loneragan 1964 Ashbed and nutrients in the growth of seedlings of Karri (*Eucalyptus diversicolor* F.v.M.). *Journal of the Royal Society of Western Australia* 47: 75-80.

Lourandos, H. 1977 Stone tools, settlement, adaptation: a Tasmanian example. In Wright, R.V.S. (ed) *Stone tools as cultural markers: change, evolution, complexity*, pp 219-224. Australian Institute of Aboriginal Studies, Canberra.

Lourandos, H. 1980 Change or stability?: hydraulics, hunter-gatherers, and population in temperate Australia. *World Archaeology* 11: 245-263.

Lourandos, H. 1983a 10,000 years in the Tasmanian highlands. *Australian Archaeology* 16: 39-47.

Lourandos, H. 1983b Intensification: a late Pleistocene-Holocene sequence from south-western Victoria. *Archaeology in Oceania* 18: 81-94.

Lourandos, H. 1985 Intensification and Australian prehistory. In Price, T.D., and J.A. Brown (eds) *Prehistoric hunter-gatherers*, pp 385-423. Plenum Press, New York.

Lourandos, H. 1997 Continent of hunter-gatherers. New Perspectives in Australian prehistory. Cambridge University Press, Cambridge.

Lovis, W.A. 1976 Quarter sections and forests: an example of probability sampling in the northeastern woodlands. *American Antiquity* 41: 364-372.

Lowry, D.C. 1967 Busselton and Augusta. Geological Survey of Western Australia 1:250,000 Geological Series.Bureau of Mineral Resources, Perth.

Lowry, D.C. 1979 Karst styles in Western Australia. In Saar, A., and R. Webb (eds) *WACCON: Proceedings of the Twelth Biennial Conference of the Australian Speleological Federation, Perth 1979*, pp 44-52. The Australian Speleological Federation, Broadway NSW.

Lowry, D., and J. Lowry 1968 Report on dig in Calgardup Cave 17 Feb. 1968, and Notes on the Deepdene Cave bone deposit. *The Western Caver* 8(2): 30-33.

Lumsden, J. 1971 *Elementary statistical method*. University of Western Australia Press, Perth.

Lundelius, E.L. 1960 Post Pleistocene faunal succession in Western Australia and its climatic interpretation. *Proceedings of the XXIst International Geological Congress, Norden, Part IV*: 142-153.

Lundelius, E.L. 1966 Marsupial carnivore dens in Australian caves. *Studies in Speleology* 1: 174-180.

Lundelius, E.L. 1983 Climatic implications of Late Pleistocene and Holocene faunal associations in Australia. *Alcheringa* 7: 125-149.

Lundelius, E.L. 1989 The implications of disharmonious assemblages for Pleistocene extinctions. *Journal of Archaeological Science* 16: 407-417.

Lyman, R.L. 1994 *Vertebrate taphonomy*. Cambridge University Press

McArthur, W.M. 1991 *Reference soils of south-western Australia*. Department of Agriculture, Perth.

McArthur, W.M., and E. Bettenay 1974 Development and distribution of soils of the Swan Coastal Plain, Western Australia. *Soil Publication No. 16*. CSIRO, Melbourne.

McBrearty, S., L. Bishop, T. Plummer, R. Dewar, and N. Conard 1998 Tools underfoot: human trampling as an agent of lithic artifact edge modification. *American Antiquity* 63: 108-129.

MacBryde, I. 1977 Determinants of assemblage variation in New England prehistory- environment, subsistence economies, site activities, or cultural tradition? In R.V.S. Wright (ed), Stone tools as cultural markers, pp 225-250. *Prehistory and Material Culture Series* No. 12. Australian Institute of Aboriginal Studies, Canberra.

MacBryde, I. (ed) 1978 *Records of times past*. Australian Institute of Aboriginal Studies, Canberra.

McCarthy, F.D. 1967 *Australian Aboriginal stone implements*. The Australian Museum, Sydney.

McCaw, W.L. 1986 Behaviour and short term effects of two fires in regenerate karri (*Eucalyptus diversicolor*) forest. *Department of Conservation and Land Management Report* 9.

McCorkle, V., H.H. Veeh, and D.T. Heggie 1994 Glacial-Holocene palaeoproductivity off Western Australia: a comparison of proxy records. In Zahn, R., M.A. Pedersen, M.A. Kaminski, and L. Labeyrie (eds) *Carbon cycling in the glacial ocean: constraints on the ocean's role in global change (Quantitative Approaches in Palaeooceanography)*, pp 443-479. Springer-Verlag, Berlin.

McKenzie, N.L., A.A. Burbridge, and A. Baynes no date [1998] Australian mammal map updates. Retrieved August 1998 from the World Wide Web: http: //www.calm.wa.gov.au

MacDonald, Hales, and Associates 1994 *Aboriginal heritage in the southern forests*. Unpublished report to the Australian Heritage Commission, Canberra.

McNicol, L.J. 1999 Environment at Witchcliffe Rock Shelter: a palynological investigation of the last 2 000 years, southwest Western Australia. Unpublished Honours thesis, Department of Geography and Centre for Archaeology, University of Western Australia.

McNiven, I. 1994 Technological organization and settlement in southwest Tasmania after the glacial maximum. *Antiquity* 68: 75-82.

MacPhail, M.K. 1979 Vegetation and climates in southern Tasmania since the last glaciation. *Quaternary Research* 11:306-341,

MacPhail, M.K. 1980 Regeneration processes in Tasmanian forests. *Search* 11 (6): 184-190.

MacPhail, M.K. 1984 Small-scale dynamics in an early Holocene wet sclerophyll forest in Tasmania. *New Phytologist* 96: 131-147.

MacPhail, M.K., and E.A. Colhoun 1985 Late last glacial vegetation, climates and fire activity in southwest Tasmania. *Search* 16:43-45.

Major, R.H., (ed) 1859 Early voyages to Terra Australis now called Australia: a collection of documents and extracts from early manuscript maps illustrative of the history of discovery on the coasts of that vast island from the beginning of the sixteenth century to the time of Captain Cook. Hakluyt Society, London.

Marchant, N.G. 1973 Species diversity in the southwestern flora. *Journal of the Royal Society of Western Australia*56: 23-30.

Marean, C.W. 1991 Measuring the post-depositional destruction of bone in archaeological assemblages. *Journal of Archaeological Science* 18: 677-694.

Marshall, B. 1986 An experimental evaluation of the criteria used to distinguish owl-deposited bone in archaeological cave deposits in Australia. *Australian Archaeology* 22: 104-121.

Marshall, B., and R. Cosgrove 1990 Tasmanian Devil (*Sarcophilus harrisii*) scat-bone: signature criteria and archaeological implications. *Archaeology in Oceania* 25: 102-113.

Martin, H.A. 1973 Palynology and historical ecology of some cave excavations in the Australian Nullarbor. *Australian Journal of Botany* 21: 283-316.

Matthews, P.G. 1985 *Australian Karst Index 1985*. Australian Speleological Federation, Broadway, NSW.

Meagher, S.J. 1974 The food resources of the Aborigines of the south-west of Western Australia. *Records of the Western Australian Museum* 3(1): 3-65.

Meagher, S.J., and W.D.L. Ride 1979 Use of natural resources by the Aborigines of south-western Australia. In Berndt, R., and C. Berndt (eds) *Aborigines of the West*, pp 66-80. University of Western Australia Press, Perth.

Meehan, B. 1988 The 'dinner-time camp'. In Meehan, B., and R. Jones (eds) *Archaeology with ethnography: an Australian perspective*, pp 171-181. Department of Prehistory, Research School of Pacific Studies, The ANU, Canberra.

Mellars, P. 1971 Some comments of the notion of 'functional variability' in stone-tool assemblages. *World Archaeology* 2: 74-89.

Mellars, P.A. 1976 Fire ecology, animal populations, and man: a study of some ecological relationships in prehistory. *Proceedings of the Prehistorical Society* 42: 15-46.

Mellars, P.A., and S.C. Reinhardt 1978 Patterns of Mesolithic land-use in southern England: a geological perspective. In Mellars, P.A. (ed) *The early post-glacial settlement of Northern Europe*, pp 243-294. University of Pittsburgh Press, Pittsburgh.

Merrilees, D. 1968 Man the destroyer: late Quaternary changes in the Australian marsupial fauna. *Journal of the Royal Society of Western Australia* 51: 1-24.

Merrilees, D. 1975 Ancient animals and men in caves. *Wildlife in Australia* 12: 93-97.

Merrilees, D. 1979a Prehistoric rock wallabies (*Marsupialia, Macropodidae, Petrogale*) in the far south-west of Western Australia. *Journal of the Royal Society of Western Australia* 61: 73-96.

Merrilees, D. 1979b The prehistoric environment in Western Australia. *Journal of the Royal Society of Western Australia* 62: 109-128.

Merrilees, D. 1984 Comings and goings of Late Quaternary mammals in extreme southwestern Australia. In Martin, P.S. and R.G. Klein (eds) *Quaternary extinctions*, pp 629-638. University of Arizona Press, Tucson, Arizona.

Merrilees, D., and J.K. Porter 1979 Guide to the identification of teeth and some bones of native land mammals occurring in the extreme south west of Western Australia. Western Australian Museum, Perth.

Miller, G.H, J.W. Magee, and A.J.T Jull 1997 Low-latitude glacial cooling in the Southern Hemisphere from amino-acid racemization in emu eggshells. *Nature* 385: 241-244.

Moore, G.F. 1884 *Diary of ten eventful years of an early settler in Western Australia.*1978 facsimile edition, introduction by C.T. Stannage.UWA Press, Perth.

Morse, K. 1984 First record of painted Aboriginal rock-art in a south-western Australian limestone cave. *Records of the Western Australian Museum* 11(2): 197-199.

Mount, A.B. 1979 Natural regeneration processes in Tasmanian forests. *Search* 10:180-186.

Mulvaney, D.J., and N. Green 1992 *Commandant of solitude. Diary of Commandant Collet Barker.* Melbourne University Press, Carlton, Victoria.

Mulvaney, D.J., and J. Kamminga 1999 Prehistory of Australia. Allen and Unwin, St Leonards, NSW.

Nance, J.D., and B.F. Ball 1986 No surprises? The reliability and validity of test-pit sampling. *American Antiquity* 51: 457-504.

Neill, J. 1845 Catalogue of reptiles and fish, found at King George's Sound. In Eyre, E.J., *Journals of expeditions of discovery into Central Australia* 1: 412-431. [original not seen, cited in Meagher 1974]

Newcomer, M.H., and C. Karlin 1987 Flint chips from Pincevint. In Sieveking, G. de G., and M.H. Newcomer (eds) The human uses of flint and chert. Proceedings of the Fourth International Flint Symposium, British Polytechnic 10-15 April 1983, pp 33-36. Cambridge University Press, Cambridge.

Newsome, J., and E. Pickett 1993 Palynology and palaeoclimatic implications of two Holocene sequences from southwestern Australia. *Palaeogeography, Palaeoclimatology, Palaeoecology* 101:245-261.

Nichols, O.G., and B. Muir 1989 Vertebrates of the Jarrah forest. In Dell, B., J.J. Havel, and N. Malajczuk (eds) *The Jarrah forest: a complex Mediterranean ecosystem*, pp 133-154. Kluwer, Dordrecht.

Nicholson, A., and S. Cane 1991 Desert camps: analysis of Australian Aboriginal proto-historic campsites. In Gamble, C.S., and W.A. Boismier (eds) *Ethnoarchaeological approaches to mobile campsites*, pp 263-349. International Monographs in Prehistory, Ann Arbor.

Nicholson, R.A. 1993 A morphological investigation of burnt animal bone and an evaluation of its utility in archaeology. *Journal of Archaeological Science* 20: 411-428.

Nind, S. 1831 Description of the natives of King George's Sound (Swan River Colony) and adjoining country. *Journal of the Royal Geographical Society* 6: 21-51.

Northcote, K.H., E. Bettenay, W.M. McArthur, and H.M. Churchward 1967 Dominant soils of the Perth-Albany-Esperance area. *Atlas of Australian Soils* Sheet 5. CSIRO, Melbourne.

O'Connell, J.F. 1977 Aspects of variation in Central Australian lithic assemblages. *In* R.V.S. Wright (ed) *Stone tools as cultural markers*, pp 269-281. Australian Institute of Aboriginal Studies, Canberra.

O'Connell, J.F. 1987 Alyawara site structure and its archaeological implications. *American Antiquity* 52: 74-108.

O'Connell, J.F., and B. Marshall 1989 Analysis of kangaroo body part transport among the Alyawara of Central Australia. *Journal of Archaeological Science* 16: 393-405.

O'Connor, A. 1986 An evaluation of a diatom assemblage as a biological indicator of palaeoenvironmental change in southwest Western Australia. Unpublished Honours thesis, Department of Geography, UWA.

O'Connor, S., P. Veth, and N. Hubbard 1993 Changing interpretations of postglacial human subsistence and demography in Sahul. In M.A. Smith, M. Spriggs, and B. Fankhauser (eds) *Sahul in review*, pp 95-105. Occasional Papers in Prehistory 24, Department of Prehistory, Research School of Pacific and Asian Studies, The ANU, Canberra.

Patterson, W.A., and K.E. Sassaman 1988 Indian fires in the prehistory of New England. In Nicholas, G.P. (ed) *Holocene human ecology in northeastern North America*, pp 107-135. Plenum Press, New York.

Pearce, R.H. 1977 Relationship of chert artefacts at Walyunga in southwest Australia, to Holocene sea levels. *Search* 8: 375-377.

Pearce, R.H. 1978 Changes in artefact assemblages during the last 8000 years at Walyunga, Western Australia. *Journal of the Royal Society of Western Australia* 61: 1-10.

Pearce, R.H. 1979 Analysis of some Western Australian small tool assemblages. Unpublished MA thesis, University of Western Australia.

Pearce, R.H. 1981 Artefact usage and sea-level changes. *The Artefact* 6: 30-33.

Pearce, R.H. 1982 Archaeological sites in jarrah forest, southwest Australia. *Australian Archaeology* 14: 18-24.

Pearce, R.H. 1989 Pre-colonial usage of jarrah forest by indigenous people. In Dell, B., J.J. Havel, and N. Malajczuk (eds) *The Jarrah forest*, pp 219-228. Kluwer, Dordrecht.

Pearce, R.H., and M. Barbetti 1981 A 38,000-year-old archaeological site at Upper Swan, Western Australia. *Archaeology in Oceania* 16: 173-178.

Pickett, E.J. 1997 The Late Pleistocene and Holocene vegetation history of three lacustrine sequences from the Swan Coastal Plain, Southwestern Australia. Unpublished PhD thesis, Department of Geography, University of Western Australia.

Pierce, R. 1978 The evidence of J. Ainsworth on the diet and economy of the Ballina Horde. In MacBryde, I. (ed), *Records of times past*, pp 1116-121. Australian Institute of Aboriginal Studies, Canberra.

Piperno, D.R. 1994 Phytolith and charcoal evidence for prehistoric slash-and-burn agriculture in the Darien rain forest of Panama. *Holocene* 4: 321-325.

Piperno, D.R., M.B. Bush, and P.A. Colinvaux 1990 Paleoenvironments and human occupation in late-glacial Panama. *Quaternary Research* 33: 108-116.

Pizzey, G., and R.B. Doyle 1991 *A field guide to the birds of Australia.* Revised edition. Harper Collins, North Ryde, NSW.

Porter, J.K. 1979 Vertebrate remains from a stratified Holocene deposit in Skull Cave, Western Australia, and a review of their significance. *Journal of the Royal Society of Western Australia* 61:109-117.

Prentis, M. 1984 Prelude to dispossession? First contacts between Aborigines and Europeans in the Northern Rivers region of New South Wales, 1770-1840s. *Royal Australian Historical Society Journal* 70(1):3-18.

Prentiss, W.C. 1998 The reliability and validity of a lithic debitage typology: implications for archaeological interpretation. *American Antiquity* 63: 635-650.

Price, T.D., and J.A. Brown (eds) 1985 *Prehistoric hunter-gatherers: the emergence of cultural complexity*. Academic Press, New York.

Prideaux, G., G. Gully, and L. Hatcher 1997 Palaeontology and stratigraphyu of Tight Entrance Cave, southwestern Western Australia, conference paper abstract. In Baynes, A., and J. Long (eds) Conference on Australasian Vertebrate Evolution, Palaeontology, and Systematics 1997, upublished programme of speakers, pp 52. Western Australian Museum, Perth.

Prior, J., and P. Gasson 1993 Anatomical changes on charring six African hardwoods. *International Association of Wood Anatomists Journal* 14 (1): 77-86.

Qin, B., S.P. Harrison, and J.E. Kutzbach 1998 Evaluation of modelled regional water balance using lake status data: a comparison of 6 ka simulations with the NCAR CCM. *Quaternary Sicence Review* 17:535-548.

Rayner, M.E. 1992 Application of dendrochronology, stem analysis and inventory data in the estimation of tree and stand ages in Karri forest. *Technical Report* 27. Department of Conservation and Land Management, Como.

Riek, E.F. 1967 The freshwater crayfish of Western Australia (Decapoda: Parastacidae). *Australian Journal of Zoology* 15: 103-121.

Richardson, N. 1992 Conjoin sets and stratigraphic integrity in a sandstone shelter: Kenniff Cave (Queensland, Australia). *Antiquity* 66: 408- 418.

Ringrose, T.J. 1993 Bone counts and statistics: a critique. *Journal of Archaeological Science* 20:121-157.

Rolland, N., and H.L. Dibble 1990 A new synthesis of Middle Paleolithic variability. *American Antiquity* 55: 480-499.

Ross, A. 1985 Archaeological evidence for population change in the middle to late Holocene in south-eastern Australia. *Archaeology in Oceania* 20: 81-89.

Rossen, J., and J. Olsen 1985 The controlled carbonization and archaeological analysis of SE U.S. wood charcoals. *Journal of Field Archaeology* 12: 445-456.

Roth, W.E. 1903 Notes of savage life in the early days of west Australian settlement. *Proceedings of the Royal Society of Queensland* 17: 45-69.

Roth, B.J., and H.L. Dibble 1998 Production and transport of blanks and tools at the French Middle Paleolithic site Combe-Capelle Bas. *American Antiquity* 63: 47-62.

Rotheram, I. 1983 Suppression of growth of surrounding regeneration by veteran trees of Karri (*Eucalyptus diversicolor*). *Australian Forestry* 46: 8-13.

Rowland, M.J. 1999 Holocene environmental variability: have its imacts in Australian pre-History? *The Artefact* 22: 11-48.

Rozen, K.C., and A.P. Sullivan 1989a Measurement, method, and meaning in lithic analysis: problems with Amick and Mauldin's middle-range approach. *American Antiquity* 54: 169-175.

Rozen, K.C., and A.P. Sullivan 1989b The nature of lithic reduction and lithic analysis: stage typologies revisited. *American Antiquity* 54: 169-175.

Rule, A. 1967 *Forests of Australia*. Angus and Robertson, Sydney.

Sandweiss, D.H., J.B. Richardson III, E.J. Reitz, H.B. Rollins, and K. A. Maasch 1996 Geoarchaeological evidence from Peru for a 5000 years BP onset of El Niño. *Science* 273: 1531-1533.

Schiffer, M.B. 1983 Toward the identification of formation processes. *American Antiquity* 48: 675-706.

Schmitt, D.N., and K.E. Juell 1994 Toward the identification of coyote scatological faunal accumulations in archaeological contexts. *Journal of Archaeological Science* 21: 249-262.

Schrire, C. 1980 An enquiry into the evolutionary states and apparent identity of San hunter-gatherers. *Human ecology* 8: 9-31.

Schwede, M.L. 1990 Quartz, the multi-faceted stone: a regional prehistory of the Helena River Valley on the Swan Coastal Plain of south-western Australia. PhD thesis, Department of Archaeology, University of Western Australia, Perth.

Semeniuk, V. 1986 Holocene climate history of coastal southwestern Australia using calcrete as an indicator. *Palaeogeography, Palaeoclimatology, Palaeoecology* 53: 289-308.

Serventy, D.L., and H.M. Whittel 1967 *Birds of Western Australia*. 4th ed. Lamb, Perth.

Shackleton, C.M., and F. Prins 1992 Charcoal analysis and the "Principle of Least Effort" - a conceptual model. *Journal of Archaeological Science* 19: 631-637.

Shackleton, C.M., and F. Prins 1993 Charcoal yields from dry wood samples of ten savanna trees. *South African Archaeological Bulletin* 48: 42-44.

Shackleton, N.J., and N.D. Opdyke 1973 Oxygen isotope and paleomagnetic stratigraphy of equatorial Pacific core V28-238: oxygen isotope temperatures and ice volumes on a 10^5 and 10^6 year scale. *Quaternary Research* 3:39-55.

Shackley M. 1978 A sedimentological study of Devil's Lair, Western Australia. *Journal of the Royal Society of Western Australia* 60: 33-40.

Shea, J.J., and J.D. Klenck 1993 An experimental investigation of the effects of trampling on the results of lithic microwear analysis. *Journal of Archaeological Science* 20: 175-194.

Shennan, S.J. 1988 *Quantifying archaeology*. Edinburgh University Press, Edinburgh.

Shepherd, N.C. 1981 Predation of Red Kangaroos, *Macropus rufus*, by the Dingo, *Canis familiaris dingo* (Blumenbach), in north-western New South Wales. *Australian Wildlife Research* 8: 255-262.

Shipman, P., G. Foster, and M. Schoeninger 1984 Burnt bones and teeth: an experimental study of color, morphology, crystal structure, and shrinkage. *Journal of Archaeological Science* 11: 307-325.

Simmons, I.G., G.W. Dimbleby, and C. Grigson 1981 The Mesolithic. In I.G. Simmons and M.J. Tooley (eds) *The environment in British prehistory*, pp 82-124. Cornell University Press, Ithaca, New York.

Singh, G., A.P. Kershaw, and R. Clark 1981 Late Quaternary vegetation and fire history in Australia.In Gill, A.M., R.H. Groves, and I.R. Noble (eds) *Fire and the Australian biota*, pp 23-54. Australian Academy of Science, Canberra.

Skinner, P.R. 1984 Seed production and survival of some legumes in the forests of Western Australia. *Forests Department of Western Australia Research Paper* 76.

Smart, T.L., and E.S. Hoffman 1988 Environmental interpretation of archaeological charcoal. In Hastorf, C.A., and V.S. Popper (eds) *Current paleoethnobotany*, pp. 167-205. University of Chicago Press, Chicago.

Smith, A.G., A. Whittle, E.W. Cloutman, and L.A. Morgan 1989 Mesolithic and Neolithic activity and environmental impact on the south-east fen-edge in Cambridgeshire. *Proceedings of the Prehistoric Society* 55: 207-249.

Smith, F.G. 1973 Vegetation map of Busselton and Augusta. Scale 1:250,000. Vegetation survey of Western Australia. Western Australian Department of Agriculture.

Smith, M. 1982 Review of the Thylacine (Marsupialia, Thylacinidae). In Archer, M., (ed) *Carnivorous marsupials*, pp 237-253. Royal Zoological Society of New South Wales, Sydney.

Smith, M.A. 1986 The antiquity of seedgrinding in arid Australia. *Archaeology in Oceania* 21: 13-28.

Smith, M.A. 1989 The case for a resident human population in the central Australian ranges during full glacial aridity. *Archaeology in Oceania* 24: 93-105.

Smith, M.A., L. Vellen, and J. Pask 1995 Vegetation history from archaeological charcoals in central Australia: the late Quaternary record from Puritjarra rock shelter. *Vegetation history and archaeobotany* 4: 171-177.

Smith, M.V. 1982 Late Pleistocene zamia exploitation in southern Western Australi. *Archaeology in Oceania* 17: 117-121.

Smith, M.V. 1993 Recherche à l'Espérence. Unpublished PhD thesis, University of Western Australia.

Solomon, S. 1990 What is this thing called taphonomy? In Solomon, S., I. Davidson, and D. Watson (eds) *Problem solving in taphonomy. Archaeological and palaeontological studies from Europe, Africa, and Oceania*, pp 25-33. Tempus 2, Archaeology and material culture studies in anthropology. Anthropology Museum, University of Queensland, St Lucia.

Solomon, S., and B. David 1990 Middle rangetheory and actualistic studies: bones and dingoes in Australian archaeology. In Solomon, S., I. Davidson, and D. Watson (eds) *Problem solving in taphonomy. Archaeological and palaeontological studies from Europe, Africa, and Oceania*, pp 234-255. Tempus 2, Archaeology and material culture studies in anthropology. Anthropology Museum, University of Queensland, St Lucia.

Specht, R.L. 1970 Vegetation. In Leeper, G.W. (ed) *The Australian environment*, pp 44-67. CSIRO/Melbourne University Press, Melbourne.

Startup, R., and E.T. Whittaker 1982 *Introducing Social Statistics*. Allen and Unwin, London.

Stephens, R. 1954 Fairfield and its founder: William Henry Graham. *Western Australian Historical Society, Journal and Proceedings* 4(5): 47-81.

Strahan, R. (ed) 1995 The mammals of Australia: the national photographic index of Australian wildlife. The Australian Museum with Angus and Robertson, Sydney.

Straus, L.G. 1990 The archaeology of the Pleistocene-Holocene transition in southwest Europe. In L.G. Straus, B.V. Eriksen, J.M. Erlandson, and D.R. Yesner (eds) *Humans at the end of the Ice Age - the archaeology of the Pleistocene-Holocene transition*, pp 83-100. Plenum Press, New York.

Straus, L.G., B.V. Eriksen, J.M. Erlandson, and D.R. Yesner (eds) 1990 *Humans at the end of the Ice Age - the archaeology of the Pleistocene-Holocene transition*. Plenum Press, New York.

Sturman, A., and N. Tapper 1996 *The weather and climate of Australia and New Zealand*. Oxford University Press Australia, South Melbourne.

Stuiver, M., and P.J. Reimer 1993 Extended 14C database and revised CALIB radiocarbon calibration program. *Radiocarbon* 35: 215-230.

Sullivan, A.P., and K.C. Rozen 1985 Debitage analysis and4 archaeological interpretation. *American Antiquity* 50: 755-779.

Sullivan, S. 1978 Aboriginal diet and food-gathering methods in the Richmond and Tweed River valleys, as seen in early settler records. In MacBryde, I. (ed), *Records of times past*, pp 101-115. Australian Institute of Aboriginal Studies, Canberra.

Temple, R.B., and M. Barbetti 1981 Revisions to radiocarbon dates from the Sydney University Radiocarbon Laboratory. *Australian Archaeology* 14: 28-29.

Theunissen, R., J. Balme, and W. Beck 1998 Headroom and human trampling: cave ceiling-height determines the spatial patterning of stone artefacts at Petzkes Cave, northern New South Wales. *Antiquity* 72: 80-89.

Thomas, I. 1993 Late Pleistocene environments and Aboriginal settlement patterns in Tasmania. *Australian Archaeology* 36:1-11.

Thomas, I. 1994 Ethnohistoric uses for the use of fire by Tasmanian Aborigines: a preliminary investigation of the evidence. *Research Paper No. 2*, Department of Geography and Environmental Studies, University of Melbourne, Melbourne.

Thomas, I. 1995 Models and prima-donnas in southwest Tasmania. *Australian Archaeology* 41: 21-23.

Thompson, G.B. 1994 Wood charcoals from tropical sites: a contribution to methodology and interpretation. In Hather, J.G., (ed) *Tropical archaeobotany*, pp. 9-33. One World Archaeology v. 22. Routledge, London.

Thorton, J.D., G.C. Johnson, and N.-K. Nguyen 1996 Revised natural durability rating for the outer heartwood of mature Australian timbers in ground contact. Proceedings 25th Forest Products Research Conference, November 1996. CSIRO, Melbourne.

Tilman, D. 1983 Some thoughts on resource competition and diversity in plant communities. In Kruger, F.J., D.T. Mitchell, and J.V.M. Jarvis (eds) *Mediterranean-type ecosystems: the role of nutrients*, pp 322-336. Springer-Verlag, Berlin.

Tindale, N.B. 1959 Ecology of primitive Aboriginal man in Australia. In Keast, A., R.L. Crocker, and C.S. Christian (eds) *Biogeography and ecology in Australia*, pp 36-51. Monographiae biologicalae v. 8, W. Junk, The Hague.

Tingay, A. 1985 Contemporary views of the voluntary conservation movement on the use of fuel reduction burns as a land management technique. In Ford, J.R. (ed) *Fire ecology and management in Western Australian ecosystems*, pp 215-218. Western Australian Institute of Technology, Perth.

Torrence, R. 1992 What is Lapita about obsidian? A view from the Talasea sources. *In* Galipaud, J.-C. (ed), *Potérie Lapita et Peuplement: actes du Colloque Lapita, Nouméa, Nouvelle Calédonie*, pp 111-126. O.R.S.T.O.M., Noumea.

Turney, C.S.M., M.I. Bird, L.K. Fifield, R.G. Roberts, M.A. Smith, C.E. Dortch, R. Grün, E. Lawson, G.H. Miller, J. Dortch, R.G. Cresswell, and L.K. Ayliffe 2001 Breaking the radiocarbon barrier and early human occupation at Devil's Lair. *Quaternary Research.*

Tyler, M.J. 1985 Quaternary fossil frogs from Skull Cave and Devil's Lair in the extreme south-west of Western Australia. *Records of the Western Australian Museum* 12: 233-240.

Underwood, R.J. 1978 Natural fire periodicity in the Karri (*Eucalyptus diversicolor* F. Muell.) forest. *Research Paper 41*. Forests Department of Western Australia, Perth.

van Huet, S., R. Grün, C.V. Murray-Wallace, N. Redvers-Newton, and J.P. White 1998 Age of the Lancefield megafauna: a reappraisal. *Australian Archaeology* 46: 5-11.

Vancouver, G. 1801 A voyage of discovery to the north Pacific Ocean and round the world, performed in the years 1790, 1791, 1792, 1793, 1794 and 1795 in the Discovery, sloop of war and armed tender Chatham, under the command of Captain George Vancouver. John Stockdale, London.

Veth, P. 1993 Islands in the interior. The dynamics of prehistoric adaptations within the arid zone of Australia. *International Monographs in Prehistory. Archaeological Series 3*. Ann Arbor, Michigan.

Villa, P. 1982 Conjoinable pieces and site formation processes. *American Antiquity* 47: 276-290.

Vlamingh, W. de 1696-1697 Extract from a journal of a voyage made to the unexplored South Land in the years 1696 and 1697. In Major, R.H., (ed) [1859] *Early voyages to Terra Australis now called Australia: a collection of documents and extracts from early manuscript maps illustrative of the history of discovery on the coasts of that vast island from the beginning of the sixteenth century to the time of Captain Cook*, pp 120-133. Hakluyt Society, London.

Walker, P.L. 1978 Butchering and stone tool function. *American Antiquity* 43: 710-715.

Walsh, M.R. 1998 Lines in the sand: competition and stone selection on the Pajarito Plateau, New Mexico. *American Antiquity* 63: 573-593.

Walshe, K. 1994a Tasmanian devils, *Sarcophilus harrisii*, and human occupation. Estimating the cultural component. In Sullivan, M., S. Brockwell, and A. Webb (eds) *Archaeology in the North. Proceedings of the 1993 Australian Archaeological Association Conference*, pp 375-380. Northern Australian Research Unit, ANU, Darwin.

Walshe, K. 1994b A taphonomic analysis of the vertebrate material from Allen's Cave: implications for Australian arid zone archaeology. Unpublished PhD thesis, Department of Archaeology and Anthropology, ANU.

Walters, I. 1988 Fire and bones: patterns of discard. In Meehan, B., and R. Jones (eds) *Archaeology with ethnography: an Australian perspective*, pp 215-221. Department of Prehistory, Research School of Pacific Studies, The ANU, Canberra.

Walton, D.W. (ed) 1988 *Zoological catalogue of Australia. Mammalia.* Australian Government Printing Services, Canberra.

Ward, D. 1998 Fire flogging, measles, and grass – Nineteenth century land use conflict in south-western Australia. An essay on human ecology. January 1998. Unpublished report, Department of Department of Conservation and Land Management, Western Australia.

Ward, D., and G. van Didden 1997 Reconstructing the fire history of the jarrah forest of south-western Australia. Unpublished report to Environment Australia under the Regional Forest Agreement, December 1997.

Ward, G.K., and S.R. Wilson 1978 Procedures for compaing and combining radiocarbon age determinations: a critique. *Archaeometry* 20: 19-31.

Wardell-Johnson, G., and P. Horwitz 1996 Conserving biodiversity and the recognition of heterogeneity in ancient landscapes: a case study from south-western Australia. *Forest Ecology and Management* 85: 219-238.

Wasson, R.J. 1989 Desert dune-building, dust-raising, and palaeoclimate in the Southern Hemisphere during the last 280,000 years. In Donnelly, T.H., and R.J. Wasson (eds) *CLIMANZ 3*, pp 123-137. CSIRO Division of Water Resources, Canberra.

Wasson, R.J., and T.H. Donnelly 1991 Palaeoclimatic reconstructions for the last 30,000 years in Australia - a contribution to prediction of future climate. Division of Water Resources Technical Memorandum 91/3. CSIRO, Canberra.

Watson, J.P.N. 1979 The estimation of the relative frequencies of mammalian frequencies: Khirokitia 1972. *Journal of Archaeological Science* 6: 127-137.

Wells, F.E., and C. Bryce 1984 A guide to the common molluscs of south-western Australian estuaries. Western Australian Museum, Perth.

Wells, P.E. and G.M.W. Wells 1994 Large-scale reorganization of ocean currents offshore Western Australia during the Late Quaternary. *Marine Micropaleontology* 24: 157-186.

White, J.P., and J.F. O'Connell 1982 *The prehistory of Australia and Sahul*. Academic Press, Sydney.

White, J.P., N. Modjeska, and I. Hipuya 1977 Group definitions and mental templates. *In* R.V.S. Wright (ed) *Stone tools as cultural markers*, pp 380-390. Australian Institute of Aboriginal Studies, Canberra.

Whitehouse, S.J.O. 1977 The diet of the Dingo in Western Australia. *Australian Wildlife Research* 4: 145-150.

Willcox, G.H. 1974 A history of deforestation as indicated by charcoal analysis of four sites in eastern Anatolia. *Anatolian Studies* 24: 117-133.

Williams, J.E., and R.J. Whelan, and A.M. Gill 1994 Fire and environmental heterogeneity in southern temperate forest ecosystems: implications for management. *Australian Journal of Botany* 42: 125-137.

Williams, M.A.J., D.L. Dunkerley, P. De Deckker, A.P. Kershaw, and T. Stokes 1993 *Quaternary environments*. Edward Arnold, London.

Williamson, K. 1979 Stream caves in the coastal limestone of moist south-west Australia. In Saar, A., and R. Webb (eds) *WACCON: Proceedings of the Twelth Biennial Conference of the Australian Speleological Federation, Perth 1979*, pp 60-67. The Australian Speleological Federation, Broadway NSW.

Wilson, T.B. 1835 *Narrative of a voyage round the world*. Sherwood, Gilbert, & Piper, London.

Witter, D. 1990 The taphonomy of meat storage. In Solomon, S., I. Davidson, and D. Watson (eds) *Problem solving in taphonomy. Archaeological and palaeontological studies from Europe, Africa, and Oceania*, pp 256-259. Tempus 2, Archaeology and material culture studies in anthropology. Anthropology Museum, University of Queensland, St Lucia.

Woodroffe, H.E., Jr, B.G. Thom, and J.M.A. Chappell 1985 Development of widespread mangrove swamps in mid-Holocene times in northern Australia. *Nature* 317:711-713.

Woodward, B.H. 1909 Extinct marsupials of Western Australia. *Geological Magazine, London* 6: 210-212.

Woodward, B.H. 1910 Fossil marsupials of Western Australia. *Records of the Western Australian Museum and Art Gallery* 1: 9-10.

Wright, H.E., Jr 1993 Introduction. In Wright, H.E., Jr, J.E. Kutzbach, T. Webb III, W.F. Ruddiman, F.A. Street-Perrott, and P.J. Bartlein (eds) *Global climates since the Last Glacial Maximum*, pp 1-4. University of Minnesota Press, Minneapolis.

Wright, R.V.S. 1977 Introduction and two studies. In R.V.S. Wright (ed) *Stone tools as cultural markers*, pp 1-3. Australian Institute of Aboriginal Studies, Canberra.

Wright, R.V.S. 1992 Doing multivariate archaeology and prehistory: handling large data sets with MV-ARCH. Second edition (2nd impression) MV-ARCH, Balmain, NSW.

Wyrwoll, K.-H. 1979 Late Quaternary climates of Western Australia: evidence and mechanisms. *Journal of the Royal Society of Western Australia* 62:129-142.

Yassini, I., and G.W. Kendrick 1988 Middle Holocene ostracodes, foraminifers, and environments of beds at Point Waylen, Swan River Estuary, south-western Australia. *Alcheringa* 12: 107-121.

Yellen, J.E. 1991 Small mammals: post-discard patterning of !Kung San faunal remains. *Journal of Anthropological Archaeology* 10: 152-192.

Appendices

Appendix 1 Ethnographic plant and animal resources in south-western Australia

Table A1.1 Plant foods and other resources in the Darling Botanical District, south-western Australia

Family	Scientific name	Common name	Nyoongar names	Recorded Aboriginal or other use in south-western Australia	Ethnographic reference	Botanical sub-district within Darling B.D.	Vegetation type	Biogeographic references
Ferns								
Dennstaedtiaceae	*Pteridium esculentum*	bracken fern	not recorded	? Rhizomes and frond tips edible if leached; has medical uses	6	Menzies, Warren	Tall open-forest, open-forest	B, C, D, F, H, I
Gymnosperms								
Podocarpaceae	*Podocarpus drouynianus*	Emu plum	koolah	Fruit eaten	9	Warren	Tall open-forest, open-forest	C, D, F, H, I
Zamiaceae	*Macrozamia riedlei*	Zamia palm	baio, boyoo, jeeriji	Fruit buried and soaked to leach toxins, roasted, eaten	6, 8, 9, 17	All	Open-forest, woodland, scrub, heath	B, C, D, F, H
Monocotyledons								
Anthericaceae	*Caesia parviflora*	Pale grass lily	karhrh	Tuber eaten	3	All	Open-forest, woodland, swamps	C, F, H
Anthericaceae	*Caesia* spp.	grass lilys	karhrh	Tuber eaten	14	All	Open-forest, woodland, swamps	H
Anthericaceae	*Dichopogon fimbriata*	Chocolate lily	not recorded	Tuber eaten raw	4	Not Warren	Open-forest, woodland	H
Anthericaceae	*Thysanotus patersonii*	Climbing fringe lily	tjungoori	Stem and leaves heated, gound, eaten with Eucalypt seeds	6, 13	All	Open-forest, woodland, heath	C, D, H
Colchicaceae	*Burchardia umbellata*	milkmaid	kara	Tubers eaten	6	All	Woodland, heath	C, D, H
Colchicaceae	*Wurmbea* sp.	none	not recorded	? Edible tuber	12	All	Open-forest, woodland	B, H
Cyperaceae	*Lepidosperma gladiatum*	Coastal sword-grass	kerbein	Stem base eaten; fibres used for making string, ? seed can be ground	6	Drummond, Warren	Coastal woodland	D, F, H, I
Cyperaceae	*Mesomelaena tetragona*	Semaphore sedge	not recorded	Fibres used for string	6	All	Sedgeland	B, D, F, H, I
Haemodoraceae	*Haemodorum sp.*	blood roots	gwerdin, mean, quaradine	Root beaten, mixed with mould, cooked, eaten	2, 4, 6	All	Woodland, heath	C, D, H
Haemodoraceae	*Haemodorum paniculatum*	none	madge	Root eaten	6, 7	All	Woodland, heath	C, D, H
Haemodoraceae	*Haemodorum simulans*	none	mutta	Spicy bulb eaten	6, 13	Not Warren	Woodland, heath	C, D, H
Haemodoraceae	*Haemodorum spicatum*	none	bhin, koolung, mirn	Spicy root eaten raw or cooked	1, 4, 7, 9, 14	All	Woodland, heath	C, D, F, H
Juncaginaceae	*Triglochin procera*	water ribbon	not recorded	? Edible tuber	12	Warren	Swamps	H
Liliaceae	*Dianella revolta*	flax lily	not recorded	? Edible fruit; tuberous root edible	12	Drummond	Open-forest, woodland, scrub, heath	B
Orchidaceae	*various*	orchids	not recorded	Tubers eaten	4	All	Tall open-forest, open-forest, woodland, heath, swamps	B, C, D, F, H, I
Orchidaceae	*Burnettia nigericans*	Red beak orchid	djubag	Tuber eaten raw or roasted	4, 6	All	Open-forest, woodland, scrub	D, G, H
Orchidaceae	*Lyperanthus serratus*	Rattle beak orchid	gar	Tuber eaten	4, 6	Drummond, Menzies, Warren	Tall (Jarrah) open-forest, open-forest, woodland, heath	D, F, G, H
Orchidaceae	*Prasophyllum fimbria*	Finged leek orchid	djubak	Roots eaten	14	All	Open-forest, woodland, sedgeland	G, H
Orchidaceae	*Prasophyllum sp.*	leek orchids	chokern, naank, tuboc	Young roots eaten	15	All	Tall open-forest, open-forest, woodland, heath, swamps	C, H

Family	Scientific name	Common name	Nyoongar names	Recorded Aboriginal or other use in south-western Australia	Ethnographic reference	Botanical sub-district within Darling B.D.	Vegetation type	Biogeographic references
Orchidaceae	*Pteristylis recurva*	Jug orchid	not recorded	Tubers eaten	6	All	Open-forest, woodland, scrub	D, G, H
Orchidaceae	*Thelymitra* spp.	sun orchids	not recorded	Tubers eaten	6	All	Open-forest, woodland, scrub	D, F, G, H
Restionaceae	*Anarthria scabra*	sword grass	not recorded	? Leaf base can be eaten, seed canbe ground?	12	Warren	Open-forest, woodland, sedgeland	H
Restionaceae	*Restio tremulus*	none	not recorded	? Seed can be ground	12	Warren	Open-forest, woodland, sedgeland	A, H
Typhaceae	*Typha* spp.	bulrushes	jetta, quirting, yandijut, yangeti	Root eaten raw or punded and cooked	6, 7, 9, 14	All	Sedgeland	B, C, D, F, H, I
Typhaceae	*Typha domingensis*	bulrush	not recorded	Root eaten	6	All	Sedgeland	A, C, D, F
Xanthorrhoeaceae	*Xanthorrhoea gracilis*	slender blackboy	mimidi	Nectar eaten, base of flower spike eaten?	6	All	Open-forest, woodland	A, D, H
Xanthorrhoeaceae	*Xanthorrhoea preissii*	blackboy	balga	Nectar, gum eaten in cakes; leaf-base burnt for scent; leaves used for thatch	4, 9, 14	All	Open-forest, woodland, scrub-heath	A, C, D, F, H
Dicotyledons								
Aizoaceae	*Carpobrotus* spp.	pigface	majerak, main, manbidi	Fruit eaten	4, 14	Drummond, Warren	Coastal heath	H
Aizoaceae	*Carpobrotus virescens*	Stout pigface	bain	Fruit eaten fresh or dried	6	Drummond, Warren	Coastal heath	C, H
Aizoaceae	*Tetragonia* sp.	pigface	not recorded	? Edible leaves	12	Drummond, Warren	Coastal heath	H
Apiaceae	*Apium prostratum*	sea celery	not recorded	? Edible leaves	12	Drummond, Warren	Coastal heath	C, H
Apiaceae	*Platysace* spp.	none	eringo, youck	Root eaten	10, 11	All	Open-forest, woodland	H
Apiaceae	*Platysace cirrosa*	Kahno	kahno, koorda	Tuber eaten	3	Dale	Open-forest, woodland	C
Apiaceae	*Platysace deflexa*	none	yug	Tuber eaten raw or roasted	4	Menzies	Open-forest, woodland	E
Apiaceae	*Platysace maxwelli*	Round yam	karno	Roots eaten raw or cooked	13	Dale, Menzies	Open-forest, woodland	C
Asteraceae	*Brachysema aphyllum*	Ribbon pea	not recorded	Nectar eaten	13	Drummond	Scrub	C
Asteraceae	*Olearia axillaris*	Beach rosemary	not recorded	? herb, eaten for iron content?	12	Drummond, Warren	Coastal heath	H, I
Casuarinaceae	*Allocasuarina fraseriana*	sheoak	not recorded	Seeds may be sucked to relieve thirst	4	All	Open-forest, woodland	A, B, F, H
Chenopodiaceae	*Enchyloeana tomentosa*	samphire	not recorded	? Fruit eaten	12	All	Woodland, coastal heath	A
Chenopodiaceae	*Rhagodia baccata*	berry saltbush	not recorded	Fruit eaten	6	Drummond, Warren	Coastal heath	D, F, H
Dasypogonaceae	*Acanthocarpus preissii*	Prickle lily	not recorded	? Fruit eaten	12	Drummond, Warren	Coastal heath	A, H
Dioscoreaceae	*Dioscorea hastifolia*	Native yam	adtjikoh, warran	Tuber pounded, cooked, eaten	6, 7, 8, 9, 14, 16	Drummond	Open-forest, woodland	C, D
Droseraceae	*Drosea* spp.	none	boon	Tuber eaten	9	All	Open-forest	F, H

Family	Scientific name	Common name	Nyoongar names	Recorded Aboriginal or other use in south-western Australia	Ethnographic reference	Botanical sub-district within Darling B.D.	Vegetation type	Biogeographic references
Epacridaceae	*Astroloma* spp.	none	gordin, mul	Fruit eaten	4	All	Open-forest, woodland, scrub	C, H
Epacridaceae	*Astrolama ciliatum*	Moss-leaved heath	not recorded	Fruit eaten	3	All	Open-forest, woodland, scrub	C, F
Epacridaceae	*Astrolama pallidum*	Kick bush	not recorded	Fruit eaten	3	All	Open-forest, woodland, scrub	C, F
Epacridaceae	*Astroloma prostratum*	Native cranberry	not recorded	Fruit eaten	4	Menzies, Warren	Open-forest, woodland	H
Epacridaceae	*Astrolama serratifolium*	Kondrung	murrumburru	Fruit eaten	13	Drummond, Dale	Open-forest, woodland, scrub	C
Epacridaceae	*Leucopogon parviflorus*	Bearded heath	not recorded	Fruit eaten	3	Warren	Coastal scrub	C, H
Epacridaceae	*Leucopogon verticillatus*	Tassel bush	not recorded	Very small fruit probably eaten	6	Warren	Tall open-forest, open-forest	B, D, F, H
Epacridaceae	*Styphelia tenuiflora*	Common pin heath	not recorded	Seed coating eaten	6	Drummond, Dale	Open-forest, woodland, heath	D, H
Lauraceae	*Cassytha* spp.	Dodder laurels	not recorded	? Fruit eaten	6	All	Open-forest, woodland, scrub, heath	C, D, F, H
Loranthaceae	*Amyema fitzgeraldii*	Southern mistletoe	not recorded	Fruit eaten	13	Dale	Woodland	C
Loranthaceae	*Amyema linophyllum*	Slender mistletoe	not recorded	? Fruit eaten	6	Drummond	Coastal woodland	C
Loranthaceae	*Amyema miqueli*	Stalked mistletoe	not recorded	? Fruit eaten	6	All	Open-forest, woodland, others?	H
Loranthaceae	*Amyema preissii*	Wireleaf mistletoe	not recorded	? Fruit eaten	6	Not Warren	Open-forest, woodland, others?	F, H
Loranthaceae	*Lysiana casuarinae*	mistletoe	nyilla-nyilla	Fruit eaten	6	Drummond	Coastal woodland	C, D
Loranthaceae	*Nuytsia floribunda*	Christmas tree	mungah	Young suckers peeled, eaten; nectar eaten; gum eaten	10	All	Woodland, low woodland, scrub-heath	B, C, H, I
Mimosaceae	*Acacia acuminata*	Jam wattle	galyang	Seeds gound into meal to make quonert (unleavened bread)	4, 10	All	Woodland	C, H
Mimosaceae	*Acacia cyclops*	Red-eyed wattle	not recorded	Seeds ground into flour; gum eaten; grubs taken from trunk	6	Drummond, Warren	Coastal woodland	D, F, H, I
Mimosaceae	*Acacia microbotrya*	Manna wattle	kunart, manna	Gum eaten or ground for making unleavened bread	4, 7, 8, 10, 14	Drummond, Dale	Woodland	C, H
Mimosaceae	*Acacia saligna*	Golden wreath wattle	wuanga	Seeds gound into meal to make quonert	10	All	Woodland	F, H
Myoporaceae	*Myoporum insulare*	none	not recorded	? Edible berries	12	Warren	Open-forest, woodland	H
Myrtaceae	*Baekea camphorosmae*	Camphor myrtle	kurren	Scent used to treat headaches	14	All	Open-forest	B, C, H
Myrtaceae	*Eucalyptus calophylla*	Marri	mnkar, ngumbit	Sweet gum eaten; flowers soaked to make drink	7, 14	All	Tall open-forest, open-forest, tall woodland, woodland	B, H, I
Myrtaceae	*E. cornuta* ?	yate (more than one sp.)	yate	Sap eaten; seeds gound into meal to make quonert	10	Menzies, Warren	Woodland	C, F, H, I
Myrtaceae	*E. loxophleba*	York gum	doatta, goatta, twotta	Gum eaten	4, 7, 14	All	Woodland	C, H

Family	Scientific name	Common name	Nyoongar names	Recorded Aboriginal or other use in south-western Australia	Ethnographic reference	Botanical sub-district within Darling B.D.	Vegetation type	Biogeographic references
Myrtaceae	*E. wandoo*	Wandoo	wandoo	Nectar eaten; bark colour indicates tree hollow contains water	6, 7	All	Woodland	B, D, F, H
Myrtaceae	*E. gomphocephalla*	Tuart	duart, duotta	Nectar eaten	3	Drummond	Tall woodland	A, C, I
Pittosporaceae	*Billardiera* spp.	billardieras	not recorded	Fallen fruit eaten	4	All	Woodland	F, H
Pittosporaceae	*Billardiera lehmanniana*	none	gurup, kurup	Fruit eaten	4	Not Warren	Woodland?	H
Pittosporaceae	*Marianthus* sp.	none	not recorded	? Edible fruit	12	Not Warren	Open-forest, woodland	F, H
Pittosporaceae	*Pronaya fraseri*	Elegant pronaya	nurjin	Fallen fruit eaten	4	Not Warren	Open-forest, woodland	F, H
Pittosporaceae	*Sollya* spp.	bluebells	kuruba, naman	Fruit eaten	6, 14	All	Open-forest, woodland	D, H
Pittosporaceae	*Sollya heterophylla*	Australian bluebell	gumug	Fallen fruit eaten	4, 6	All	Open-forest, woodland	C, D, F, H
Proteaceae	*Banksia* spp.	banksias	mungat	Nectar eaten	4, 5, 9, 10	All	Open-forest, woodland, scrub-heath	F, H, I
Proteaceae	*Banksia attenuata*	Candle banksia	biara	Nectar eaten	14	All	Open-forest, woodland, scrub-heath	B, C, H, I
Proteaceae	*Banksia grandis*	Bull banksia	mangaitch, mangyt, mungat	Nectar eaten or fermented for drink; smouldering cones carried for warmth	14, 15, 16	All	Open-forest, woodland	B, C, H, I
Proteaceae	*Banksia sphaerocarpa*	Round-fruited banksia	nugoo	Nectar eaten	13	All	Scrub-heath	C
Proteaceae	*Dryandra* spp.	dryandras	dunjin, birnd	Nectar eaten	4	All	Scrub-heath	H
Proteaceae	*Dryandra fraseri*	Orange honeypot	budjan, butyak	Nectar eaten	14	Drummond, Dale, Menzies	Scrub-heath	C
Proteaceae	*Dryandra nivea*	Couch honeypot	putjarn	Nectar eaten	3	All	Scrub-heath	C
Proteaceae	*Dryandra sessilis*	Parrotbush	pudjak	Nectar eaten	6	Warren	Open-forest, woodland, scrub-heath	D, F, H, I
Proteaceae	*Hakea* spp.	hakeas	dolgar	Gum eaten; nectar eaten	14	All	Open-forest, woodland, scrub-heath	B, H, I
Proteaceae	*Persoonia ellipticus*	Spreading snottygobble	not recorded	? Fruit eaten	12	Warren	Open-forest, woodland	H
Proteaceae	*Persoonia longifolia*	Snottygobble	not recorded	? Fruit eaten	12	All	Open-forest, woodland	A, B, H
Proteaceae	*Persoonia saccata*	Pouched snottygobble	cadgeegurrup	Fruit eaten	6	Drummond, Warren	Woodland	B, D, F, H
Proteaceae	*Xylomelum occidentale*	Woody pear	dumbung	?	14	All	Open-forest, woodland, scrub-heath	H, I
Ranunculaceae	*Clematis microphylla*	Small-leaved clematis	taaruk	Tuberous roots roasted and pounded into paste	6	All	Coastal woodland, scrub	C, D, F, H
Ranunculaceae	*Clematis pubescens*	Western clematis	not recorded	Tuberous roots eaten	3	Menzies, Warren	Tall open-forest, forest	C, H
Rhamnaceae	*Trymaliam floribundum*	Creek hazel, soapbush	not recorded	Placed in pools to drug animals; stems perhaps used for spears	6	Menzies, Warren	Tall open-forest, open-forest	A, B, C, D, F, H
Rutaceae	*Eriostemon spicatus*	"pepper and salt"	not recorded	Spicy leaves	12	Drummond, Warren	Open-forest, woodland	B, H

Family	Scientific name	Common name	Nyoongar names	Recorded Aboriginal or other use in south-western Australia	Ethnographic reference	Botanical sub-district within Darling B.D.	Vegetation type	Biogeographic references
Santalaceae	*Exocarpos odoratus*	none	dtulya	Fruit eaten	4, 10, 14	Uncertain	Coastal woodland	D, F, H
Santalaceae	*Exocarpos sparteus*	Broom ballart	chuck, djuk	Fruit eaten	4, 10, 14	All	Coastal woodland, scrub-heath	D, F, H, I
Santalaceae	*Santalum acuminatum*	Quandong	dumbari, jawirli, worinj	Fruit eaten; kernel eaten?	4, 6, 10	Drummond, Dale	Woodland	C, D
Santalaceae	*Santalum spicatum*	Sandalwood	poilyenum, willarak	Seeds eaten	4, 9	Avon	Woodland	C
Solanaceae	*Solanum symonii*	none	not recorded	? Edible fruit	12	Warren	Coastal scrub	H
Thymelaeaceae	*Pimelia* spp.	pimeleas	not recorded	Fibres used for string	6	All	Tall open-forest, open-forest	C, D, F, H, I
Thymelaeaceae	*Pimelia ferruginea*	none	not recorded	? Edible fruit	12	Warren	Coastal scrub	H
Zygophyllaceae	*Nitraria billardieri*	Nitre bush, wild grape	not recorded	? Edible fruit	12	Warren	Coastal scrub	H

Ethnographic references:
1 = Anon. 1834*
2 = Backhouse 1843*
3 = Bindon 1996
4 = Bird and Beeck 1988
5 = Bunbury 1930*
6 = Daw *et al.* 1997
7 = Drummond 1842*
8 = Grey 1841*
9 = Hammond 1933*
10 = Hassell 1936*
11 = Irwin 1835*
12 = G. Magyar, Boranup Eco-Tours, pers. comm.
13 = Meagher 1974
14 = Moore 1884*
15 = Nind 1831*
16 = Roth 1903*
17 = Stokes 1846*
*cited in Meagher 1974

Biogeographic references:
A = Beard 1981
B = Bell & Heddle 1989
C = Bindon 1996
D = Daw *et al.* 1997
E = Department of Environment and Heritage 1999
F = R. Hearn, Department of Conservation and Land Management, pers. comm.
G = Hoffman & Brown 1992
H = Hopper *et al.* 1992
I = Smith 1973

Table A1.2 Animal foods and other resources in south-western Australia

Family	Scientific name	Common name	Nyoongar names	Aboriginal procurement of animal or animal products	Ethnographic reference	Botanical sub-district within Darling B.D.	Associated vegetation formation (if any)	Biogeographic references
Insects								
Formicidae	not identified	ants	biddit, budjin, bulolo, kardagut, kwalak	eggs eaten	7, 8, 12	na	na	na
		native bee	not recorded	honey eaten	9	na	na	na
Hemiptera	not identified	"manna"	womela	insect deposit on leaves of York gum	4	na	na	na
not known	not identified	insect (not identified)	not known	found under edible white scale (meenah) on "blue gum" leaves	7	na	na	na
Hepialidae	*Bardistus cibarius*	Bardi grub	bardi	found under bark or in Xanthorrhoea trunks	7, 8, 10	All	Open-forest, woodland, scrub, heath	See Table A1.1
not known	not identified	large grub (not identified)	changut	found in rotten trees, bulrushes	12	na	na	na
not known	not identified	large grub (not identified)	wulgang	found in Xanthorrhoea trunks	10	na	na	na
Molluscs								
Hyriidae	*Westralunio carteri*	fresh-water mussel	marel, yinbi	not eaten as supposedly toxic, but cf C.E. Dortch et al. (1984)	3, 10	All	Streams in all formations	D, N
Crustaceans								
Parastacidae	*Cherax quinquecarinatus*	jilgie	not recorded	trapped or caught by hand	3, 13	Drummond	Streams in all formations	K
Parastacidae	*Cherax preisii*	koonak	not recorded	trapped, caught by hand, or excavated from burrow	3, 12, 13	Menzies, Warren	Streams in all formations	K
Parastacidae	*Cherax tenuimanus*	marron	not recorded	trapped, caught by hand, or excavated from burrow	3, 12, 13	Drummond, Warren	Streams in all formations	K
Fish								
Carangidae	*Caranx* sp.	trevally	madawick	speared in shallows (marine or estuarine)	11	Drummond, Warren	Marine and estuarine waters near all formations	F
Cheilodactylidae	*Dactylophora nigricans*	Dusky morwong	toorjenong	not stated	11	Drummond, Warren	Marine and estuarine waters near all formations	F
Gonorynchidae	*Gonorhynchus greyi*	Beaked salmon, sand fish	pining, waunugur	not stated, perhaps rarely eaten	11	Drummond, Warren	Marine and estuarine waters near all formations	F
Hemiramphidae	*Hemiramphus* spp.	garfishes	yillah, yellin	not stated	7, 10	Drummond, Warren	Marine and estuarine waters near all formations	F
Kyphosidae	*Kyphosus sydneyanus*	Silver drummer, buffalo bream	memon (1), muddier	not stated	11	Drummond, Warren	Marine and estuarine waters near all formations	F
Monocanthidae	*Nelusetta ayraud*	Chinaman leatherjacket	tabeduck	speared in deep water in rocky places (marine)	11	Drummond, Warren	Marine and estuarine waters near all formations	F
Mugilidae	*Aldrichetta forsteri*	Yelloweye mullet	knamler, kwullah, kalkada, ngamler	speared along sandy shores (marine or estuarine)	7, 10, 11	Drummond, Warren	Marine and estuarine waters near all formations	F
Mugilidae	*Mugil cephalus*	Sea mullet	merrong, mirrong	speared in estuaries, marine fish	7, 10, 11	Drummond, Warren	Marine and estuarine waters near all formations	F
Odacidae	*Haletta semifasciata*	Blue-weed whiting	ianont, worogut	speared along sandy shores from estuary banks	11	Drummond, Warren	Estuarine waters near all formations	F
Platycephalidae	various spp.	flatheads	nolkah	not stated	7	Drummond, Warren	Marine and estuarine waters near all formations	F
Plotosidae	*Cnidoglanus macrocephalus*	Estuary catfish or cobbler	karal-ya, moyor	not stated	7, 10	Drummond, Warren	Marine and estuarine waters near all formations	F

Family	Scientific name	Common name	Nyoongar names	Aboriginal procurement of animal or animal products	Ethnographic reference	Botanical sub-district within Darling B.D.	Associated vegetation formation (if any)	Biogeographic references
Pomatomidae	*Pomatomus saltator*	Tailor	billah	not stated	7	Drummond, Warren	Marine and estuarine waters near all formations	F
Scorpididae	*Scorpis aequipinnis*	Sea sweep	memon (2)	speared from rocky shores (marine)	11	Drummond, Warren	Marine waters near all formations	F
Sillaganidae	*Sillaginodes punctata*	King George Whiting	chondelar	speared in shallows (marine or estuarine)	11	Drummond, Warren	Marine and estuarine waters near all formations	F
Sparidae	*Acanthopagrus* spp.	bream	jillbah	not stated	7	Drummond, Warren	Marine and estuarine waters near all formations	F
Sparidae	*Chrysophrys auratus*	Snapper	yolkah, ijrap	not stated	7, 10	Drummond, Warren	Marine and estuarine waters near all formations	F
Sphyraenidae	*Sphyraena obtusata*	Striped seapike	kordong	speared in shallows (marine or estuarine)	11	Drummond, Warren	Marine and estuarine waters near all formations	F
not identified	not identified	shark (not identified)	kwillah, munda, bugor, madjit	not stated	7, 10	Drummond, Warren	Marine and estuarine waters near all formations	F
not identified	not identified	small fish (not identified)	wappi	taken by sweeping river pools with branches	10	na	na	na
Herpetofauna								
not identified	not identified	frog (not identified)	goya, guya	collected by hand from swamps and lakes	11	na	na	na
not identified	not identified	frog (not identified)	tdunjar	collected by hand from swamps and lakes	10	na	na	na
not identified	not identified	frog (not identified)	tuk	collected by hand from swamps and lakes	10	na	na	na
not identified	not identified	frog (not identified)	wurgyl	collected by hand from swamps and lakes	10	na	na	na
not identified	not identified	frog (not identified)	youangur	collected by hand from swamps and lakes	10	na	na	na
Cheluidae	*Chelodina oblonga*	Long-necked tortoise	not stated	collected by hand from swamps and lakes	5, 10	All	Swamps	C, E, I
Boidae	*Morelia spilota imbricata*	Carpet snake	wackul, wakel	not stated	12	All	Tall open-forest, open-forest, woodland, scrub, heath	C, E, I
Elapidae	*Echiopsis curta*	Bardick	bardick	not stated	11	All	Tall open-forest, open-forest, woodland, scrub, heath	C, E, I
Elapidae	*Notechis coronatus*	Crowned snake	werr	not stated	9	All	Tall open-forest, open-forest, woodland, scrub, heath	C, E, I
Elapidae	*Notechis scutatus orientalis*	Tiger snake	norn	not stated	11, 12	All	Tall open-forest, open-forest, woodland, scrub, heath	C, E, I
Elapidae	*Pseudonaja affinis*	Dugite	dubyt, karbarda, tookyte, tornock	not stated	10, 11	All	Tall open-forest, open-forest, woodland, scrub, heath	C, E, I
Elapidae	*Rhinoplocephalus bicolour*	Mueller's snake	torkite	not stated	11	All	Tall open-forest, open-forest, woodland, scrub, heath	C, E, I
Elapidae	not identified	snake (not identified)	wango	not stated	10	na	na	na

Family	Scientific name	Common name	Nyoongar names	Aboriginal procurement of animal or animal products	Ethnographic reference	Botanical sub-district within Darling B.D.	Associated vegetation formation (if any)	Biogeographic references
Pygopodidae	*Lialis burtonis*	Burton's snake-lizard	kerrygura, william lunger	not stated	10, 11	All	Tall open-forest, open-forest, woodland, scrub, heath	C, E, I
Scincidae	*Ergenia kingii*	King skink	wandie	found among rocks and in hollow trees	12	All	Tall open-forest, open-forest, woodland, scrub, heath	C, E, I
Scincidae	*Tiliqua rugosa*	Bobtail lizard	youern	found in long grass	12	All	Tall open-forest, open-forest, woodland, scrub, heath	C, E, I
Varanidae	*Varanus gouldii*	Gould's monitor	carta, munnaar	caught by hand; eggs cooked & eaten	8, 12	Drummond, Dale, Menzies	Open-forest, woodland, scrub, heath	C, E, I
Birds								
Accipitridae	*Aquila audax*	Wedge-tail eagle	yelga, waldja, jandu, gudap, ngulor	not stated	7, 8, 10, 12	All	Tall open-forest, open-forest, woodland, scrub, heath	C, J
Accipitridae	*Haliaeetus leucogaster*	White-bellied sea-eagle	not recorded	Eggs collected	7	Drummond, Warren	Marine shores near all formations	J
Accipitridae	not identified	hawks (not identified)	not recorded	not stated	7, 12	All	na	na
Anatidae	*Cygnus atratus*	Black swan	cooljaik, guijak	caught in breeding season; eggs taken	8, 10, 12	All	Open waters near all formations	J
Anatidae	not identified	ducks (not identified)	bunnit-jee	brought down with boomerangs	7, 12	All	na	na
Cacatuidae	not identified	cockatoos (not identified)	ye-ahbi, karak, manyt, ngolak, jakkal-yakkal	boomerangs used; wounded birds used as decoys	5, 7, 10	All	Open-forest, woodland	C, J
Columbidae	*Phaps chalcoptera*	Bronze-wing pigeon	moorhait, wodta	common in summer when acacia seeds ripen	5, 10, 12	All	Open-forest, woodland, scrub	J
Cracticidae	*Gymnorhina tibicen*	Australian magpie	not recorded	not stated	7, 13	All	Tall open-forest, open-forest, woodland	C, J
Dromaiidae	*Dromaius novaehollandiae*	Emu	wait, wayji	speared during laying season	12, 13	All	Open-forest, woodland, scrub, coastal heath	C, J
Loriidae	not identified	lorrikeets (not identified)	not recorded	brought down with boomerangs	7	All	Tall open-forest, open-forest, woodland	C, J
Otididae	*Ardeotis australis*	Australian bustard	bibilyer	not stated	7, 10	Dale?	Open-woodland, open-scrub, grassland	J
Phasianidae	not identified	quails (not identified)	not recorded	not stated	12	All	Heath, swamp	C, J
Psittacidae	not identified	parrots (not identified)	not recorded	brought down with boomerangs	7, 12	All	Tall open-forest, open-forest, woodland	C, J
Mammals								
Muridae	not identified	native mouse (not identified)	nuji	not stated	10	na	na	na
Muridae	not identified	native mouse (not identified)	djilyur	not stated	10	na	na	na
Dasyuridae	*Antechinus flavipes*	Yellow-footed antechinus	mardo	not stated	10	All	Tall open-forest, open-forest, woodland, scrub	B, C, E, I, L, M
Peramelideae	*Perameles ?*	bandicoot (not identified)	quoint	not stated	1	na	na	na
Peramelideae	*Perameles bougainville*	Western barred bandicoot	not known*	tracked to nest; flushed from cover with fire	2	Outside Darling*	Woodland, scrub, open heath	A, L, M, N

Family	Scientific name	Common name	Nyoongar names	Aboriginal procurement of animal or animal products	Ethnographic reference	Botanical sub-district within Darling B.D.	Associated vegetation formation (if any)	Biogeographic references
Perameliodeae	*Isoodon obesulus*	Short-nosed bandicoot	kundi, quenda	not stated	10	All	All	B, C, E, G, I, L, M
not identified	not identified	possum (not identified)	ballawara	not stated	10	na	na	na
not identified	not identified	possum (not identified)	dyinda	not stated	10	na	na	na
Pseudocheiridae	*Pseudocheirus peregrinus*	Ring-tail possum	nworra	not stated	12	All	Tall open-forest, open-forest, low forest	B, C, E, G, I, L, M
Phalangeridae	*Trichosurus vulpecula*	Brush-tail possum	comal, kumal	smoked out of nest and caught by hand	7, 10, 12	All	Tall open-forest, open-forest, low forest, woodland	B, C, E, G, I, L, M
Potoroidae	*Potorous tridactylus*	Long-nosed potoroo	garlgyte	not stated	10	Menzies, Warren	Tall open-forest, open-forest, heath	B, C, E, G, I, L, M
Potoroidae	*Bettongia penicillata*	Brush-tailed bettong	walyo, woail, woile, woylie	speared in the nest (leaves on the ground)	1, 10	All	Open-forest, woodland, scrub	B, C, E, G, I, L, M
Potoroidae	*Bettongia lesueur*	Burrowing bettong	boodie, burdi	not stated	10	All	Open-woodland, grassland	B, C, E, G, I, L, M
Macropodidae	*Setonix brachyurus*	Quokka	kwaker, kwogger	stalked or flushed from shelter with fire or dogs	10	Drummond, Menzies, Warren	Tall open-forest, open-forest, low forest	B, C, E, G, I, L, M
Macropodidae	? *Lagorchestes hirsutus*	wallaby (not identified)	wurak	stalked or driven from habitat	10	na	na	na
Macropodidae	*Petrogale lateralis*	Black-footed rock-wallaby	not known*	speared when basking or flushed from shelter with dogs	2, cf. 6	Outside Darling*	Open scrub; shelters in rocks	A, L, M
Macropodidae	*Macropus eugenii*	Tammar	tammar	stalked or flushed from habitat with fire or dogs	12, 14	All	Open-forest, woodland, scrub; shelters in thickets	B, C, E, G, I, L, M
Macropodidae	*Macropus irma*	Brush wallaby	gurhra, ingwahra	stalked or driven from habitat	10, 12, 14	All	Open-forest, woodland	B, C, E, G, I, L, M
Macropodidae	*Macropus fuliginosus*	Western grey kangaroo	yangor, yongar	stalked or driven; trapped in pits	7, 10, 12, 14	All	Open-forest, woodland, scrub	B, C, E, G, I, L, M
Macropodidae	*Macropus fuliginosus* ?	kangaroo, male	kubit	stalked or driven; trapped in pits	10	na	na	na
Macropodidae	*Macropus fuliginosus* ?	Kangaroo, female	warru	stalked or driven; trapped in pits	10	na	na	na
Macropodidae	not identified	wallaby (not identified)	banggap	stalked or driven; trapped in pits	10	na	na	na
Canidae	*Canis familiaris dingo*	Dingo	dwerda-yuggyn, durda mokyn	not stated; puppies sometimes eaten	5, 7, 8, 12	All	Open-forest, woodland, scrub, heath	B, C, E, G, I, L, M
Cetacea	*Physeta catodon*	Sperm whale	mimanga (species not determined)	stranded specimens eaten on the beach	7, 9, 10	Drummond, Warren	Beaches near all formations	E, H
Cetacea	*Pseudorca crassidens*	False killer whale	mimanga (species not determined)	stranded specimens eaten on the beach	9	Drummond, Warren	Beaches near all formations	E, H
Cetacea	*Tursiops truncatus*	Bottle-nose dolphin	kwillen-ah, warranang	stranded specimens eaten on the beach	7, 9, 10	Drummond, Warren	Beaches near all formations	E, H

*animal has fossil record in Darling Botanical District; not recorded there as Aboriginal food

Ethnographic references:
1 = Anon. 1834*
2 = Burbridge et al. 1988
3 = Dortch & Kelly 1997
4 = Drummond 1843*
5 = Grey 1841
6 = Gould 1863
7 = Hammond 1933
8 = Hassell 1938*
9 = Meagher 1974
10 = Moore 1884*
11 = Neill 1845
12 = Nind 1831*
13 = Roth 1903*
*cited in Meagher 1974

Biogeographic references:
A = Burbridge *et al.* 1988
B = Christensen 1992
C = Christensen *et al.* 1985
D = Halse & Blyth 1992
E = How *et al.* 1987
F = Hutchins & Swainston 1983
G = McKenzie *et al.* nd
H = Meagher 1974
I = Nichols & Muir 1989
J = Pizzey and Doyle 1991
K = Riek 1967
L = Strahan 1995
M = Walton 1988
N = Wells & Bryce 1984

Appendix 2 Assessment of archaeological potential of Leeuwin-Naturaliste Region cave and rock shelters and their floor deposits

Table A2.1: Archaeological potential of Leeuwin-Naturaliste Region cave entrances and rock shelters
Table continues on following pages.

Terminology

Entrance type:
Note that any entrance type may lead to deeper caverns or passages, not investigated in the current survey.
Fissure: low entrance probably formed by slow growth and a series of small collapses of a subterranean chamber or passage towards the surface; floor is sometimes inclined (cf. Bastian 1964).
Cavern: high entrance probably formed by intersection and exposure of a subterranean cavern by sudden widening or collapse of doline or cliffs; floor is often flat.
Overhang: Simple cliff overhang, e.g. in doline (therefore part of former chamber wall) or in limestone outcrop (therefore never part of cave system).
Pothole: Sinkhole, narrow vertical entry. Only potholes less than 2 m height were inspected, as the maximum length for access if ropes are not used.
Accessibility: Easy: walk in. Moderate: Uneven surface, may need to use hands. Difficult: climb in, must use hands.
For people to visit caves or rock shelters inside dolines they have to descend the sides of the doline. In most cases this descent is only moderately difficult.
Light levels: Good: equivalent to being in full shade at midday. Medium: equivalent to twilight. Poor: Artificial light required at midday.
Area under shelter: Large: more than 100 m². Medium: between 10 m² and 100 m². Small: less than 10 m².
Floor deposit: Type of sediment (sand, boulders, flowstone, etc).

Karst Index no.	*Name (if any)*	*Entrance type and ease of access*	*Light level*	*Area under shelter*	*Floor deposit*	*Arch'l excavation potential*
Augusta		**No. inspected caves: 3**				
Au- 1	Deepdene Cave	Wide fissure, horizontal, easy	Medium	Large	Boulders	Poor
Au- 9	Old Kudardup Cave	Wide fissure, declining slope, moderate	Medium	Large	Boulders and sand	Good - artefacts, hand-stencils
Au- 10		Wide fissure, declining slope, moderate	Medium	Medium	Boulders and sand	Moderate
Au- 17		Pothole, not found, difficult access?	na	na	na	na
Au- 22-25	Deepdene Gorge	Many caves, landowner refused access to gorge	na	na	na	na
Witchcliffe		**No. inspected caves: 49**				
Wi- 1		Overhang in doline, horizontal deposit, easy	Good	Small	Sand	Poor
Wi- 2	Green Cave	Wide fissure, declining slope, moderate	Medium	Large	Boulders	Poor
Wi- 5	Midgie Hollow	Fissure in doline, short descent, moderate	Medium	Medium	Sand	Moderate
Wi- 6	Goanna Cave	Cavern in doline, horizontal floor, easy	Good	Large	Sand	Good
Wi- 7	Skittle Cave	Cavern in doline, horizontal floor, moderate	Good	Large	Sand and boulders	Moderate
Wi- 8		Shallow dolines, easy	Good	None	Boulders and sand	Poor - offers no shelter
Wi- 10		Fissure, steep descent, difficult	Medium	Medium	Boulders	Poor
Wi- 11		Short pothole, steep descent, difficult	Poor	Medium	Boulders and sand	Poor
Wi- 13	Golgotha Cave	Cavern at base of cliff, declining slope, moderate	Medium	Large	Boulders	Poor
Wi- 16		Enormous cavern in doline, horiz. floor, easy	Good	Large	Sand	Good
Wi- 17	Mordang Dar	Fissure in shallow doline, steep declining slope, difficult	Poor	Large	Boulders and sand	Poor
Wi- 21	Giants Cave east entrance ("21A")	Cavern in doline, horizontal floor, steep descent into doline, difficult	Medium	Large	Sandy	Moderate
Wi- 21	Giants Cave doline, east cavern ("21B")	Cavern in doline, horizontal floor, steep descent into doline, difficult	Good	Large	Sandy	Moderate
Wi- 22	Giant's Cave west entrance	Fissure, steep descent, difficult	Medium	Large	Boulders	Poor
Wi- 25		Overhang in doline, easy	Good	Small	Sand	Poor
Wi- 28		Overhang in doline, moderate	Good	Medium	Sand	Poor

Karst Index no.	*Name (if any)*	*Entrance type and ease of access*	*Light level*	*Area under shelter*	*Floor deposit*	*Arch'l excavation potential*
Wi- 29		Overhang in doline, moderate	Good	Medium	Sand	Poor
Wi- 33	Tunnel Cave, west entrance	Short pothole, steep descent, moderate	Poor	Large	Boulders and sand	Moderate (near site)
Wi- 35		Fissure in doline, steep descent to horiz. floor, difficult	Poor	Medium	Sand	Poor
Wi- 37	Orchid Cave	Cavern in shallow doline, declining slope, easy	Medium	Medium	Sand	Moderate
Wi- 38	Mammoth Cave, east entrance	Cavern in cliff face, stream outflow, horizontal entry, easy	Medium	Large	Sand and flowstone, boulders	Good
Wi- 39	Mammoth Cave, west entrance	Short pothole	Poor	Small	Boulders	Poor
Wi- 42	Terry Cave	Cavern in small doline, horiz. floor, easy	Medium	Medium	Boulders over sand	Moderate
Wi- 43	Terry Cave 2nd entrance	Small doline	Good	None	Boulders and sand	Poor
Wi- 45		Fissure in doline, steep descent to horiz. floor, moderate	Medium	Medium	Boulders and sand	Moderate
Wi- 46		Fissure in doline, steep descent to horiz. floor, moderate	Poor	Medium	Boulders	Poor
Wi- 49	Calgardup Cave	Short pothole, boulder pile	Medium	Large	Sand	Good (has stream)
Wi- 51	Ruddocks Cave	Fissure, stream inflow, horizontal floor, easy	Medium	Medium	Sand	Moderate (had artefact)
Wi- 53	Zamia Nut Cave	Narrow fissure, steep descent to horizontal floor	Poor	Large	Sand	Poor
Wi- 55		Fissure in gully, steep descent, moderate	Poor	Medium	Boulders	Poor
Wi- 60	Nannup Cave	Low fissure in doline, horizontal entry, moderate	Poor	Small	Sand	Poor
Wi- 61	Devil's Lair	Fissure in doline, horizontal entry, easy	Poor	Large	Sand and flowstone	Good
Wi- 62	Crystal Cave	Wide fissure, steep descent to horiz. floor, moderate	Poor	Large	Sand	Moderate
Wi- 67	Acoustic Cave	Fissure, steep descent, moderate	Medium	Medium	Boulders	Poor
Wi- 69		Cavern in doline, moderate	Medium	Medium	Sand	Moderate
Wi- 70		Overhang, easy	Good	Small	Sand	Poor
Wi- 84		Fissure, very steep descent, dificult	Medium	Small	Boulders	Poor
Wi- 93	Arnor Cave	Fissure, horizontal entry, easy	Medium	Medium	Sand	Moderate
Wi- 95		Short pothole in doline, steep descent to horiz. floor, difficult	Poor	Medium	Sand	Moderate
Wi- 97	Tunnel Cave, east entrance	Cavern in doline, horizontal floor, easy	Good	Large	Sand	Good
Wi- 99		Overhang in doline, steep descent, moderate	Good	Small	Sand	Moderate
Wi- 100		Cavern in doline, horizontal floor, easy	Good	Medium	Sand, much removed by large slump	Moderate
Wi- 107	Fisherman's Cave	Cavern in cliff, short ascent, easy	Good	Large	Sand, much disturbed	Moderate
Wi- 108		Low fissure in cliff, short ascent, moderate	Medium	Medium	Sand	Poor
Wi- 110		Short pothole, steep descent, difficult	Poor	Medium	Boulders and sand	Poor
Wi- 116		Short pothole, vertical descent, difficult	Poor	Medium	Boulders and sand	Poor
Wi- 122	Pentorafice	Fissure, steep descent to horizontal floor, moderate	Poor	Large	Boulders and sand	Moderate
Wi- 123		Low overhang, easy	Good	Small	Sand	Poor
Wi- 148		Overhang, moderate	Good	Small	Sand	Poor
	Margaret River	**No. inspected caves: 6**				
MR- 1	Witchcliffe Cave ("1A")	Cavern, former stream outflow, short ascent, easy	Poor	Large	Sand and flowstone	Moderate (nr site)
	Witchcliffe Rock Shelter ("1B")	Cavern, former stream outflow, horizontal floor, easy	Medium	Medium	Sand	Good (site)
MR- 3	Rainbow Cave	Cavern, horizontal entry, easy	Good	Medium	Sand	Good (site)
MR- 4	Wallcliffe Cave	Fissure, steep descent to horiz. floor through squeeze, difficult	Poor	Large	Sand	Poor
MR- 9	Foxhole Cave	Fissure, steep descent to horizontal floor, moderate	Poor	Medium	Sand	Poor

Karst Index no.		Name (if any)	Entrance type and ease of access	Light level	Area under shelter	Floor deposit	Arch'l excavation potential
MR-	10		Low fissure, steep descent to low & narrow passage, moderate	Poor	Medium	Sand	Poor
MR-	11		not found	na	na	na	na
MR-	17	"Rabbit's Lair"	Fissure, steep descent to horizontal floor, moderate	Medium	Medium	Boulders and sand	Moderate (but no artefacts found)
		Cowaramup	**No. inspected caves: 3**				
Co-	1	Quinninup Lake Cave	Cavern in dune face, stream outflow, horizontal entry, easy	Poor	Large	Sand	Good (contains stream, nr lake)
Co-	7	Cowaramup Cave	Fissures, stream outflow, steep descent, moderate	Poor	Large	Sand	Moderate
Co-	8	Meekadorabee Cave	Fissure, stream outflow, horizontal entry behind waterfall, easy	Poor	Medium	Rock floor	Poor
Co-	9		not found	na	na	na	na
		Yallingup	**No. inspected caves: 21**				
Ya-	6(?)	"Mt Duckworth Doline"	Cavern in doline, horizontal entry, moderate	Medium	Large	Sand	Moderate
Ya-	3	Seven Sisters Cave	Short pothole, steep descent, difficult	Medium	Large	Sand	Poor
Ya-	11		Fissure, steep descent, moderate	Medium	Medium	Sand	Moderate
Ya-	13	Warrigal Cave	Overhang in cliff, stream outflow, horizontal entry, easy	Good	Large	Sand, slumped	Moderate
Ya-	14 to 22	Cape Naturaliste sea caves	Caverns in sea-cliff, variable ease of access	All good	All small	Beach sand, storm or wave deposit	Poor
Ya-	40	Barbilla Cave	Cavern in shallow doline, horizontal entry, easy	Good	Large	Sand	Good
Ya-	42 to 48	Cape Naturaliste sea caves	Caverns in sea-cliff, variable ease of access	All good	All small	Beach sand, storm or wave deposit	Poor

Appendix 3 Devil's Lair Harris Matrix

The Devil's Lair Harris Matrix (cf. Barker 1982) is drawn using the Harris Matrix editing program ArchEd, version 1.4.1 (cf. Mutzel *et al.* 1999). It can be viewed here on the enclosed compact disc, in any of three versions: the original ArchEd (.hm) file; a "bitmap" (.bmp) version accessible to most imaging programs; and an "autocad" (.dxf) version accessible to many design programs.

Computer files
Devil's Lair Harris Matrix.hm
Devil's Lair Harris Matrix.bmp
Devil's Lair Harris Matrix.dxf

References
Barker, P. 1982 *Techniques of archaeological excavation.* Batsford, London
Mutzel, P., C. Hundack, I. Pouchkarev, and S. Thome 1999 ArchEd. Computer program, version 1.0. Max Planck Institut für Informatik, Saarbrücken.

Appendix 4 Assessment of radiocarbon age estimates

Statistical testing and the mean pooled age

Ward and Wilson (1978) show how to identify statistically similar radiocarbon age estimates. This method serves as a preliminary means of identifying the age estimates that can be used in more sophisticated comparisons (cf. Holdaway and Porch 1996).

A pooled mean may be calculated from any group of age estimates by the formula (Ward and Wilson 1978, Formula 1):

$$A_p = (\sum_1^n A_i / E_i^2) / (\sum_1^n 1 / E_i^2)$$

where A_p is the pooled mean age, and A_i and E_i the ith age estimate and its associated error (as in: radiocarbon age is $A \pm E$). Calculating A_p allows one to calculate the test-statistic T and test the Null Hypothesis that the age estimates are effectively the same (Ward and Wilson 1978, Formula 2):

$$T = \sum_1^n (A_i - A_p)^2 / E_i^2$$

T is compared to the chi-squared distribution with $n - 1$ degrees of freedom. Calculating the probability of the Null Hypothesis (p H0) for each set of possible groupings of the same age estimates helps in assessing their reliability (Tables 6.3 and 6.7). Where the Null Hypothesis can be rejected (p H0 < 0.05), one may remove estimates that are unacceptable for further analysis because of reasons such as contamination. Where the Null Hypothesis can be accepted (p H0 > 0.95), the variance or error of the pooled age can be calculated by an additional formula (Ward and Wilson 1978, Formula 3). Unfortunately, p (H0) is rarely this high, and so the method of Holdaway and Porch (1996) is used to show the probable age range of the pooled age.

Estimating age range: the normalised probability curve

To identify the age range of a deposit with multiple radiocarbon age estimates, I identify sets of age estimates that cannot be distinguished from one another (see above; Ward and Wilson 1978). In many cases, applying this method shows that the Null Hypothesis, that the estimates are identical, can neither be accepted nor rejected. One cannot identify a pool of statistically identical estimates, from which to derive their variance or error (cf. Ward and Wilson 1978). Instead, Holdaway and Porch (1996) suggest that the age range of a set of estimates that can neither be distinguished nor pooled is indicated by the minimum and maximum ages of the youngest and oldest estimates in the set.

The simplest method of doing this is to identify the youngest and oldest possible ages suggested by the error margins on each estimate in the group. However, where the estimates are inverted, and there are no reasons to suspect contamination or laboratory error, one cannot identify the correct estimate. The alternative method is to construct a normalised probability curve which indicates the probability that a set of estimates is correct for any given time period. In the analyses used here, the smallest practicable period is 100 years. In the text, I provide graphs showing the normalised probability curve for all the radiocarbon estimates considered together within each deposit at Devil's Lair and Tunnel Cave (Figures 6.5, 6.11). Below, I provide similar graphs for each group of age estimates, as determined on stratigraphic grounds and from analysis of their statistical similarity (Tables 6.3, 6.7). These graphs for individual groups allow calculation of the probability that the estimates are correct. To infer an age range at a given probability level (say 0.95), one identifies the x-intercepts of the 0.025 and 0.975 probability levels on the y-axis of each graph. For more precise calculation, one may refer to the tables of summed probabilities on which the curves are based, but to illustrate the method, I show only the curves themselves.

The normalised probability curves are calculated as follows (Holdaway and Porch Appendix 3; Holdaway pers. comm.):

$$P(R) = \{1/(s\sqrt{2\pi})\} . e^{-(R-U)^\wedge 2/2 s^\wedge 2}$$

where $P(R)$ is the probability of age R; s is the reported standard deviation of the age estimate; U is the reported mean; and e is a constant, the base of the natural logarithm (~2.718). $P(R)$ is calculated for each period of 200 years and summed for all age estimates; the resulting summed totals form the y-values of a normalised probability curve with years BP as the x-axis. Periods or layers with only a single age estimate are not plotted.

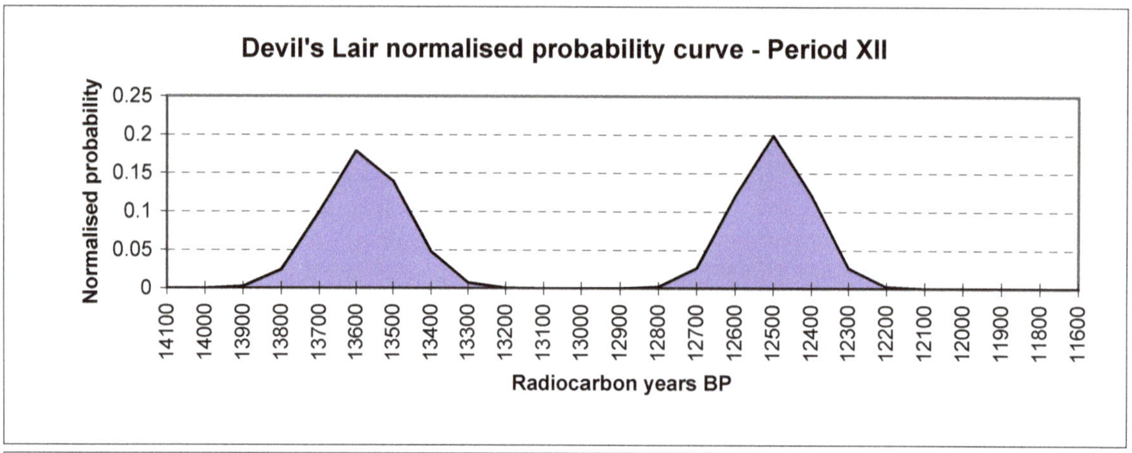
Devil's Lair normalised probability curve - Period XII
Normalised probability
0.25
0.2
0.15
0.1
0.05
0
14100 14000 13900 13800 13700 13600 13500 13400 13300 13200 13100 13000 12900 12800 12700 12600 12500 12400 12300 12200 12100 12000 11900 11800 11600
Radiocarbon years BP

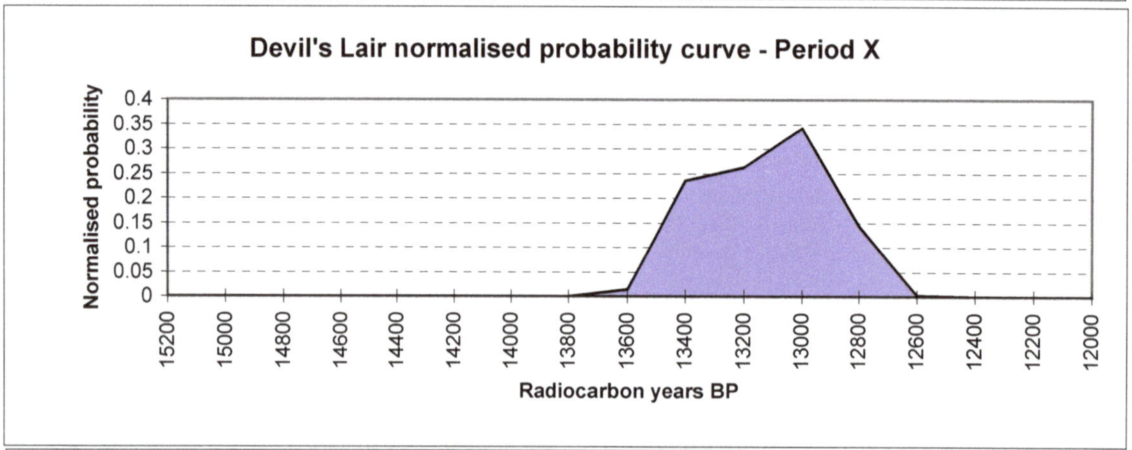
Devil's Lair normalised probability curve - Period X
Normalised probability
0.4
0.35
0.3
0.25
0.2
0.15
0.1
0.05
0
15200 15000 14800 14600 14400 14200 14000 13800 13600 13400 13200 13000 12800 12600 12400 12200 12000
Radiocarbon years BP

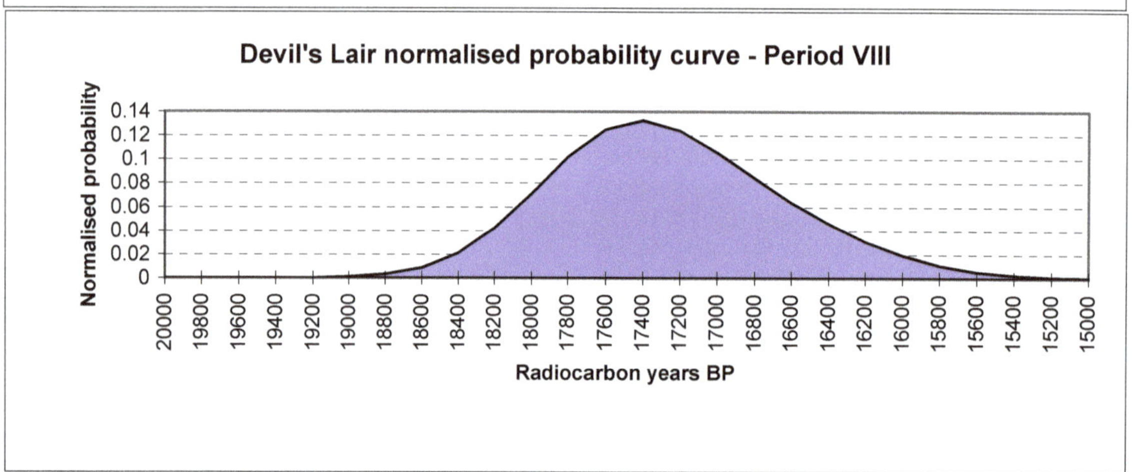
Devil's Lair normalised probability curve - Period VIII
Normalised probability
0.14
0.12
0.1
0.08
0.06
0.04
0.02
0
20000 19800 19600 19400 19200 19000 18800 18600 18400 18200 18000 17800 17600 17400 17200 17000 16800 16600 16400 16200 16000 15800 15600 15400 15200 15000
Radiocarbon years BP

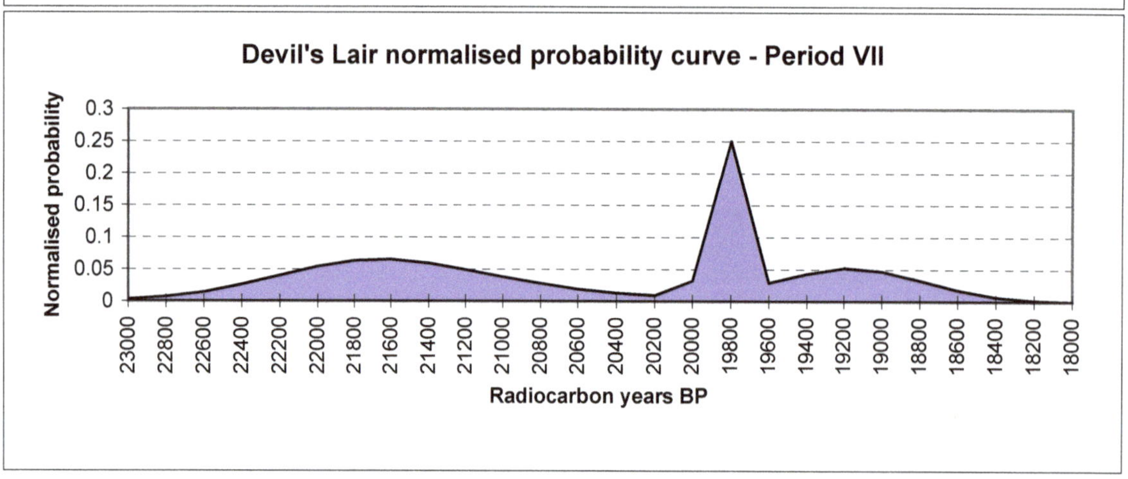
Devil's Lair normalised probability curve - Period VII
Normalised probability
0.3
0.25
0.2
0.15
0.1
0.05
0
23000 22800 22600 22400 22200 22000 21800 21600 21400 21200 21000 20800 20600 20400 20200 20000 19800 19600 19400 19200 19000 18800 18600 18400 18200 18000
Radiocarbon years BP

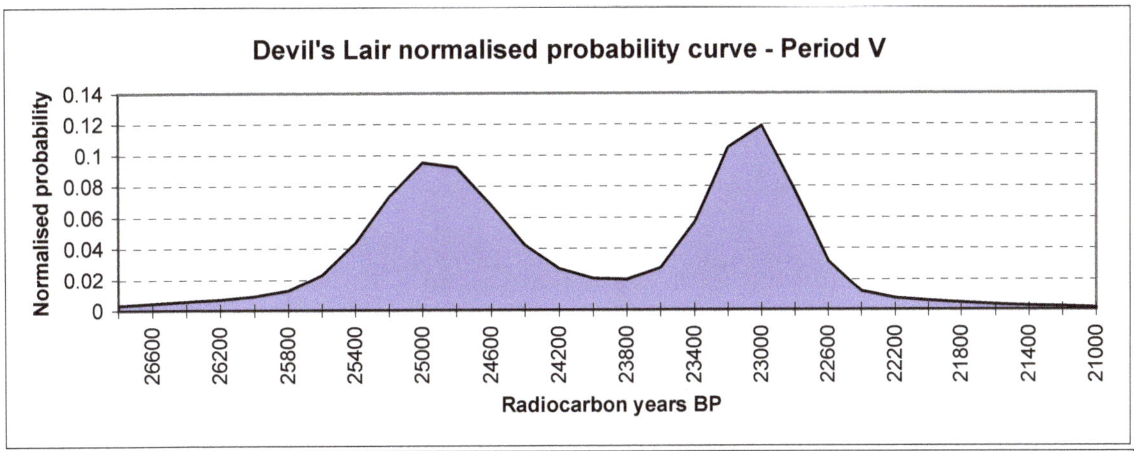
Devil's Lair normalised probability curve - Period V
Normalised probability
0.14
0.12
0.1
0.08
0.06
0.04
0.02
0
26600
26200
25800
25400
25000
24600
24200
23800
23400
23000
22600
22200
21800
21400
21000
Radiocarbon years BP

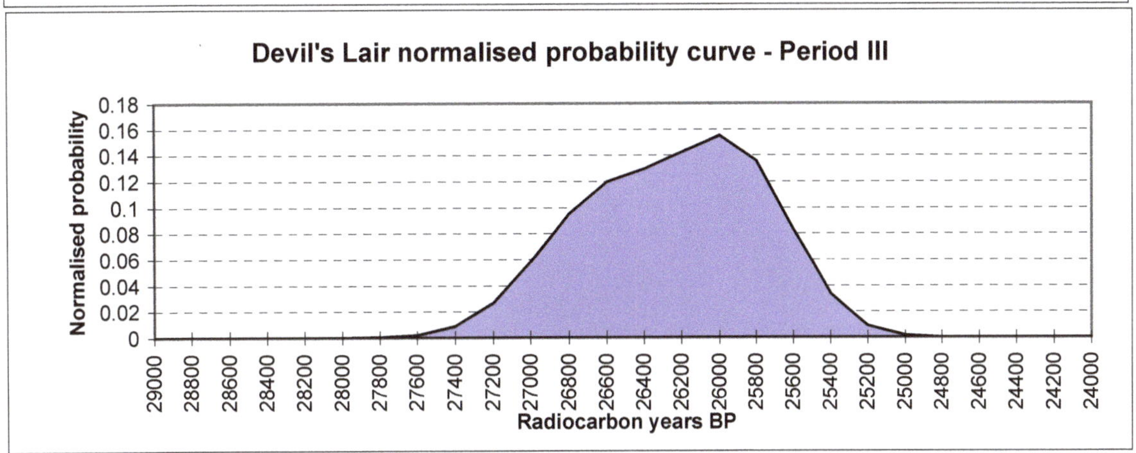
Devil's Lair normalised probability curve - Period III
Normalised probability
0.18
0.16
0.14
0.12
0.1
0.08
0.06
0.04
0.02
0
29000
28800
28600
28400
28200
28000
27800
27600
27400
27200
27000
26800
26600
26400
26200
26000
25800
25600
25400
25200
25000
24800
24600
24400
24200
24000
Radiocarbon years BP

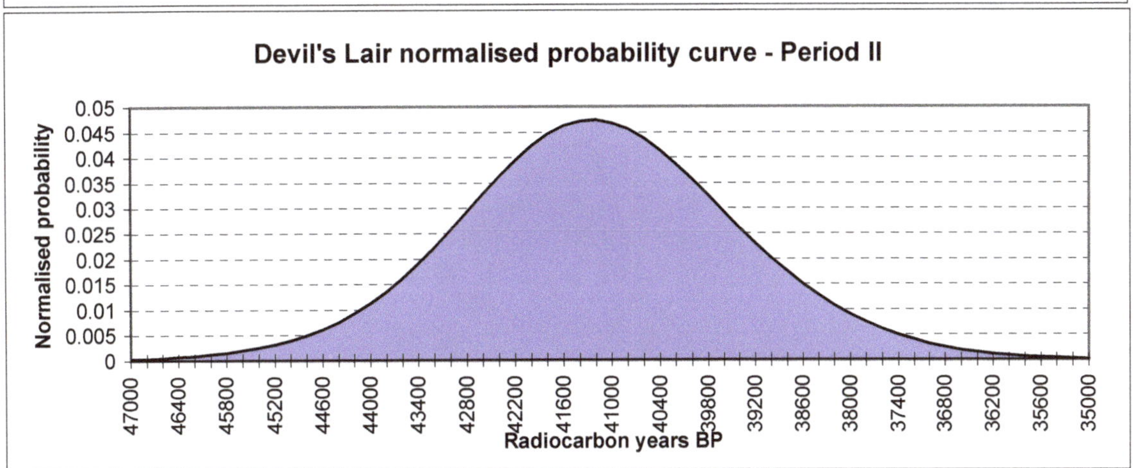
Devil's Lair normalised probability curve - Period II
Normalised probability
0.05
0.045
0.04
0.035
0.03
0.025
0.02
0.015
0.01
0.005
0
47000
46400
45800
45200
44600
44000
43400
42800
42200
41600
41000
40400
39800
39200
38600
38000
37400
36800
36200
35600
35000
Radiocarbon years BP

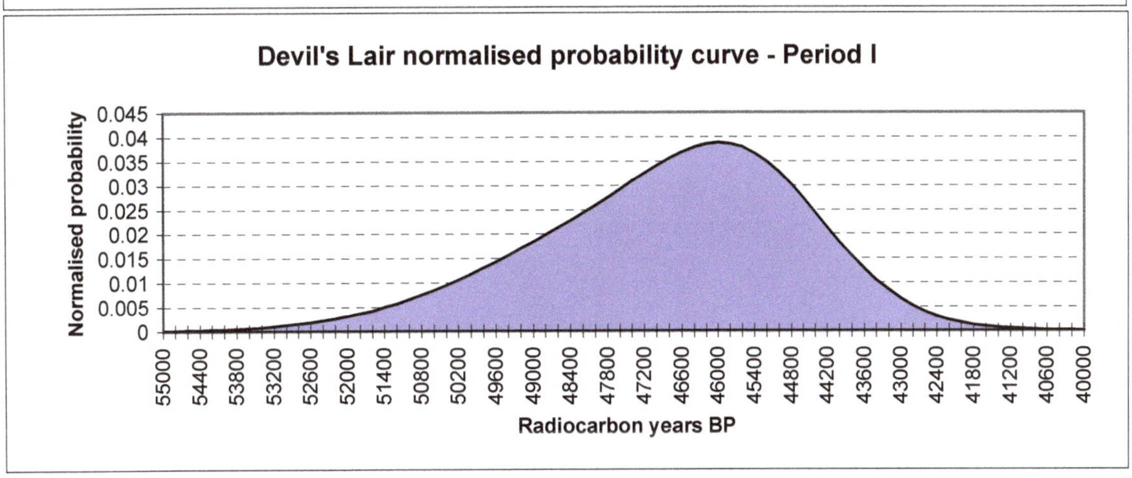
Devil's Lair normalised probability curve - Period I
Normalised probability
0.045
0.04
0.035
0.03
0.025
0.02
0.015
0.01
0.005
0
55000
54400
53800
53200
52600
52000
51400
50800
50200
49600
49000
48400
47800
47200
46600
46000
45400
44800
44200
43600
43000
42400
41800
41200
40600
40000
Radiocarbon years BP

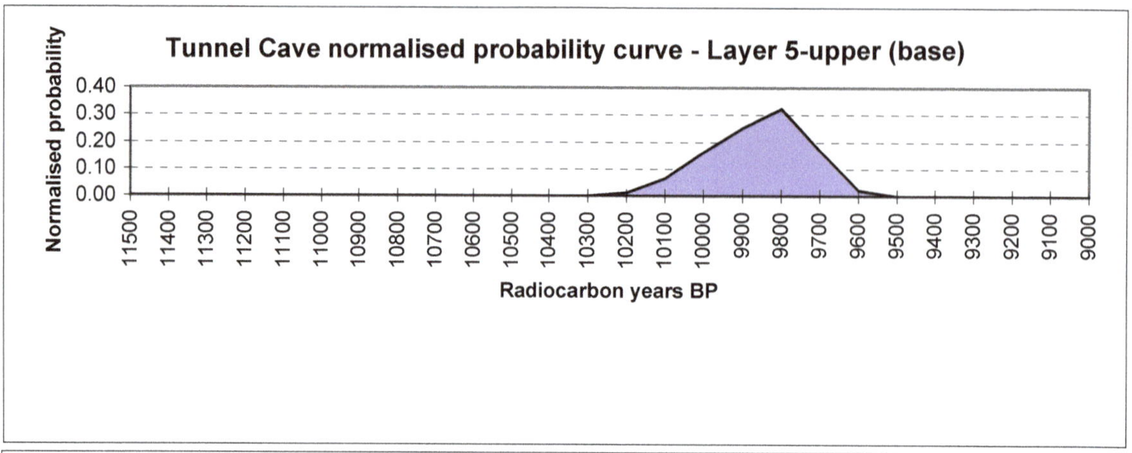
Tunnel Cave normalised probability curve - Layer 5-upper (base)
Normalised probability
0.40
0.30
0.20
0.10
0.00
11500 11400 11300 11200 11100 11000 10900 10800 10700 10600 10500 10400 10300 10200 10100 10000 9900 9800 9700 9600 9500 9400 9300 9200 9100 9000
Radiocarbon years BP

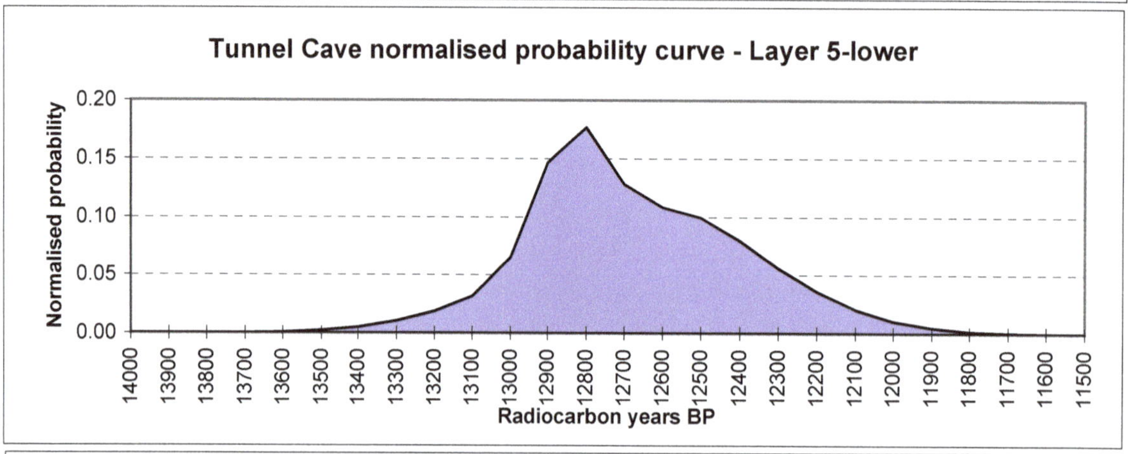
Tunnel Cave normalised probability curve - Layer 5-lower
Normalised probability
0.20
0.15
0.10
0.05
0.00
14000 13900 13800 13700 13600 13500 13400 13300 13200 13100 13000 12900 12800 12700 12600 12500 12400 12300 12200 12100 12000 11900 11800 11700 11600 11500
Radiocarbon years BP

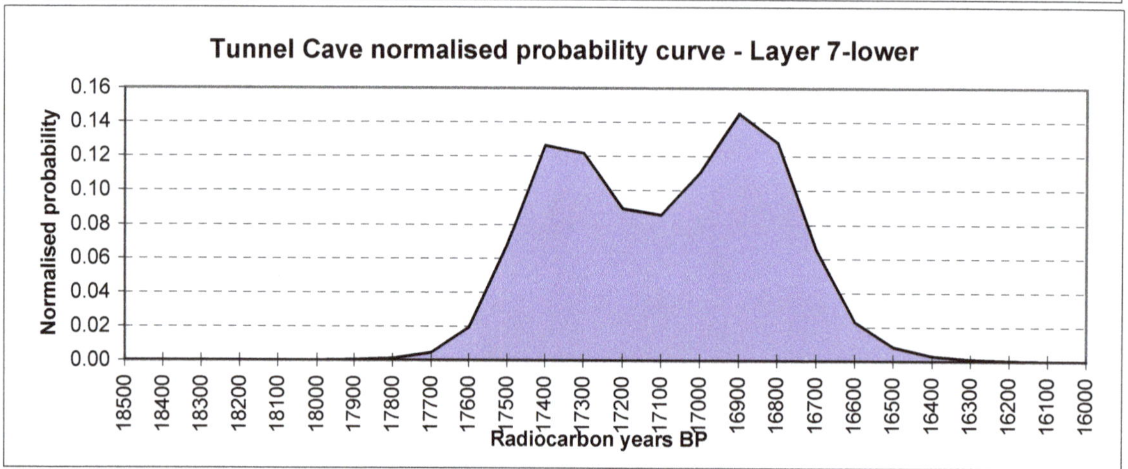
Tunnel Cave normalised probability curve - Layer 7-lower
Normalised probability
0.16
0.14
0.12
0.10
0.08
0.06
0.04
0.02
0.00
18500 18400 18300 18200 18100 18000 17900 17800 17700 17600 17500 17400 17300 17200 17100 17000 16900 16800 16700 16600 16500 16400 16300 16200 16100 16000
Radiocarbon years BP

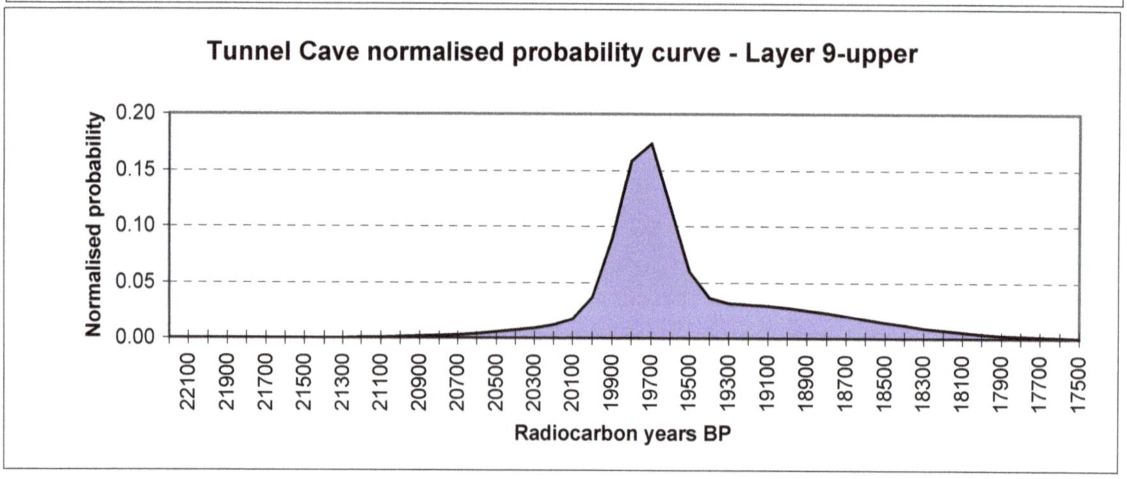
Tunnel Cave normalised probability curve - Layer 9-upper
Normalised probability
0.20
0.15
0.10
0.05
0.00
22100 21900 21700 21500 21300 21100 20900 20700 20500 20300 20100 19900 19700 19500 19300 19100 18900 18700 18500 18300 18100 17900 17700 17500
Radiocarbon years BP

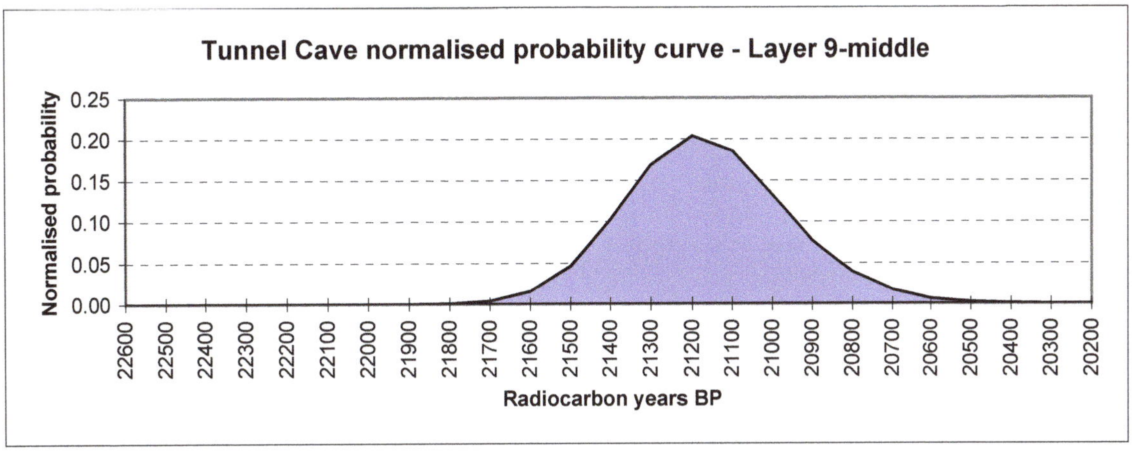
Tunnel Cave normalised probability curve - Layer 9-middle
Normalised probability
0.25
0.20
0.15
0.10
0.05
0.00
22600
22500
22400
22300
22200
22100
22000
21900
21800
21700
21600
21500
21400
21300
21200
21100
21000
20900
20800
20700
20600
20500
20400
20300
20200
Radiocarbon years BP

Appendix 5 Photographs of Tunnel Cave hearths

Tunnel Cave hearths are circular and comprise ash and charcoal layers that are thickest in the centre, pinching out at the edges, and hence referred to as lenses (Photo 1). They are of various thicknesses: the ash lens is up to 15 cm thick (*e.g.*, Feature 12); the charcoal lens is up to 5 cm thick; and the scorched layer is up to 2 cm thick (Photos 2, 3). The charcoal lens indicates the cooler coals; the ash lens indicates the hot upper part of the hearth (cf. Photo 6). The charcoal lens extends a few centimetres beyond the ash lens, so that in plan it appears as a black ring around the white ash layer (Photo 4).

Thick ash layers suggest very large fires; thin lenses of grey or dark brown sediment within the larger ash lenses suggest repeated firing episodes (*e.g.*, Features 5D, 12-15). These fine lenses may indicate lenses of "cooler" burning created by building fires (with new charcoal lenses) on existing ash beds. Future researchers might examine soil micro-morphology (cf. Courty *et al.* 1989, cf. Photos 8, 9) to assess this suggestion of re-firing. All these features are adjacent to one another and cover the squares G10 and G11. They appear to be part of an enormous complex of thick ash and charcoal lenses, largely unexcavated, and almost certainly indicating large or prolonged hearth-fires (Photos 2, 5, cf. Photo 7). Soil movements in square G11 appear to have caused layer 7 hearth sediments to slump, losing their original shape, but not their stratigraphy (Photo 5).

Relatively thin hearths F1, F4, F5A-C, F8A-C, F9, F10, F11, F17, F18, and F19 suggest single or brief occupations (see section diagrams, Photos 1-3). Hearths F17-F19 seem to be less well preserved, perhaps because of greater age or greater compaction.

Photograph 1 Overview of Tunnel Cave test-excavation showing hearths in section in north and east walls of trench

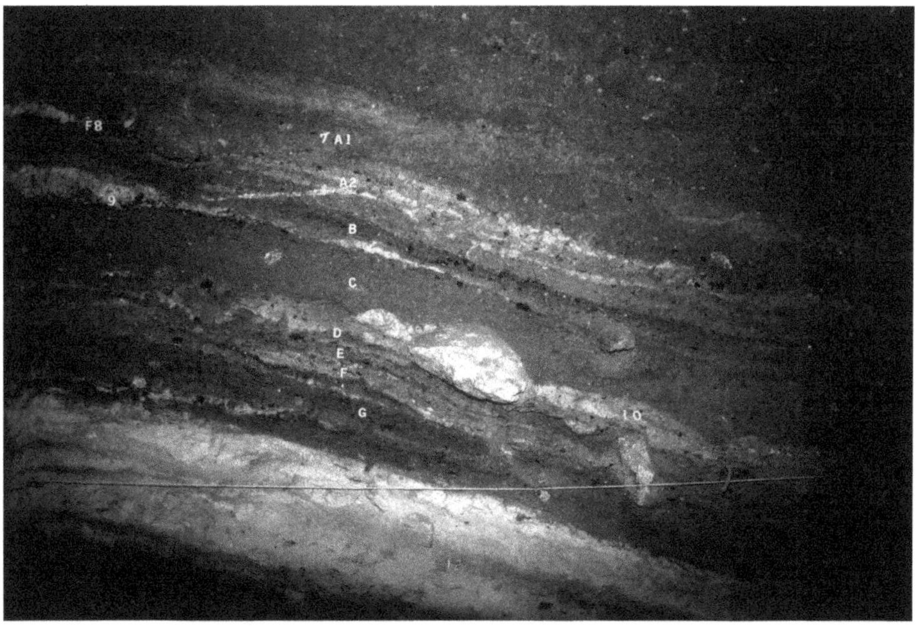

Photograph 2 Variety of hearth thicknesses: hearth features in layer 7 (width of photo is approximately 1 m)

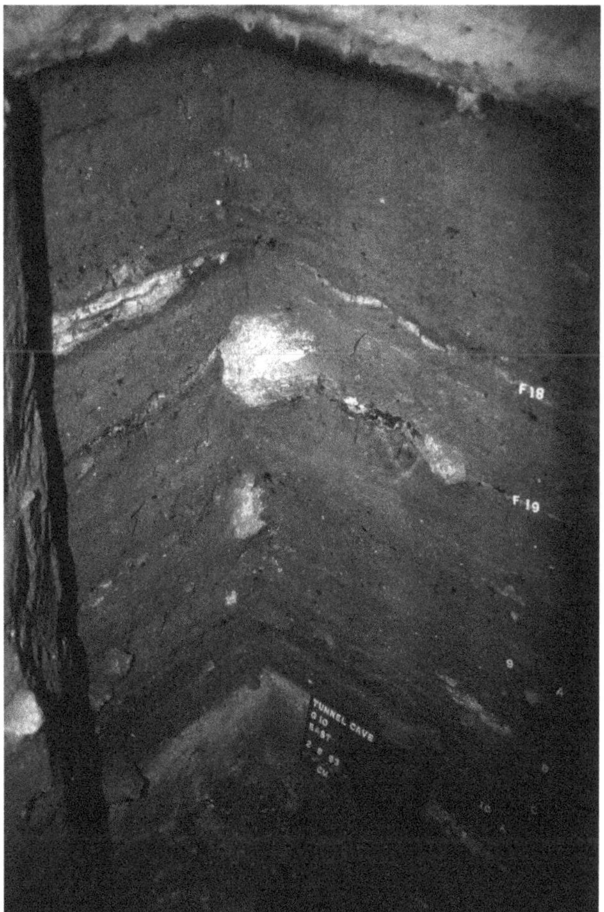

Photograph 3 Thin hearths in section: Tunnel Cave Features 17, 18, and 19

Photograph 4 Tunnel Cave hearth (F8) in plan view
The ash and charcoal lenses of Feature 8 occupy the lower left hand quarter of the picture.

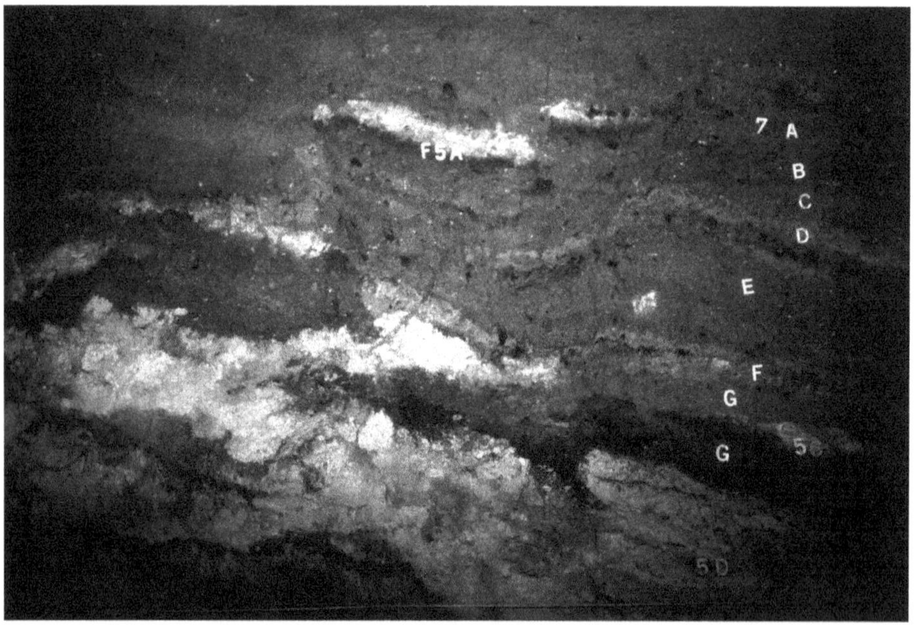

Photograph 5 Large convoluted hearth lenses in the hearth complex at the base of layer 7-lower.
Note dipping and "faulted" hearth lenses, presumably caused by subsidence beneath these compacted layers; interfingering ash lenses and sand layers, suggesting multiple hearth construction episodes; and faint dark lines in ash lenses suggesting re-firing of hearths.

Photograph 6 Hearth in six-month old campsite.
Note ash and charcoal lenses, and underlying black scorched zone, which still stands out from surrounding dark humic sediment. Campfire made on deflated dune sands at Boranup, 5 km south of Tunnel Cave, used for three or four days in August 1995, cleared from leaf litter and photographed in January 1996. Scale in centimetre blocks.

Photograph 7 Ash and charcoal lenses in large bonfire, made at Boranup in November 1995, photographed in January 1996.

Additional photographs cited in the text

Photograph 8 Water-borne sedimentation in action at Tunnel Cave.
During one severe storm in June 1993, fieldworkers had to make a bank of sandbags to prevent sand and forest litter washing into the excavation (on right).

Photograph 9 Patterns of water-borne sedimentation at Tunnel Cave.
Before excavation began in April 1993, a wash of reddish sand (lighter shade in this photograph) was discernible on the surface of the deposit. This sand sheet proved to be 1 to 2 cm thick. Ranging poles 1 m long in centre of picture indicate position of future excavation.

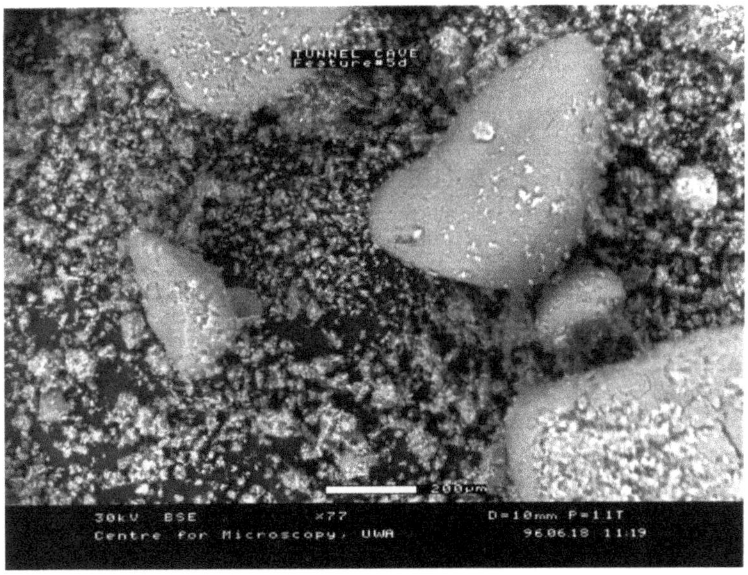

Photograph 10 SEM photomicrograph at × 77 of sediment in Tunnel Cave hearth F5D.
The large quartz grains are rounded, possibly because of extreme heating; the fine carbonaceous sediment derives from the ash lens. The white scale bar is 200 μm (0.2 mm) long.

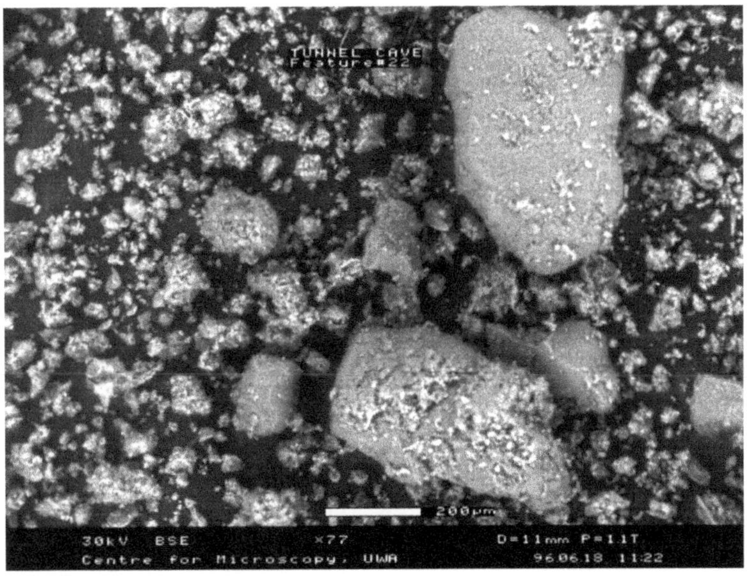

Photograph 11 SEM photomicrograph at × 77 of sediment in Tunnel Cave F22 - not a hearth.
Here the quartz grains are sub-angular, suggesting no heating, and the carbonaceous sediment is not as fine as that in Photograph 8, suggesting that it is not as comminuted by heating. The white scale bar is 200 μm (0.2 mm) long.

Appendix 6 Analysis of stone artefacts

Stone artefact terminology used in this study

The following notes are intended not as a complete review of stone artefact definitions but as clarification of terms used in the text, some of which have multiple meanings in the literature. For example, terms used here for debitage types (cf. Prentiss 1998) are equivalent to the terms of Sullivan and Rozen (1985; see also Rozen and Sullivan 1989a, 1989b). The terms flake, proximal broken flake, distal broken flake, and fragment correlate to their terms complete flake, broken flake, fragment, and debris. Sullivan and Rozen's 1985 simplified debitage classification is based on the same characteristics of flaked stone artefacts as other classifications, which are preferred here (Amick and Mauldin 1989; Ensor and Roemer 1989; Prentiss 1998).

No ground stone artefacts were identified at any of the sites analysed here, so only flaked stone artefacts are discussed below. The various techniques for detaching flakes from cores considered here include direct percussion, bipolar percussion, and pressure (Cotterell and Kamminga 1990, Hiscock and Mitchell 1993, Inizan et al. 1992).

Broken flakes are either proximal (bulbar end), distal, or longitudinal. Cores are identified from the absence of a positive bulb of percussion and the presence of one or more flake scars with negative bulbs of percussion. They can have single or multiple platforms, depending on the number of striking directions. Hammerstones are usually unmodified cobbles, convenient to hold in the hand. They can be identified by abrasion or pecking on the parts that struck the core. Other percussors, e.g. wood or bone, have not been identified in the sites analysed here.

The cores and the few flakes characteristic of bipolar technique are recognisable by having two or more bulbar surfaces; bulbs and flakelet scars at both ends; "crushing" or reduction of the platform to a sharp edge by many fine fractures; or negative flake scars on opposing surfaces.

Pressure flaking may be indicated by small, thin conchoidal flakes, and definitely indicated if such flakes have a overhanging, curved lip rather than a bulb along the proximal ventral edge. These flakes are produced by pressure from a soft object (e.g., wood or bone). They are initiated with a bending fracture that starts well inside the core from the point of applied force. Bending fractures also form on stone edges that are scraped along soft materials, e.g., hides (Kamminga 1982).

A fragment is simply a piece with no bulb, ring crack, platform, eraillure, or bulbar surface, but still identifiable as flaked stone or at least an artefact by its dorsal scars.

Almost any stone artefact with a sharp edge could have been used as a tool, and expedient or informal tools are recognised from use-wear such as microscopic abrasions, flake scars, or residues acquired during cutting or scraping (Cotterell and Kamminga 1990). Formal tools are those flakes or fragments that have been further modified by retouch or grinding. Retouch is made by percussion or pressure flaking and is distinguished from edge-damage (acquired in use or in deposition) by its size and regularity (Kamminga 1982) or style, e.g., denticulate, notched, serrated, etc (Inizan et al. 1992). A scraper, as recognised here, is any flake or fragment, not necessarily of pre-determined shape, with retouch (Cotterell and Kamminga 1990). Dortch and Dortch (1996) tentatively identify one artefact from Devil's Lair as an adze, a flake with steep retouch on one edge, that is intended to be hafted to a wooden handle and used as a chisel (McCarthy 1967).

Other tools identified at Devil's Lair include notched and denticulated pieces (Dortch 1984: 47, 55). The descriptions of these pieces as "ad hoc" (Rolland and Dibble 1990), or "adventitious" (Flenniken and White 1985) suggest that their incidence in a deposit would be random. However, it can relate to tool longevity and raw material supply: when raw material is easily available one may expect to find fewer retouched or well-used tools, because there is less requirement to make a stronger, longer-lasting, or more versatile edge (Walker 1978). The re-use and progressive reduction of simple flake tools such as scrapers can indicate the quality or the scarcity of a raw material (Dibble 1987, Hiscock 1996).

Studying debitage in relation to the *chaîne opératoire* and raw material conservation

A major goal of the analyses reported in this study is the interpretation of prehistoric conservation of raw material from debitage, since debitage makes a large proportion of all artefacts from the sites analysed here. Debitage studies can indicate aspects of tool manufacture, use, and discard, and thus allow one to compare treatment of raw material and potentially, intensity of human occupation, at different times in the same site.

The origins for debitage encompass all parts of the *chaîne opératoire*:

All stone knappers inadvertantly produce numerous minute flakes, broken flakes, and fragments, none of which are useful for making tools. The quantity of these flakes and fragments is in inverse proportion to the quality of the stone, as measured by toughness, hardness, absence of flaws; the quality of the core, as indicated by its size, shape, departure from the ideal platform angle for that rock, and extent of overhang on the platform edge; and the skill and intentions of the knapper, which direct the force and angle of the blow, and hence the speed and direction of the fracture (Cotterell and Kamminga 1979, 1990; Crabtree 1972; Hiscock 1986; Inizan *et al.* 1992; Prentiss 1998).

Fragments or breakages in the debitage piece also derive from fracturing of the piece as it is detached from the core, or if very thin, as it hits the ground. This breakage results in broken flakes (proximal or distal ends), longitudinal broken flakes, and fragments. Transversely broken flakes deriving from debitage are often indistinguishable from fragments deriving from breakage after deposition (see below) but post-depositional breakage is rarely the cause of longitudinally broken flakes. The fracture in these usually passes through the thickest, most robust part of the flake, near the point of impact.

The eraillure flake is an additional type of unintended flake, produced from the ventral surface, but its production relates more to the quality of material than flaking technique.

All such unintended pieces are technically *breakage* (cf. Inizan *et al.* 1992), but since they have bulbs of percussion, platforms, margins, conchoidal surfaces, etc, they are virtually impossible to distinguish from actual knapping products such as small retouch chips. I cannot evaluate the intentions behind flakes and fragments, and treat them all as debitage.

Every strike with a hammerstone may result in flaking or breakage of the hammerstone itself: pieces from hammerstones may be recognised if the hammerstones are of a different stone to the core or nucleus being flaked, especially if it is a tough rather than brittle stone, like basalt (Cotterell and Kamminga 1990). Since prolonged use of a hammerstone weakens its structure, one can find that they eventually produce their own small quantities of debitage.

Core preparation often includes very light flaking of a platform edge, to remove overhang and improve the fracture of the main flake removal (Flenniken and White 1985, Prentiss 1998). Tiny chips from overhang removal can form an abundant residue around stone-knappers. Probably only the larger ones are recoverable (Newcomer and Karlin 1987).

Several forms of fine debitage from tool edges are considered here. *Retouch flakes* or *flakelets* are usually small because they give the precise shape to a tool or tool edge whose general form exists already as a blank. Retouch flakes may show traces on their dorsal faces of earlier retouch flake removals and tend to have more feather terminations (Cotterell and Kamminga 1990, Newcomer and Karlin 1987). Because they are often broken (Sullivan and Rozen 1985), they are potentially difficult to distinguish from step-terminated overhang removal flakes. However, they are additionally distinguished by short length, remaining feather terminated ends, and dorsal scars of similar form. Retouch flakes from a sharp flake edge are similar in form to transverse re-sharpening flakes (Cornford 1986: Figures 29.1: 1-3; Newcomer and Karlin 1987: Figure 4.4).

Use-flakes or flakelets are the products of *edge-fracture*, which arises from use (Grace 1996, Kamminga 1982). These flakes can have the same characteristics as knapping products, e.g. bulbs of percussion and flake scars on the dorsal faces. In other cases they are distinguished from knapping products by having a bending-initiated fracture, which arises from use of a stone artefact on soft or yielding material, and more often when used in a direction parallel to the tool edge (Lawrence 1979). Tools used on hard materials, in a direction transverse to the tool edge, or with a chopping motion tend to produce use-wear flakes with conchoidal fracture and step or hinge terminations. When such conchoidal flakes have abundant residues or large step-terminations on their dorsal faces and platforms one can infer that they derive from use or tool rejuvenation.

Re-sharpening flakes (transverse type) are effectively similar to retouch flakes, with the added feature of use-wear or ancient retouch, because they are from the retouch of tools that have been used or even re-sharpened and rejuvenated on another occasion. They are a type of tool rejuvenation flake. On their dorsal faces, they may bear traces of residues, use-fractures, or earlier, resharpening flake scars. Their short and broad form is similar to that of retouch flakelets.

Post-depositional breakage has several causes, including people walking on artefacts, re-working of deposits, and breakage of thin edges or thin artefacts during knapping. Fragments and broken flakes broken after deposition are hard to distinguish from those broken in knapping, but pieces broken by a snap, or with edges chipped and broken by snaps, are probably broken after deposition. Snaps are caused by pressure applied to a thin edge on both sides of the snap, either accidentally (e.g. when a flake is trodden on or moved in a deposit) or from use of the edge. Snaps deriving from use are usually aligned in one direction, are regular, and located on one side of the edge. Snaps that are irregular along an edge, in terms of direction, size, or location, are likely to derive from accidental damage to the flake in the deposit. Ideally, a snap is distinct from a step terminated end of a flake or a straight break along a flaw. However, as either of these indicate breakage in knapping (e.g. from retouch, overhang removal) or tool-use, the separation of snaps from other breaks indicates the degree of post-depositional breakage relative to knapping- or use-breakage.

Trampling and post-depositional movement of artefacts can also snap or flake their edges. The edge-chips themselves are usually less than 2 or 3 mm (Shea and Klenck 1993), so would probably be not be found in these deposits, which were excavated and sieved with 3 and 5 mm mesh sieves. However, the incidence of chipped edges serves to reinforce the interpretation of the degree of artefact damage (see main section on *Post-depositional alteration*, below).

Repeated use or knapping of a core eventually renders it too small to hold and knap further by direct percussion. The knapper at this stage has limited options: discard the piece, or hold it on an anvil and pound it with a hammerstone. This bipolar technique produces hundreds of tiny fragments of stone, of which many are useless because they are too small to hold in the fingers or too irregular. Bipolar technique favours the production of longitudinally broken flakes because the compressive forces on the flake act longitudinally in two directions instead of one (Cotterell and Kamminga 1990). However, a knapper can usually obtain a few moderate size flakes with sharp edges, so the bipolar technique is useful for obtaining a few flakes from a small nucleus. Use of the bipolar technique implies that natural nuclei or cores are small. The latter probably owe their small size to continual flaking.

All other factors held equal, small-sized artefacts imply poor raw material supply. Poor supply implies either that people are obtaining the stone less often (e.g., because people are occupying the site or its immediate region for longer periods), or that sources of large nuclei are inaccessible. If comparisons of fine debitage types are made within the raw material types, they can indicate intense or prolonged use of tools and cores as follows:

1. more flakes and broken flakes relative to fragments;

2. more step and hinge-terminated flakes from platform preparation and core and tool exhaustion, relative to other debitage;

3. more hammerstone fragments;

4. more use-, retouch-, or tool rejuvenation flakes;

5. more bipolar flakes.

In contrast the presence of more snapped artefacts and snapped or chipped artefact edges will indicate more damage from trampling of artefacts or re-working of a deposit. In layers with large numbers of such damaged flakes, proximal and distal broken flakes are as much attributable to post-depositional damage as debitage techniques.

Summary of attributes

The following list summarises the attributes searched for on every stone artefact selected from Tunnel Cave and Devil's Lair:

Retouched and retouch flakelets:

On tool: flake scars on edge show conchoidal or bending initiation, feather termination, symmetrical shape, and are adjacent to or overlap scars of similar size and shape.

On debitage flakelet or fragment: flake has conchoidal or bending initiation, feather termination, symmetrical shape, may show use-wear on dorsal face or platform. Flake scars on dorsal face show conchoidal or bending initiation, feather termination, symmetrical shape, overlap scars of similar size and shape to each other and the flake in question.

Use-fractures and use-flakelets:

General: presence of multiple overlapping traces of use-wear.

On tool: shown on edge.

On debitage flake or fragment: shown on intersection of platform and dorsal face.

Types of use-wear:

Residues: coloured staining not attributable to soil carbonates, adhering to artefacts as commonly seen on non-artefactual material in the same sites.

Use-fractures - on relatively thick edge:

Conchoidal: from chopping or scraping hard materials, e.g. hardwood, bone.

Bending: possibility of retouch.

Feather termination: possibility of retouch or post-depositional damage.

Hinge or step termination: from chopping, scraping, or cutting hard materials.

Use-fracture - on relatively thin edge (consider alternative possibility of post-depositional damage):

Conchoidal: from chopping, scraping, or cutting hard materials, e.g. hardwood, bone.

Bending: scraping soft materials; light scraping or cutting hard materials.

Feather termination: scraping or cutting of soft and hard materials.

Hinge or step termination: scraping or cutting of hard materials.

Precision flaking:

Suggested by elongation of flakes, presence of a focal platform, and prevalence of flakes struck from the core either along an axis between parallel crests, or along a crest.

Platform preparation:

Flaked or facetted platforms, proximally truncated flake scars on cores and flake dorsal surfaces (also interpretable as platform rejuvenation).

Overhang removal:

Multiple dorsal scars of short length and size and with step or hinge terminations.

Core rotation:

On core: multiple platforms.

On flake, broken flake, or fragment: multiple flaking directions as inferred from flake axis and direction of flake scars on dorsal face.

Heat-treatment before flaking:

Discolouration (e.g., whitening of exterior surfaces of flake or fragment) of the dorsal side of a flake only: i.e., no discolouration of interior.

Bipolar technique:

On cores, flakes or indeterminate pieces: flake scars originating from opposing ends that are too obtuse to have been flaked by direct percussion; presence of wedging-initiated fractures; flake scars originating from opposite sides of the same end.

Flaking of poor quality material or exhausted cores:

Many step and hinge-terminated fractures indicate continued flaking or core to the limits of its useful life.

Flaking of cortical surfaces:

Presence of cortex. On fossiliferous chert there is no identifiable cortex, on quartz look for rounded abraded surfaces indicating artefact is flaked from river pebble.

Post-depositional alteration of artefacts:

Burning and potlidding from being incorporated into a hearth (including burning and potlidding of bulbar surfaces). Burnt chert artefacts are white or matt grey, and have a chalky texture, friable edges, or potlidded surfaces. Unburnt chert artefacts are harder, glossier, and coloured yellow-grey, brown-grey, or red-brown. Potlidding on flakes' bulbar surfaces indicates burning after flaking;

Fracture of edges when lying in the deposit, due to severe water-rolling, trampling, soil movement: identified from presence of small conchoidal fractures and snaps on fine edges, often random in appearance and of different sizes; such fractures may expose interior surfaces of different colour from outer, patinated or burnt surfaces of artefacts;

Damage during excavation and analysis: recognised by 'fresh' fractures (very clean unpatinated surfaces exposed) or fracturing and then later breakage inside the protective plastic sample bag or aluminium foil wrapping.

Appendix 7 Statistical comparison of stone artefact attributes

Kolmogorov-Smirnov (K-S) tests

Shennan (1988: 55-61) details the K-S test, used here for distinguishing differences between the distribution of observations in two different categories, for example, are the numbers of burnt artefacts significantly different from the numbers of unburnt artefacts throughout a deposit? The test is applicable to observations measured at the ordinal scale or higher.

Counts of observations are converted to proportions of their category total. If there are 174 artefacts in the category "burnt chert artefacts" at Tunnel Cave, and 26 of them are in layer 5-lower, the proportion of burnt stone artefacts in that layer is 26/174, or 0.149. For each of the two categories being compared, one calculates and then accumulates the proportions of observations in each layer, as in the example below (Table A7.1). The right hand column of this table shows the differences between the cumulative proportions; the largest difference shows whether the two categories are significantly different. For a maximum difference between proportions to be significant, at p (H0) > 0.05, it must exceed the figure obtained by the formula:

$$1.36 \sqrt{[(n_1 + n_2)/(n_1 n_2)]}$$

(Shennan 1988: 61)

where n_1 and n_2 refer to the total number of observations in each category.

Table A7.1 below shows a typical calculation formulated in a computer spreadsheet for the comparison of burnt and unburnt artefacts.

A requirement of the test is that the two categories are independent, a condition that is argued in the example above: any increase in the number of burnt artefacts is assumed to be caused by more hearth-building, and not an increase in the production or discard rate of all artefacts. In other K-S comparisons (see Tables 7.2-7.23), independence between the two categories is more evident.

Analysis of variance (ANOVA)

ANOVA is a method of determining from multiple samples of interval-scale data whether there are significance differences between the samples (Lumsden 1971, Startup and Whittaker 1982). It is therefore very useful for analysing variation in measurements over time, as represented by samples of stone artefacts from successive layers or periods, assuming that samples are normally distributed and selected randomly. These criteria are met at Tunnel Cave and Devil's Lair, the two sites analysed here (for example, see Figure A7.1). As these artefacts are small, many measurements tend towards zero (*e.g.*, many artefacts' maximum dimension is less than 1 cm) so the measurements are logged to approximate the normal distribution.

Table A7.1 Calculation of a K-S test result in a computer spreadsheet program
The table compares burnt and unburnt chert artefacts at Tunnel Cave."D max" is the largest difference between the cumulative proportions; "required D max" is obtained from the formula given in the text. Cells in the lower section contain algorithms to calculate results from the data in the upper section.

Layer	Number of burnt chert artefacts	Number of unburnt chert artefacts	Cumulative proportions of burnt chert artefacts	Cumulative proportions of unburnt chert artefacts	Difference
1	0	2	-	-	-
2	0	1	-	-	-
3	0	1	-	-	-
5-upper	0	3	-	0.001	0.001
5-lower	26	223	0.149	0.110	0.040
6	2	41	0.161	0.136	0.025
7-upper	25	251	0.305	0.324	0.019
7-lower	120	749	0.989	0.937	0.051
8	0	21	0.989	0.952	0.036
9-upper	2	35	1.000	0.971	0.029
9-middle	0	35	1.000	0.993	0.007
9-lower, 10	0	21	1.000	1.000	0.000
Total	174	1383	-	-	-
				Observed D max:	0.051
				Required D max (at p (H0) > *0.05):*	0.109
				Result:	Can't reject *H0*

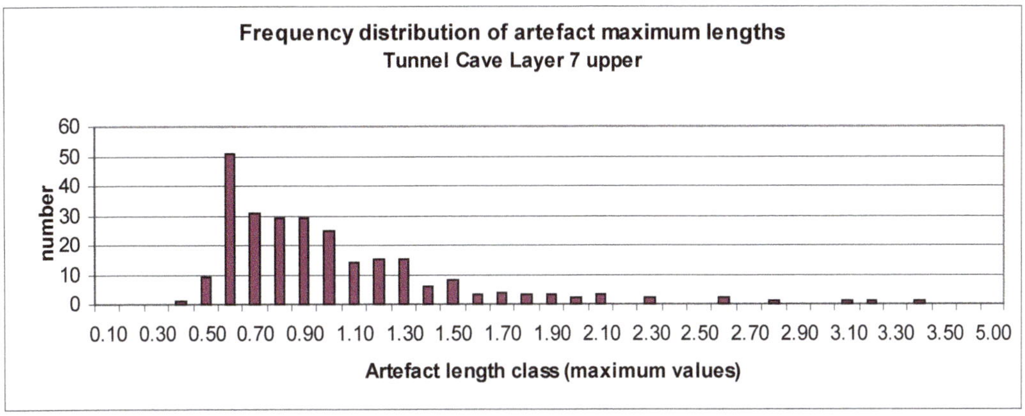

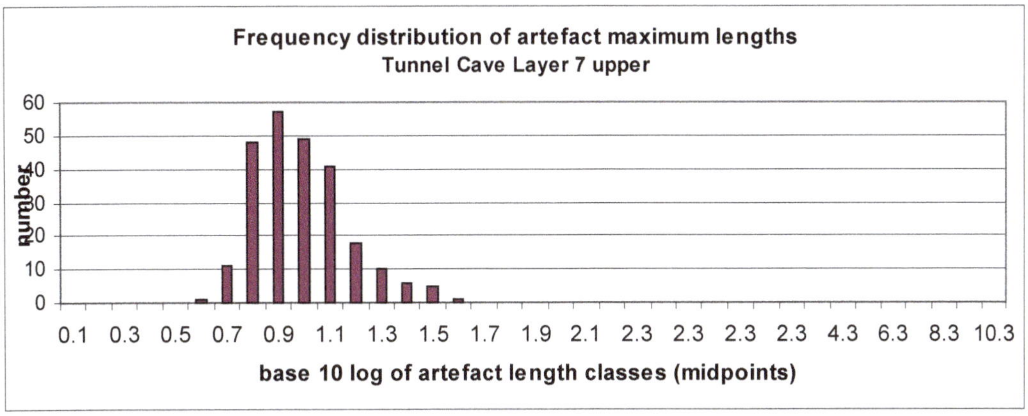

The lower diagram shows that logged artefact length classes provide an approximately normal distribution. The diagrams represent chert artefacts only.

Figure A7.1 Frequency diagrams for raw and logged measurement classes

The basis of ANOVA is to determine variance between samples and variance within samples, calculate their ratio, and compare it with the *F*-distribution for a given significance level and the calculated degrees of freedom. Where the variance between samples is greater than variance within them, one can identify the source of significant variation by carrying out a series of *T*-tests comparing every possible pairing of samples.

The variance estimates are calculated from the sums of squares of deviations from the mean ("sums of squares", or SS). The Total Sum of Squares (TSS) is the sum of squared deviations, calculated from all scores in all samples (see below); the Within Sum of Squares (WSS) from summing the sums of squares within each sample; and the Between Sum of Squares (BSS) from the sum of variances from each sample, minus the variance of all scores taken together. All the sums of squares then lead to an estimation of variance (σ^2) within and between the samples, by dividing the SS by the corresponding degrees of freedom (*d.f.*).

Formulae from Lumsden (1971: 117) and Startup and Whittaker (1982: 121-123):

$$\text{TSS} = \Sigma X^2 - (\Sigma X)^2/N$$

$$\text{WSS} = \sum_{i=a}^{k} \left[\Sigma X_i^2 - (\Sigma X_i)^2/n\right]$$

$$\text{BSS} = \sum_{i=a}^{k} \left[(\Sigma X_i)^2/n\right] - (\Sigma X)^2/N$$

$$\sigma^2_{(W)} = \text{WSS}/d.f. \qquad \sigma^2_{(B)} = \text{BSS}/d.f.$$

Degrees of freedom within samples is given by $k - 1$, where k = number of samples, and between samples, by $N - k$.

The presentation of ANOVA is the summary table (Table A7.2), as shown in the example below (see Appendix 8 for data). These summary tables were written in a spreadsheet program (MS-Excel version 5.0) for each ANOVA, and the results are summarised also in Tables 7.12-7.13 and 7.26-7.27.

Table A7.2 Example of ANOVA summary table

ANOVA - logged values Tunnel Cave chert artefact maximum length				
Source	SS	Degrees freedom	Variance estimate	Variance ratio
Between layers	0.74	8	0.09	2.33
Within layers	54.13	1356	0.04	
Total layers	54.87	1364		
Sum between & within	54.87	1364	*F*-probability:	0.02
Check	OK	OK	Result:	Reject H0

The "check" shown in the table compares TSS and total degrees of freedom with the sums of the corresponding values for between and within samples - these should be the same. The comparison is written into the spreadsheet program to give the answer, either "OK" or "Check analysis!". The *F*-probability and the result are calculated from the *F*-distribution, as given by the program MS-Excel, and from an algorithm written in MS-Excel. *H0* is assessed at $\alpha = 0.05$ and degrees of freedom for within and between samples, as calculated above, and as required for the *F*-test.

Where *H0* can be rejected, one may identify the source of variation by carrying out *t*-tests on every pair of means. Certainly, *t*-tests could be done without carrying out ANOVA, but ANOVA allows firstly, a decision about whether total variation in the group of samples is significant, and secondly, a quick way of carrying out the *t*-test (Lumsden 1971: 123), as follows:

$$t = (X_1 - X_2) / \sqrt{[\sigma^2_{(W)} (1/n_1 + 1/n_2)]}$$

t is then evaluated against the two-tailed t-distribution with $n_1 + n_2 - 2$ degrees of freedom. Here, this procedure is carried out by the computer spreadsheet program (MS-Excel v. 5.0). Significant results of two-tailed *t*-tests, where *H0* is rejected, are listed in Tables 7.12-7.13 and 7.26-7.27. Lumsden (1971) cautions that comparisons of *t*-probabilities are potentially misleading where there are few definite ideas about conditions affecting the data (as is the case here), because the *F*-test does not provide enough protection against the possibility that one may be unknowingly comparing the most extremely different means from several samples. He recommends conducting multiple analyses with different data sets that test the same hypothesis, and to hold results of each analysis lightly, until general tendencies are seen. Hence I have only relied on ANOVA results where ANOVA on several different measurements of stone artefacts suggested the same thing.

Appendix 8 Measurements and attributes of stone artefacts

The compact disc enclosed in the back cover of this volume contains two spreadsheets containing all of the raw data used in calculations based on Devil's Lair and Tunnel Cave stone artefacts. Other spreadsheets were constructed from these to obtain the Period or layer sub-totals and statistical tests discussed in Chapter 7. The disc may also be obtained on application to the author. The spreadsheets are formatted in MS-Excel version 5.0 for personal computers. In the Devil's Lair spreadsheet, some columns are hidden for ease of viewing; these are intended for construction of sub-totals. Users of the spreadsheets may view the hidden columns by "unlocking" them with the password "change" (Tools Menu, choose Protection, choose Unprotect sheet, enter password).

Abbreviations used in the spreadsheets are given in Table A8.1.

Table 8.1 Abbreviations used in Devil's Lair and Tunnel Cave stone artefact spreadsheets

Flake or scar initiation	
conc.	conchoidal
bend.	bending
bip.	bipolar
init.	initiation
Flake or scar termination	
feath.	feather
retr. h.	retroflex hinge
term.	termination
Flake shape	
contr.	contracting
exp.	expanding
obliq.	oblique
Other	
"1" in cell	feature present
frac.	fracture
ret.	retouch
foss. chert	fossiliferous chert

Computer files

Appendix 8 Devil's Lair artefacts.xls
Appendix 8 Tunnel Cave artefacts.xls

Appendix 9 Tunnel Cave stone tools

Nine stone artefacts at Tunnel Cave are recognised as tools or used pieces. Only one of these artefacts is retouched (J0085/2; see drawing below). Unlike Devil's Lair there are no notched flakes at Tunnel Cave. The sole retouched piece is a scraper made on a large longitudinally broken flake, which has retouch around the entire perimeter except along the thick edge formed by splitting of the flake during knapping. The retouched margins show traces of bending- and conchoidally-initiated fractures on both sides.

All the other pieces shown here are informal tools: flakes or broken flakes with macroscopically visible traces of use-wear (J0074/8, J0074/'9, J0075/4, J077/3, J0078/79, J0078/1, J0114/2. Also illustrated below is a chert artefact (J0078/2) that initially appeared to be a tool, but is in fact a core flaked by bipolar technique in two directions, giving it characteristic step-terminated scars running across the piece and truncated at their upper ends by bipolar scars[1].

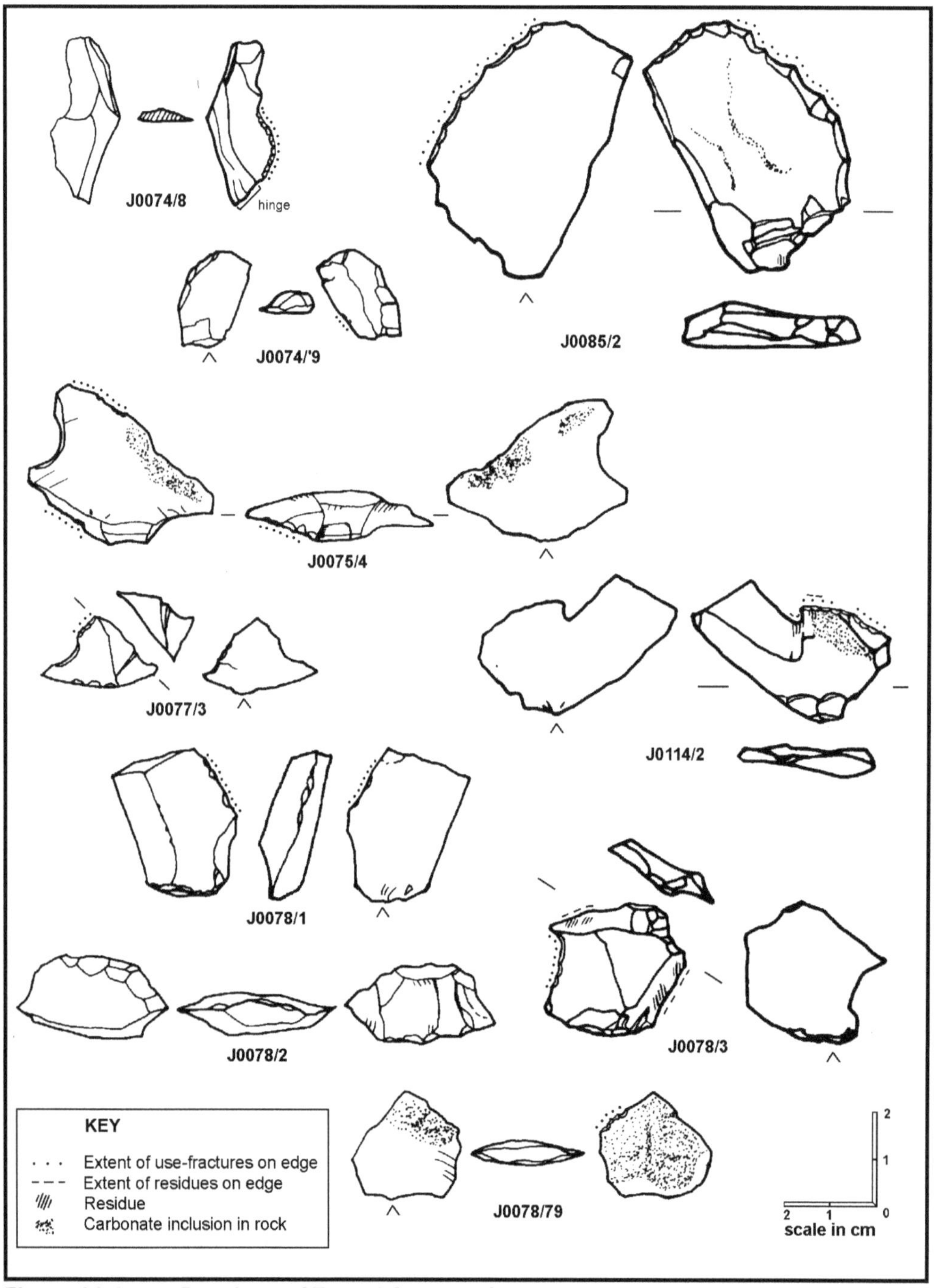

Figure A9.1 Drawings of Tunnel Cave artefacts identified as tools

[1]I acknowledge the advice of J. Kamminga for this interpretation

Appendix 10 Experimental flaking of quartz nodules and pebbles

The aim of this procedure was to identify problems and advantages of different flaking techniques as applied to quartz, and to obtain a large sample of quartz artefacts made by bipolar technique, for comparison to prehistoric assemblages. The number of artefacts actually recognisable as produced by bipolar technique would indicate the proportion of small quartz artefacts in archaeological deposits that might be expected to result from use of bipolar technique.

Method

In June 1997, I obtained about 3 kg of polycrystalline quartz nodules, each one 10-15 cm across, and several river-pebbles of crystal quartz, from the E. de C. Clarke Museum, Geology Department, UWA. All the quartz came from Kambalda in the Western Australian Goldfields (*c*.800 km east of Perth). I attempted to flake the nodules and pebbles by direct freehand percussion until they became too small to hold in the hand. I then proceeded to bipolar percussion, i.e., holding a fragment of quartz on an anvil and pounding it with a hammerstone.

Results

When flaked the polycrystalline quartz nodules broke transversely with the crack seeeming to start at the point of percussion. In isotropic and homogenous rock a good conchoidal fracture would have been initiated. Instead, the nodule broke, probably along the nearest plane of weakness, producing only fragments not even recognizable as core fragments. Under direct percussion, every fragment soon broke down to a fragment too small to flake by direct percussion. However, I had obtained some triangular fragments 5-8 cm high suitable for bipolar cores. Several long flakes and blades produced - many thin, potentially useful flakes, as well as irregular lumps and fragments. The cores produced good flakes until about 3-4 cm high when they became too small to hold between forefinger and thumb. In bipolar-flaking the first fragment I produced one thin, narrow blade-like flake (about 2 by 0.5 cm) and several equally thin but shorter flakes. All the bipolar cores were easily recognizable as such: flake scars originating from the base and sometimes from the top, crushing on a thin-edged top, and a flat base. However, none of the half-dozen good quality, thin flakes showed bipolar attributes. Among other debitage, many pieces would be doubtful artefacts in locations where natural quartz is abundant. Many of the small flakes and fragments produced from the polycrystalline nodules looked like crystal quartz.

The crystal quartz water-worn pebbles were up to 5 cm long. Taking the largest, I attempted percussion flaking but only produced one short thick flake by flaking off the end of the pebble and leaving no more suitable platform angles. The material is superior - absolutely glass-like. Bipolar-flaking the same pebble core, I obtained many fine flakes of which about 40% were recognizable as bipolar flakes under the microscope - more than with the polycrystalline quartz. Irregular and lumpy fragments, some very small, were also produced. I obtained many more blade-like flakes and thin flakes from the crystal quartz than the polycrystalline quartz. I retained the crystal quartz bipolar core and one large flake, as their glassy surfaces showed several bipolar attributes well.

Discussion

Bipolar technique was needed to flake all the polycrystalline quartz and all the crystal quartz pebbles, which were less than five centimetres diameter. Yet 60% of the flakes and fragments produced by bipolar technique showed no attributes of the technique.

Aboriginal people occupying south-western Australian archaeological sites also flaked quartz of similar quality and nodule size (Dortch and Kelly 1997), and produced some artefacts recognisable as bipolar (e.g., Cocks 1993, Coventry 1998, Dortch this thesis, Dortch and Kelly 1997, Gardos 1997). Many more of the quartz artefacts at these sites probably derive from bipolar flaking. The archaeological quartz artefacts comprise bipolar fragments, cores, and flakes (up to 13% of the total number of artefacts), and also, many small flakes indistinguishable from direct percussion flakes, generally treated as such, but potentially also derived from bipolar flaking.

Acknowledgment

I thank Mike Gartell, teaching assistant at the E. de C. Clarke Geology Museum, for providing the quartz for flaking.

Appendix 11 Potential non-human bone accumulators in south-western Australian cave sites

Sarcophilus is perhaps the pre-eminent bone-accumulator in caves throughout most of the periods discussed here. It possibly disappeared from mainland Australia *c.*3000 BP (Mulvaney and Kamminga 1999, Walshe 1994a), but may have persisted in south-western Australia until 600-400 BP (Archer and Baynes 1974). It possibly survived many environmental changes as it thrives in Tasmania, despite setbacks in the 20th C from the introduction of feral dogs, land clearance, and bounty hunting (Guiler 1982). Whatever the date of its disappearance in south-western Australia it would have deposited bones in most or all of the deposits analysed here.

In Tasmania, *Sarcophilus* has an almost legendary capacity to consume any carcase that it finds in its nocturnal foraging (Green 1967, Guiler 1970). Several individuals attacking a wallaby carcase leave little more than "trampled grass, claws and tufts of hair" - everything else is chewed, fragmented, and ingested. It also hunts birds, possums, and perhaps small wallabies (Guiler 1982, Walshe 1994a). A larger species, *S. lanarius*, extant in the Late Pleistocene, hunted perhaps slightly larger prey animals. *Sarcophilus* could have carried to cave deposits the bones of live prey - predominately small animals - as well as scavenged bones of any species.

Individual *Sarcophilus* carry bones to cave deposits not in their jaws but in their digestive tracts. Almost all *Sarcophilus*-derived bone in cave deposits comes from their scats. Contrary to some researchers' statements (Lundelius 1966, Marshall and Cosgrove 1986), they leave scats in caves and rock shelters not because they den there (they prefer small burrows or hollow logs), but to mark their presence (Walshe 1994b). They may also scavenge carcases already in caves, such as human meal debris (Balme *et al.* 1978), since living *Sarcophilus* sometimes scavenge human rubbish near houses and campsites (Green 1967, Guiler 1977).

Douglas *et al.* (1966) and Marshall and Cosgrove (1986) analysed *Sarcophilus* meal and faecal debris to identify characteristics of their bone leavings. The bones they found were notched, incised, punctured, or broken into fragments generally less than 15 mm long, occasionally up to 40 mm long. Such bone fragments probably number in the millions in Australian cave deposits (*e.g.*, Balme *et al.* 1978, Bowdler 1984, Hope *et al.* 1977). On the basis of physical characteristics alone, much of the bone in cave deposits is attributable to *Sarcophilus* - even bones from animals actually hunted by people.

Canis familiaris dingo is an important potential bone-accumulator in Late Holocene deposits. Still widespread in central Australia today, dingoes were probably also common in south-western Australia until the mid-20th C (Baynes *et al.* 1975, A.P. Bussell n.d. [1834], Hammond 1933, Nind 1831). There are no known dingo fossils in Australia before *c.*3,500 BP, so this date is assumed to indicate the approximate time of the animal's arrival in Australia (Gollan 1984).

Dingoes are adept hunters and wide-ranging scavengers, therefore able to contribute bones from many animals. In the wild and as Aboriginal hunting dogs they hunt Australian animals of almost any size, but concentrate on wallabies and small kangaroos - juveniles and females rather than old males (Shepherd 1981). Besides these animals, Whitehouse (1977) also identifies remains of locusts, skinks, birds, rabbits, and rock-wallabies, and carrion in dingo stomach contents. Dingoes were kept by south-western Australia Nyoongar people (Nind 1831) and scavenged bones and food refuse from their campsites (Hammond 1933). In recent central Australian Aboriginal camps dingoes ate human food refuse and small vertebrates such as lizards (Hayden 1975).

Dingoes sometimes den in caves, to shelter, feed on prey, or raise pups (Solomon and David 1990). As they leave their scats in various locations, including their dens, and human occupation sites, they are potential bone-accumulators in caves and rock shelters. Feeding experiments show that dingoes dismember kangaroo carcases and ingest many bone fragments (Solomon and David 1990), producing bone fragments 2.5-25 mm long, occasionally up to 40 mm long, in scats. In the late Holocene sites examined here, the thousands of small bone fragments, very many less than 20 mm long, could have been left by dingoes. In sum, dingoes are proficient bone-chewers, scavengers, and hunters that have potentially contributed bone to cave and rock shelter deposits younger than 3,500 BP.

Considering other mammalian carnivores, *Dasyurus geoffroii* is probably a perennial but minor contributor of bones in cave deposits. The bone-filled scats of *Dasyurus geoffroii* (Western Quoll or Chuditch) have been found in cave deposits (Lundelius 1966). It was widespread at the time of European arrival in south-western Australia (Archer 1984), and on the basis of collection localities, would have existed in most Late Holocene and Pleistocene environments. Dasyurids like *D. geoffroii* can drag food items as large as themselves and may scavenge around human campsites (Strahan 1995). Their diet includes vertebrates smaller than themselves, insects, grubs, and even faeces of *Sarcophilus* (Walshe 1994a). However, one must assume that such avid recycling of second- or even third-hand bone fragments was too infrequent, and quolls were too small, to remove large amounts of bone material from the sites discussed here. Walshe (1994a) estimates from photographs (Lundelius 1966: Plate 28) that bone fragments in *D. geoffroii* scats are as large as 30 mm, and are mostly much smaller.

These characteristics are insufficient to distinguish scat-bones contributed by *D. geoffroii* from those contributed by other carnivores.

A poorly understood potential source of bones in caves is the activity of extinct carnivores, *Thylacoleo carnifex* (Marsupial "lion"), and *Thylacinus cyanocephalus* (Thylacine). There is uncertain evidence for the existence of *Thylacoleo* anywhere in Australia when people began to occupy the Leeuwin-Naturaliste Region sites. (Gillespie *et al.* 1978, van Huet *et al.* 1998). In south-western Australia, its remains have been found only at Mammoth Cave, in a non-occupation deposit older than 37,000 BP (Merrilees 1984). Although it was in its day Australia's largest non-human carnivore, its bone-accumulating potential in the sites analysed here is entirely unknown.

The other extinct large carnivore, *Thylacinus*, does have well-dated and younger fossils in south-western Australia. *Thylacinus* became extinct in mainland Australia sometime after 3000 BP, perhaps because of competition from the Dingo which had arrived before 3500 BP (Archer 1984). It survived in Tasmania, which had no dingoes, until placed under a government bounty and hunted to extinction in the 20th C (Smith 1984).

Thylacinus possibly did deposit bones of its prey in caves, but the lack of detailed studies of its habits prohibits certain conclusions (Smith 1984). It hunted wallabies and possibly kangaroos and small animals, and scavenged carrion, perhaps also human meal debris. It had jaws powerful enough to break bone in its prey, and it probably consumed large portions of its kills, so it could have left scats containing bone fragments in caves (Walshe 1994a). Scat-bone is probably all that one would find from *Thylacinus* in caves and rock shelters since it probably did not bring its kills to these places (Smith 1984). Bones left by *Thylacinus* in caves and rock shelters would therefore derive from scats, but there are no adequate observations from which to derive criteria for distinguishing them from bone fragments left by other animals (Lundelius 1966).

Mammalian carnivores smaller than the Chuditch (all dasyurids other than the Chuditch and listed in Table 2.8) are eliminated from further consideration, as they could have gnawed bones (Ride 1970) but not removed or contributed them.

Raptorial birds, especially owls, have probably accumulated bones in cave deposits throughout the period considered here. I consider owls are the major raptors accumulating bones in caves: unlike owls, raptors such as *Aquila audax* (Wedge-Tailed Eagle) and *Accipiter cirrhocephalus* (Collared Sparrow-Hawk) which accumulate bone-piles and perhaps even went into Devil's Lair (Baird 1986), do not use caves regularly. Marine birds, such as sea-eagles and gulls, may also accumulate fish bones and sea-shells, but these birds are not commonly observed to use caves, especially inland sites like Devil's Lair and Tunnel Cave. Marine birds probably could have only accumulated bones at Witchcliffe Rock Shelter and Rainbow Cave, and even there, their contribution was probably minor.

Owls known to roost in caves and inhabiting the Leeuwin-Naturaliste Region species include *Tyto novaehollandiae* (Masked Owl), *Tyto alba* (Barn Owl), and *Ninox novaeseelandiae* (Boobook Owl; Pizzey and Doyle 1991). At least one of these species operated at the sites studied here. At Devil's Lair, the bones of lorrikeets, parrots, quails, and bronze-wing pigeons make up most of the bird bones from the site and indicate that the most likely predator was *T. alba* (Baird 1986). The same owl may have roosted in the other sites. At Rainbow Cave, Jackson (1992) attributes a large proportion of mammal bones to owl predation.

Owls modify bones in distinctly different ways from mammalian carnivores. *Tyto novaehollandiae* (Masked owl) hunts animals weighing up to 2 kg, which includes murids, dasyurids, bandicoots, and possums (Marshall 1986, Walshe 1994b). Like all owls it regurgitates undigested material, including bones, in pellets. Whole bones up to 45 mm long remain intact in the pellets (Walshe 1994a); larger bones are broken, fractured, or fissured in feeding (Walshe 1994b). *T. alba* and *Ninox novaeseelandiae*, which hunt smaller prey, ingest bones less than 20 mm long. Owl-deposited bones are therefore largely intact or only fissured, and tend to be from small animals.

Prehistoric reptilian carnivores potentially depositing bones in sites have been considered by Balme *et al.* (1978) and Merrilees (1979a). There is little documentation of modern reptiles accumulating bones in caves or rock shelters and I therefore presume their role is small. With the possible exception of the extinct python *Wonambi naracoortensis* (cf. Merrilees 1979a), reptiles would have difficulty in transporting bones of most species analysed in this study.

Appendix 12 Statistical treatment of faunal data

Principle Components Analysis (PCA)

In PCA, the vectors that condense the variation are called components. Components are scaled eigenvectors, equations that describe a proportion of the variation in a table of data. Every eigenvector has an associated eigenvalue, which denotes the size of (percentage of variation represented by) each component. Each successive eigenvalue represents the maximum of information about variation in the table, after the variation accounted for by the preceding eigenvalue has been removed. PCA also gives component loadings, which are values of each variable (taxon) and object (layer or phase) on each component. Variables and objects are thus graphed with the components as the axes, as a visual representation of the original table of data (Figures 8.2, 8.7, 8.13, and 8.14). The results of PCA are given in Appendix 12.

Diversity

Diversity is a concept used widely in archaeology and ecology (Leonard and Jones 1989). Diversity includes two main concepts: *richness* and *evenness*. Richness is the number of species in a sample (an archaeological layer); evenness is how well or evenly are all species represented in that sample (in an even sample, the number of individuals in each of several species is the same). Richness and evenness summarise the distribution of taxa in a sample, and indicate whether its structure is broadly similar to other samples, even if it lacks species that are present in the others. Table A12.1 shows several formulae for calculating these measures.

The concept *heterogeneity* (H) summarises richness and evenness in one statistic. However, richness and evenness are such different qualities that heterogeneity is a rather nebulous concept (cf Brobowsky and Ball 1989). I use H only to calculate evenness (J), according to the method of Pielou (1977, cited in Brobowsky and Ball 1989). The second part of the formula for H approximates to Simpson's Index (λ), so similar results are obtained from J and λ when analysing samples of similar size (H_{max} is the log of the number of categories: Kintigh 1989). For comparison, the often-used Shannon-Weaver Index (H') is given below, with its corollary J'. H' and J' are not applied in this analysis because they are most heavily dependent on sample size (Brobowsky and Ball 1989, Grayson 1984).

As in all statistical analyses of abundance, sample size dependency is a problem. Deposition of small parts of relatively large organisms inside a small excavation area will tend to be irregular, creating fluctuating sample sizes. All diversity measures are potentially correlated with sample size (Cruz-Uribe 1989). Therefore, I follow Grayson (1984) and test for rank-order correlations (using Spearman's co-efficient, r_s) between the diversity measures for the layers in a site, and sample sizes for the layers.

I also test for rank-order correlation between the proportions of the most abundant taxon in each sample or layer and the sample size. This correlation indicates that the representation of taxa is dependent on sample size (Grayson 1984). Its absence indicates that the quantity of identifed bone does not determine the quantity of the taxa. Where diversity measures are correlated with sample size, but the quantity of the most abundant taxon is not, sample size is probably an important but not dominant influence on diversity measures (cf. Grayson 1984).

Table A12.1 Diversity measures

S = number of taxa in sample; N = MNI or other estimate of numbers in sample

n_i = MNI in taxon i; p_i = proportion of taxon i's MNI in sample MNI, ie n_i / N

Diversity measure	Formula	Reference	Additional ref.
Richness (dl)	$(S - 1)/\log_e N$	Margalef (1958: cited in Brobowsky & Ball 1989, Formula 1)	Cruz-Uribe (1989)
Richness (R)	$S/\sqrt{N}$	Menhinick (1964: cited in Brobowsky & Ball 1989, Formula 3)	Frankel & Holdaway (1993)
Evenness, reciprocal of Simpson's Index λ	$1 / \sum_{i=1}^{S} p_i^2$	MacArthur (1972, cited in Grayson 1984)	As "diversity": Grayson (1984)
Heterogeneity (H)	$1 - \sum_{i=1}^{S} \{[n_i (n_i - 1)]/[N (N - 1)]\}$	Pielou (1977, cited in Brobowsky & Ball 1989, Formula 24)	Frankel & Holdaway (1993)
Heterogeneity, or Shannon & Weaver Index (H')	$-\sum_{i=1}^{S} p_i \log p_i$	Shannon & Weaver (1949, cited in Brobowsky & Ball 1989, Formula 25)	As "diversity": Cruz-Uribe (1989), Grayson (1984)
Evenness, J	H/H_{max}	Pielou (1977, cited in Brobowsky & Ball, Formula 18)	Frankel & Holdaway (1993), Kintigh (1989)
Evenness, J'	H'/H'_{max}	Pielou (1977, cited in Brobowsky & Ball, Formula 19)	Kintigh (1989)

As sample size may well be correlated with representation of taxa, I do not rely solely on diversity measures. I also compare species distributions by ordering the frequencies of taxa in each layer from highest to lowest, and applying a two-sided Smirnov test on each possible pair of samples' cumulative relative frequencies (cf. Grayson 1984). This comparison is a qualitative and quantitative alternative to diversity measures, showing whether layers have different taxonomic distributions.

The formulae in Table A11.1 are calculated here either by means of a computer program, Diverse 3.0 (Frankel and Holdaway 1993), or by algorithms that I wrote in Microsoft Excel 5.0. The most appropriate estimate is the one whose results are least correlated with sample size. A significance test from Startup and Whittaker (1982: 160) is used to confirm the Spearman's r_s value, as follows:

A test statistic is calculated from the formula:

$$r_s \sqrt{[(n - 2) / (1 - r_s^2)]}$$

If the Null Hypothesis is true, the statistic is distributed as Student's t with n - 2 degrees of freedom. If the statistic exceeds the critical t-value at n - 2 degrees of freedom and a given significance level (assuming α at 0.05), the Null Hypothesis can be rejected.

Appendix 13 Results of PCA

The following pages give tables of PCA results for each of the four sites, Devil's Lair, Tunnel Cave, Witchcliffe Rock Shelter, and Rainbow Cave. The analysis of each site generates a page of results, the output of the program package MV-Arch (Wright 1992). The first paragraph of each page gives the output file name and other information; the second gives the source of the data and the programs used to transform the original data into a covariance matrix.

The next few lines show the eigenvalues for each component and the percentage of variation that they account for (on their own and cumulatively). Each column of figures relates to an eigenvalue and principle component: the left-hand one is the first and most important in accounting for variation, the next column to the right is the second and second-most important, and so on. Below these lines, the two tables show component scores (loadings) on variables (species) and on objects (layers or periods in which the species were counted). The scores are organised into columns corresponding to each component. They may be compared to the scattergrams shown above (Figures 8.2, 8.7, 8.13, and 8.14). In these scattergrams, I only graph the first two components (PC1 and PC2). The number of variables and objects is small, so the first two components describe most of the variation.

PCA - Devil's Lair species

```
C:\DATA\OUT048.DOC   Analysis by MV-ARCH program BIGPCA3
Data file name C:\DATA\DLSPS.TRA
Date 19/Jun/99   Time 10:37

C:\DATA\DLSPS.TRA    Made from file C:\DATA\DLSPS.CSV into MV-NUTSHELL file by
IMPEXP
 COVARIANCE MATRIX

Sum of PCA eigenvalues = 9

Eigenvalues
       5.36  2.28  0.78  0.58

%
      56.93 24.26  8.29  6.21

CUM %
      56.93 81.18 89.47 95.67

PCA variable loadings                  (Species)
   1: -0.35  0.54 -0.16 -0.12          Isoodon
   2: -0.88  0.83  0.25  0.07          Peramales
   3: -0.13 -0.35 -0.29  0.45          Pseudocheirus
   4:  0.96 -0.17 -0.08  0.13          Trichosurus
   5: -0.65 -0.47 -0.06  0.01          Potorous
   6:  1.18  0.04  0.20  0.10          B. pencillata
   7:  0.96  0.45 -0.23  0.10          B. lesueur
   8: -0.18 -0.82  0.25 -0.24          Setonix
   9:  0.84  0.17  0.26 -0.32          Petrogale
  10: -0.10  0.03  0.59  0.41          M. eugenii
  11:  0.21 -0.16  0.19  0.05          M. fuliginosus

PCA object scores                      (Period)
   1: -1.96  0.23 -0.35 -0.02          Period I
   2: -0.79 -0.27  0.72 -0.09          Period II
   3:  0.06 -0.44  0.00  0.32          Period III
   4:  0.43 -0.04 -0.20  0.03          Period IV
   5:  0.19 -0.10 -0.15  0.33          Period V
   6:  0.39  0.31 -0.02  0.03          Period VI
   7:  0.48  0.42  0.05 -0.10          Period VII
   8:  0.22  0.67 -0.00 -0.23          Period VIII
   9:  0.27  0.45  0.12 -0.08          Period IX
  10:  0.25  0.17  0.06  0.20          Period X
  11:  0.13 -0.58  0.02  0.10          Period XI
  12:  0.32 -0.82 -0.23 -0.49          Periods XII & XIII
```

PCA - Tunnel Cave species

```
C:\DATA\OUT030.DOC   Analysis by MV-ARCH program BIGPCA3
Data file name C:\DATA\TCSPSMNI.TRA
Date 03/Jun/99   Time 15:49

C:\DATA\TCSPSMNI.TRA    Made from file C:\DATA\TCSPSMNI.CSV into MV-NUTSHELL
file by IMPEXP
 COVARIANCE MATRIX

Sum of PCA eigenvalues = 11

Eigenvalues
       5.56  2.26  1.03

%
      52.82 21.53  9.81

CUM %
      52.82 74.35 84.16

PCA variable loadings         (Species)
   1:  0.41  0.05 -0.44       Isoodon
   2:  1.37  0.09  0.23       Peramales
   3: -0.25 -0.22  0.13       Pseudocheirus
   4:  0.00 -0.12 -0.12       Trichosurus
   5: -0.02  0.95  0.28       Potorous
   6: -0.03  0.41  0.28       B. pencillata
   7:  0.43  0.53 -0.29       B. lesueur
   8: -1.32 -0.17 -0.24       Setonix
   9:  0.92 -0.78  0.25       Petrogale
  10: -0.73  0.09  0.57       M. eugenii
  11:  0.37  0.43 -0.26       M. fuliginosus

PCA object scores             (Layer)
   1:  0.40  0.08  0.12       9-lower & 10
   2:  0.62 -0.85 -0.57       9-middle
   3:  0.78  0.12  0.22       9-upper
   4:  0.42  0.05  0.26       8
   5:  0.44  0.13  0.06       7-lower
   6:  0.34  0.09  0.00       7-lower
   7:  0.54  0.12  0.05       6
   8:  0.19  0.15 -0.13       5-lower
   9: -0.73  0.05  0.49       5-upper
  10: -1.23 -0.89  0.22       3
  11: -0.82  0.15 -0.33       2
  12: -0.93  0.79 -0.39       1
```

PCA - Witchcliffe Rock Shelter species

```
C:\DATA\OUT054.DOC    Analysis by MV-ARCH program BIGPCA3
Data file name C:\DATA\WRSSPS.TRA
Date 08/Jul/99    Time 20:17

C:\DATA\WRSSPS.TRA     Made from file C:\DATA\WRSSPS.CSV into MV-NUTSHELL file by
IMPEXP
 COVARIANCE MATRIX

Sum of PCA eigenvalues = 33

Eigenvalues
      12.45  6.83  4.85  4.37


%
      37.17 20.40 14.49 13.05

CUM %
      37.17 57.57 72.06 85.12

PCA variable loadings                  (Species)
   1:  1.89 -0.52  0.31 -0.46          Isoodon
   2: -1.05  0.37  0.51 -1.41          Pseudocheirus
   3:  1.99  0.09  0.02  0.55          Trichosurus
   4:  0.31  0.79  1.02 -0.13          Potorous
   5: -1.48  0.55 -0.59  0.89          B. pencillata
   6: -1.19 -0.95  1.32  0.41          Setonix
   7:  0.17  0.55  0.43 -0.14          Petrogale
   8:  0.22  2.01 -0.01 -0.08          M. eugenii
   9:  0.21  0.50  1.09  0.94          M. fuliginosus

PCA object scores                      (Layer)
   1:  0.20 -0.28  0.37 -0.16          1
   2:  2.45 -0.82 -0.97 -0.12          2
   3:  0.06  0.14  0.61 -0.17          3-upper
   4: -1.11 -0.46  0.36  0.39          F5
   5:  0.82  0.23 -0.16  1.01          3-lower
   6: -1.19 -0.79 -0.61  1.02          F5
   7:  0.03  0.53  0.32  0.29          4-upper
   8: -0.75  1.30 -1.33 -0.65          F6
   9:  0.49  0.68  0.46  0.01          4-middle
  10: -0.04 -1.29  0.43 -1.06          F8
  11:  0.56  0.62  0.67  0.27          4-lower
  12: -0.11  0.65  0.22 -0.55          4-lower & 5
  13: -1.40 -0.51 -0.36 -0.28          5
```

PCA - Rainbow Cave species

```
C:\DATA\OUT057.DOC   Analysis by MV-ARCH program BIGPCA3
Data file name C:\DATA\RCSPS3.TRA
Date 10/Aug/99   Time 16:40

C:\DATA\RCSPS3.TRA    Made from file C:\DATA\RCSPS3.CSV into MV-NUTSHELL file by
IMPEXP
 COVARIANCE MATRIX

Sum of PCA eigenvalues = 42

Eigenvalues
      21.84  6.59  5.19

%
      51.96 15.67 12.35

CUM %
      51.96 67.64 79.99

PCA variable loadings         (Species)
   1:  2.34  0.47  0.19       Isoodon
   2:  1.45 -0.82 -0.40       Pseudocheirus
   3: -0.48  1.21 -1.70       Trichosurus
   4: -3.26 -0.13  0.18       Potorous
   5:  1.79  0.04  0.03       B. pencillata
   6: -0.23  0.09  1.19       Setonix
   7:  0.07  0.20 -0.13       Petrogale
   8:  0.35 -0.25  0.11       M. eugenii
   9:  0.13  2.02  0.80       M. fuliginosus

PCA object scores             (Spit)
   1:  0.53  0.88 -0.53       2
   2:  0.13  1.00  0.11       3
   3: -0.17  0.57 -0.43       4
   4:  0.03  1.22  1.43       5
   5:  1.19  0.01 -0.86       6
   6:  1.33 -0.53  0.20       7
   7: -0.73 -0.92  0.86       8
   8:  1.51 -0.66  0.01       10
   9: -2.71 -0.75  0.32       12
  10: -2.50  0.10 -1.02       14
  11:  1.40 -0.93 -0.11       15
```

Appendix 14 Faunal data

The compact disc enclosed in this volume contains two spreadsheet files tabulating the tooth and jaw specimens from which I have estimated vertebrate NISPs and MNIs at Tunnel Cave and Witchcliffe Rock Shelter (Tables 8.12, 8.13, 8.21, 8.22). Comments on use of the disc are given in Appendix 8.

MNI and NISP estimates obtained from the Devil's Lair and Rainbow Cave published faunal lists are not reproduced here (see Balme *et al.* 1978, Lilley 1993). Table A14.1 gives the meanings of abbreviations used in the tabulations.

For example, a left-hand mandible in which a deciduous premolar and the first three molars are retained would be written thus in the "Left" column for the appropriate species:

MAN(dP1,M1-3)

In the tabulations, rows represent spits. Italicised numbers indicate sub-units of spits, such as hearth sediment collections. These sub-units are totalled within the tabulation, as can be seen from the row headings.

Table A14.1 Abbreviations used in recording faunal specimen data

Column headings			
uw	unburnt whole jaw	bw	burnt whole jaw
uf	unburnt fragment of jaw	bf	burnt fragment of jaw
ut	unburnt tooth	bt	burnt tooth
uwt	unburnt specimens' total weight (g)	bwt	burnt specimens' total weight (g)
unid	unidentified		
Specimen information			
I	Incisor	PRE	Premaxilla (with teeth present)
C	Canine	MAX	Maxilla (with teeth present)
P	Premolar	MAN	Mandible (with teeth present)
M	Molar	number	position of tooth in jaw
u or l	upper or lower tooth	1/2	position of tooth uncertain (i.e., 1 or 2)
d	deciduous tooth	1 - 4	indicates presence of all teeth between tooth 1 and tooth 4
j or juv	juvenile		

Computer files

Appendix 14 Tunnel Cave fauna.xls
Appendix 14 Witchcliffe Rock Shelter fauna.xls

Appendix 15 Procedures of charcoal fragment identification

This appendix deals with all aspects of charcoal identification: reference sample preparation; tabulations of reference and published sample characteristics; archaeological sample selection and preparation; and a tabulation of archaeological sample characteristics.

Reference sample preparation

The first step in any palaeontological analysis is preparation of a reference collection for comparison with archaeological specimens. A reference collection comprises modern specimens assumed to have the same identifying characteristics as prehistoric ones. The appearance of the reference specimens should approximate that of the archaeological specimens, yet without losing information. Thus whole defleshed bones are reference specimens for archaeological fragmented bones, and charcoal reference specimens must be burnt, as wood shrinks, vitrifies, and distorts in burning (February 1992, Prior and Gasson 1993, Rossen and Olson 1985, Smith *et al.* 1998).

Published identification keys are designed for wood samples, so my colleague in charcoal analysis, S. Burke, and I collected 50 dry wood samples representing 22 Leeuwin-Naturaliste Region taxa and burnt them to determine the degree of distortion. Location and other details of reference samples are given in Table A9.1.

In the reference collection Burke and I also attempt to describe variations from the identification keys in juvenile wood, as in branches and young trees, and in trees growing at the extremes of their range, e.g., the northernmost Karri forest (represented by a stand near the archaeological site of Quininup Brook: Smith 1973).

Sample selection

To avoid distortions created by burning green wood, our reference samples were collected from fallen branches beneath the desired trees. We chose branches that retained leaves or fruit so that the species culd be identified. Leaf, and fruit samples were then submitted to W. Loneragan, Department fo Botany, UWA, to "voucher" our identifications. Sections of branches 1 to 5 cm long were then cut from the branches by bandsaw. These sections were oven-burnt to simulate archaeological charcoal.

Oven burning

Burke (1997: 42-43) describes the oven burning technique for simulating archaeological charcoal. One must assume that such charcoal burnt in a low-oxygen environment (at the very base of the fire, for example), or else, it would not have survived combustion. Burke reports that the best method was to burn samples wrapped in aluminium foil (to reduce the amount of oxyen reaching them) in a sealed electric oven heated to 450° C until the samples ceased to emit smoke. Two precautions are advised: burning must be carried out in a well-ventilated room, and the oven door must be locked to prevent internal gas pressure bursting it open, which would cause an influx of oxygen and rapid combustion of the samples (cf. Smart and Hoffman 1988: 172).

Determination of distortion from burning, and analysis of juvenile wood

Burke (1997) found that burning has little effect on qualitative characteristics such as cell arrangement; however, quantitative measures were often outside the range specified by the keys. Vessel width shrank by at least 25% for Jarrah and 30% for Karri (Burke 1997: 26). Reference samples also showed more variation in cell size and structure than described by the CSIRO key and Dadswell (1972); this is probably because many specimens were from branches or young trees. Wood reference keys are based on mature trees because juveniles of different species tend to be similar to one another. The reference collection also contains samples from roots, trunks, and branches, and trees in a variety of locations, often where unfavourable conditions appear to have retarded growth and potentially altered wood anatomy.

As with burning, the age of the specimen has little bearing on qualitative characteristics, such as vessel or parenchyma distribution, but it may affect quantitative characteristics, especially vessel frequency and tangential width of vessels. While the presence of wide vessels eliminates taxa that normally have small vessels, the presence of small vessels helps very little since small vessels exist in juvenile parts of most trees (February 1993, Prior and Gasson 1993, Jane 1972). On the same principle, the presence of high rays eliminates many taxa, but the presence of short rays eliminates very few. These considerations would make it hard to distinguish a species in a very large genus such as *Eucalyptus*; fortunately for this analysis the south-western eucalypts are relatively few, and they can be separated in the identification keys (Dadswell 1972; Ilic 1996).

Table A15.1, on compact disc (enclosed with this volume or available on application to the author), gives details of the reference collection. Comments on using the disc are given in Appendix 8. Table A15.2 summarises the identifying characteristics of south-western Australian hardwoods, listed in the order used by Ilic (1996).

Table A15.2 Summary of published and observed key features of south-western Australian hardwoods

Codes

1 = Feature usually present in species according to Ilic (1996) wood key and reference charcoal.
D = Feature in reference charcoal is a deviation from Ilic (1996) wood key.
V = Variable feature according to Ilic (1996) wood key.

Where Dadswell diverges from Ilic (1996) and reference charcoal:
+ = Feature listed as present in Dadswell (1970).
- = Feature listed as absent in Dadswell (1970).

			Vessels			Vessel size (μm)			Vessel frequency			Ray structure				Ray height			Multi-seriate rays					Fibre structure		Parenchyma - abundance			Parenchyma - distribution				
	Feature, classification, and key number in CSIRO-ID (Ilic 1996):	Grain relatively straight	Exclusively solitary	In radial pairs or multiples	Marked oblique tendency	Max. tangential diameter <150	Max. tangential diameter 150-250	Max. tangential diameter >250	Less than 7.5/sq. mm	7.5-15/sq. mm	More than 15/sq.mm	Exclusively uniseriate	Mainly uniseriate, some biseriate	Mainly uniseriate, some triseriate	More than triseriate	Max. ray height 15 cells	Max. ray height 15-20 cells	Max. ray height >20 cells	Multiseriate part <7 cells	Multiseriate part 7-12 cells	Multiseriate part >12 cells	% multiseriate rays <25	% multiseriate rays >25	Wall thick.<lumen diameter	Wall. thick>lumen diameter	sparse	Fairly abundant	Abundant	Paratracheal	Vasicentric	Aliform; confluent	Apotracheal	Apotracheal-diffuse
Common name	**Scientific name**	**15**	**25**	**26**	**27**	**28**	**29**	**30**	**31**	**32**	**33**	**37**	**38**	**39**	**40**	**41**	**42**	**43**	**44**	**45**	**46**	**47**	**48**	**57**	**58**	**60**	**61**	**62**	**63**	**64**	**65**	**66**	**67**
Powderbark wandoo	*Eucalyptus accedens*		1			1					1		1	1	1	1	1		1	1			1		1	1			1			1	1
Mallet	*E. astringens*		1			1				1	1	1	1			1	1		1			1	1			1			1				1
Marri	*E. calophylla*	1		1		D	1	1	1	D	D		1			1	1		1			1	+	1	D		D+	1	1	1	1	1	1
Yate	*E. cornuta*		1		D	1	V		V	1	D		1	1		1	1	D	1			1	1	D	1	1	D		1				1
Karri	*E. diversicolor*		1		1	D+	1	V	1	V	D		1	1		D	1	1	1	1	1	1-	1	D-	1	1	1		1	1		1	V
Tuart	*E. gomphocephala*		1		1	1	V-		V	1	D		1	1	1	1	1	V	1	V	D	D	1		1	1	1	1	1	1	V-	1	1
Mountain marri	*E. haematoxylon*			1		1	1			1			1			1			1			1	1					1	1	1	1	1	1
Red tingle	*E. jacksonii*		1		1		1		1				1	1		1	1	1	1	V		1-	V+	1		1	1	1	1			1	1
Red morrell	*E. longicornis*		1							1	1		1			1			1				1		1		1	1	1			1	1
Jarrah	*E. marginata*		1		1	+	1		V	1	D		1	1	1	1			1			1	1	1		1	1	V	1	V		1	1
Bullich	*E. megacarpa*		1		1	1	+		+	1			1			V	1	V	1			1		1	+	1			1				V
Flooded gum	*E. rudis*		1			1				1			1			1			1			1		1			1		1			1	
Blackbutt	*E. patens*		1		D	V	1		1		D		1	1		1	1		1	1			1	1		V	1	1	1			1	1
Salmon gum	*E. salmonophloia*		1			1					1		1	1		1	1		1			1	1		1	1	1	1	1			1	1
Gimlet	*E. salubris*		1			1					1		1			1	1		1			1	1		1	1			1				1
Wandoo	*E. wandoo*		1			1				1	1		1	1	1	1			1	1			1		1	1			1			1	

Archaeological sample selection and preparation

Small samples were selected randomly from collection bags, which contained hundreds of fragments. I poured each bag's contents onto a sheet of plastic and took fragments from the edge of the pile until I had a sample of 10 to 30 fragments (the number varied according to that needed for the entire period). Samples include fragments of all identifiable sizes, from 2 mm to >2 cm length, suggesting no bias in fragment size. There may be some bias against woods that break into smaller fragments, which can be assessed by referring to preent-day plants at the site and the modern period.

In Table 9.1, the small collections from spits G10-8-2, G10-92-7H, and G10-100-8 were trial identifications at the start of my analysis. The three fragments from G10-8-3 are included with the modern sample on the basis of depth below surface. They are probably closer in age to it than the mid-Holocene sample.

Preparing charcoal samples requires extreme care: surfaces are easily smeared and must be split with sharp pins, preferably chisel-ended. I made such pins by rubbing the ends of dress-maker's pins on a whetstone. I also used 'Gem' brand scalpel blades (available from artists' supply shops and stationers), which are sharpened on both sides. Pressing the chisel end or scalpel blade onto the sample in line with the grain removed large areas for TLS and RLS (e.g. 1 cm^2 area). One cannot cut as large a TS, because charcoal resists breakage across the grain, but a small or uneven TS is often diagnostic anyway. With thin, elongated fragments I obtained a TS by transversely scoring one longitudinal side and snapping the piece along the line (cf. Leney and Casteel 1975). With thick fragments I rested the fragment transversely on a pin and press a double-edged blade across the grain onto the fragment above the pin. Provided one pulled the blade back at the moment of breakage, the blade would only smear part of the TS. Almost all specimens gave at least one TS, TLS, and RLS. Sometimes I obtained more than one of each section, and other times, with very small or fragile samples, I could not obtain all three sections. This variable success at obtaining sections affected identifications slightly.

All of this work was done holding specimens with very fine tweezers and viewing them with either a bright lamp and no magnification, or under a low-powered stereoscopic microscope. All sections were viewed under this microscope before further preparation, to check for broken or smeared surfaces, and to confirm that the section was correct (in splitting eucalypt wood one can inadvertently cut oblique longitudinal sections, which are of little use).

For electron microscopy I glued the split fragments to aluminium stubs (special discs about 1-2 cm diameter that fit onto a SEM stage) with nail varnish, with the section for viewing uppermost, and splutter-coated them with gold to improve conductivity. I attempted to fix all three sections to one stub. Burke (1997) used an environmental SEM, which uses a low vacuum and does not require conductive surfaces, but one can get better focus and less interference with a high vacuum SEM and gold-coated samples. In my analysis, uncoated charcoal fragments reacted to the electron beam, with very small pieces becoming charged and breaking off.

For transmitted light microscopy, with 50 specimens from spit 17, I did not prepare samples other than fixing them with sticky poster putty (e.g. "Blu-Tack") onto plastic lids, so that the correct side faced upwards. Had the specimens not been of exceptionally good quality, the results would not have been as definite; also my charcoal identification experience had greatly improved by the time I used the light microscope.

Table A15.3 (on compact disc) lists features identified in archaeological charcoal samples with codes for the stubs that the samples were mounted on. Codes in the table indicate the reliability of identifications as follows:

Code	Definition	Significance for species identification
1	feature positively identified in several places on the sample	can use feature in species identification key
(1)	feature positively identified in only one place on the sample	can use feature only if other features identified as "(1)" confirm the species identification
?	feature possibly present, but not positively identified anywhere	do not use to identify species (this tentative identification has only potential interest for any future work)

www.ingramcontent.com/pod-product-compliance
Lightning Source LLC
LaVergne TN
LVHW072252300926
849160LV00018B/1651